The
PRICE
of CARBON

DAVID D. MAENZ

The Price of Carbon
Copyright © 2017 by David Daniel Maenz

Published by One Ton Press

Tellwell Talent
www.tellwell.ca

ISBN
978-1-7752913-1-2 (Hardcover)
978-1-7752913-0-5 (Paperback)
978-1-7752913-2-9 (eBook)

To Lynn, Bryan and Kyle.

To my grandson Jackson,

Whose future was an inspiration
to the writing of this book.

Table of Contents

Part 2: Where We are Headed

Chapter 7: How We Get There – Mitigation Pathways to a Sustainable Future

Prologue

On May 1, 2016, unusually hot, dry conditions existed over much of northern Alberta. The temperature high on that date for the city of Fort McMurray was 24.6°C and well above the normal range of 2–13°C for that time of year. Over the next 3 days, temperatures would soar to nearly 33°C with wind gusts of over 70 km/hr. The previous winter had been drier than normal, with little snowfall, such that the winter snowpack had completely melted by the end of April. A wildfire under this combination of conditions had the potential for explosive growth.

The Fort McMurray fire began somewhere south of the city on May 1, and a local state of emergency was declared by 10 pm. By the evening of May 3, the situation had escalated such that the entire city was placed under a mandatory evacuation order. The images shown by the media were harrowing, with long lines of vehicles passing precariously close to raging roadside fires discharging soot and embers. Incredibly, 88,000 people were successfully evacuated with only 2 casualties as the result of a highway collision.

The intensity of the fire created its own weather system of cloud formations, wind influx and lightning. The fire continued to rage out of control, and over the next 2 months would spread over a vast area stretching into the neighbouring province of Saskatchewan. On July 4, 2016, the fire was declared to be under control.

The Fort McMurray fire was the costliest disaster in Canadian history. By May 9, approximately 2,400 structures in the city had been destroyed. Total recovery costs were estimated at over $4.5 billion with $3.6 billion in insured damage. Oil sands operations were halted such that 25% of Canada's daily oil production was lost at a cost to the Alberta economy of approximately $70 million per day.

The underlying conditions that fueled the intensity of the Fort McMurray fire cannot be assigned to any one factor. Fires in boreal forests occur naturally and are essential to forest turnover and health. An El Niño cycle led to a dry fall and winter, followed by a warm spring. Possibly, climate change could have contributed to the extreme weather conditions at the time of the fire. However, the extent of this contribution (if any) to the severity of the Fort McMurray fire cannot be ascertained.

What is certain is that emissions of greenhouse gases from the activities of man have led to an average surface temperature increase of about 1°C relative to pre-industrial times. It is also certain that future surface warming in excess of 2°C will increase the frequency, severity and damage potential of extreme weather events such as the hot, dry windy conditions in Northern Alberta that existed in early May of 2016.

The price that will be paid by future generations for continued excess emissions of greenhouse gases over the next 40 years will be extreme. As an alternative, viable, cost-effective pathways arriving at net zero-emissions in the second half of this century do exist. These pathways will limit future surface warming, and thereby minimize or avoid the damaging effects of climate change that would otherwise be imposed on future generations.

Part 1:
HOW WE GOT HERE

Chapter 1: A BRIEF HISTORY OF THE PLANET EARTH

The Latin meaning of Homo sapiens is "wise man" or "rational man." We are the only surviving species of hominids and are characterized by large well-developed brains that provide capabilities for tool use, reasoning, problem solving, advanced communication and the creation of complex social structures. Evidence of anatomically distinct Homo sapiens date back 200,000 years to the plains of Africa. For the first 199,900 years of existence, the lives of Homo sapiens, like all other animal species, had no significant effect on the atmosphere and climate of Earth. By the year 1750, the capabilities and circumstance of man lead to the onset of the Industrial Revolution whereby raw materials were extracted from the Earth in volume and combusted for energy to drive mechanical processes. By 1900, the core inventions of the Industrial Revolution and other advances in science and agriculture had placed mankind on a path toward a future of intensive fossil fuel use and other activities that would release substantial quantities of greenhouse gases into the atmosphere. Over the past 117 years, the activities of man have led to an accumulation of these gases at sufficient concentrations to elevate surface temperatures. Future surface warming is dependent on the extent of continued release and atmospheric accumulation of anthropogenic (man-made) greenhouse gases.

Over the 4.5 billion year history of Earth, the climate has fluctuated between extremes of "hot house earths", where average surface temperatures were 14°C warmer than our current climate to "snowball earths" where the entire surface of the planet was covered in snow and ice. An assessment of the triggers and processes involved in naturally occurring climate change provides an understanding of how we have transitioned to a new geological age where human activity has become the dominant influence on climate and the environment.

Our solar system began as a pre-solar nebula localized in a cold molecular cloud of hydrogen and helium gases sprinkled with particles of cosmic dust.[1] The mass of the nebula would have been slightly more than the mass of the sun, and the composition would be identical to that of the current solar system (73% hydrogen, 25% helium and 2% heavier elements). The nebula began to spin faster as material condensed and, eventually, a hot dense protostar formed at the center of a protoplanetary disk.

Gravitational collapse of the protostar continued to the point where temperatures and pressures within the core triggered the onset of hydrogen fusion reactions. The outward force of fusion reactions reached an equilibrium with the inward force of gravity, and 4.57 billion years ago the sun began to radiate.[2]

The initial burst of radiation from the newborn sun produced intense cosmic rays and solar winds. Electrons, protons, and alpha particles, along with electromagnet radiation propagated outward from the sun and fundamentally altered the dynamics of the emerging solar system. Solar winds and higher temperatures drove off gases and volatiles that comprised the early atmospheres of the inner planets. Mercury, Venus, Earth and Mars formed as rocky inner planets enriched in molten iron, nickel, silicates and other heavy elements. These planets generated internal heat from gravitational forces such that iron and other heavy metals sunk into hot liquid metallic core structures.

1.1 The Violent Beginnings of the Planet Earth

The term Hadean is derived from Hades and refers to the hellish, violent conditions over the first 600 million years after the formation of the Earth. During the Hadean Eon, the high density of asteroids, comets and smaller fragments (meteors and meteorites) that peppered the inner solar system resulted in a bombardment of Earth by massive dirty ice balls. An early collision with a very large (Mars sized) object resulted in the formation of the moon.

By 10 million years, lighter silicates had separated from molten iron, and a solid crust had formed on the surface of the planet.[3] The second atmosphere was devoid of oxygen and consisted of nitrogen, carbon dioxide (CO_2), water vapour, ammonia, and methane.[4] This atmosphere was produced by a combination of intense volcanic activity releasing entrapped gases from the molten core, plus degassing from the impacts of the comet and asteroid bombardment. With the cooling of the Earth, clouds formed, gaseous water condensed and heavy rains deposited liquid water over most of the surface. With time, ammonia, methane, and carbon dioxide levels would drop to trace (but impactful) levels.

The Archean Eon is defined by the 1.5-billion year period from the end of the Last Heavy Bombardment of the Hadean Eon to the onset of The Great Oxygenation.[3] During the Archean Eon, the interior of the Earth gradually cooled and heat flows from the core slowly diminished. Volcanic activity declined over time as did the rate of movement of tectonic plates. The atmosphere remained free of oxygen during the Archean Eon.

1.2 The Origins of Life and the Oxygenation of the Atmosphere

The mechanism for the origin of life on Earth is poorly understood. Hydrothermal vents at the bottom of the primitive oceans would have provided active conditions of high temperatures and a continuous source of inorganic chemical energy in the form of reduced elements and compounds such as ferrous iron and hydrogen sulfide. Under these conditions, a myriad of chemical reactions would have taken place, which would eventually lead to the formation of proteins from amino acids, and RNA and DNA from nucleic acids.[5] Eventually, life began in the form of a primitive single cell organism capable of metabolism and self-replication under the thermally active conditions of ocean vents. Evidence for remains of biotic life date back 4.1 billion years and imply a rapid onset of life under the conditions of early Earth.[6] Fossils of the oldest microbial life on the planet date to 3.7 billion years ago.[7] The first organism or Last Universal Ancestor was a chemoautotroph that synthesized molecules from carbon dioxide and used inorganic energy sources from ocean vents.[8] Contemporary species of chemoautotrophs can be found in what we now consider to be the hostile environment of deep ocean vents. Over time, the Last Universal Ancestor evolved such that by 3.5 billion years ago, this original organism no longer existed but had split into bacteria and archaea (prokaryotic pre-cursors to eukaryotic organisms).[*,9]

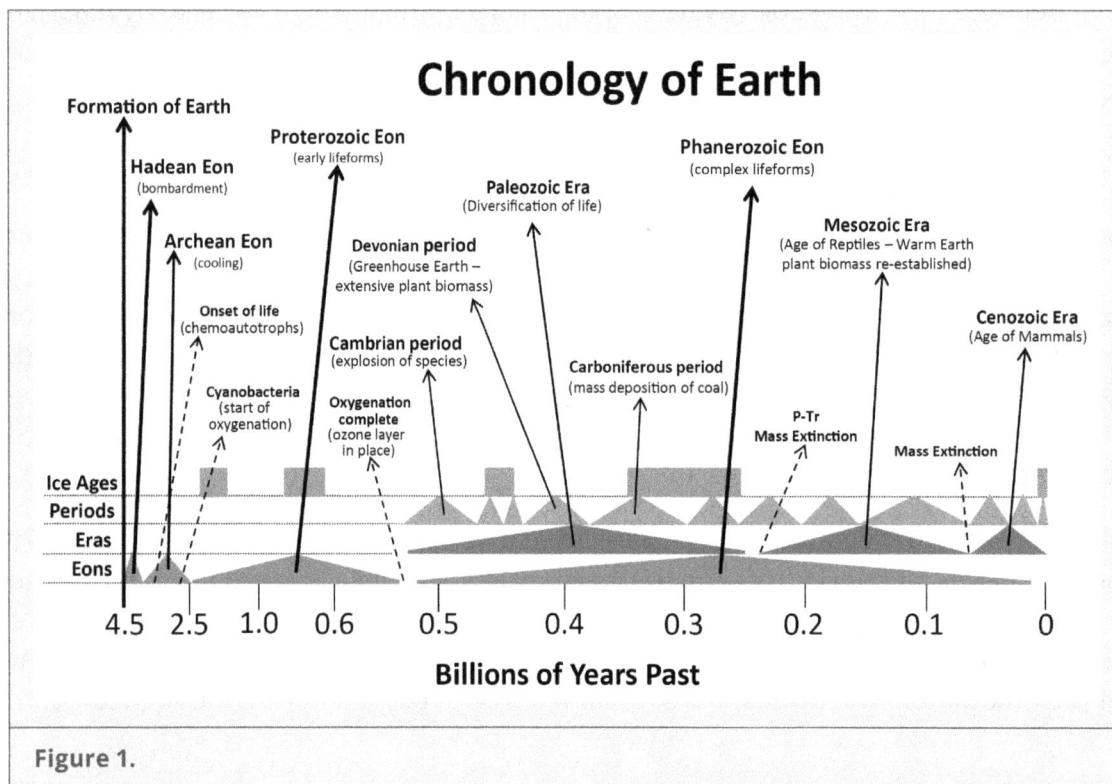

Figure 1.

* Prokaryotes are simple one-celled organisms devoid of subcellular organelles such as a nucleus and mitochondria. Eukaryotic organisms are based on more complex cells containing subcellular organelles. Eukaryotes include life forms from single-celled species to large complex multicellular animals.

As of three billion years ago, photosynthesizing cyanobacteria (blue-green algae) had come into existence and begun to flourish in the oceans.[10] These organisms used sunlight for energy, consumed CO_2 and produced oxygen. Bands of iron oxide are found in geological formations dating back 2.5–2.3 billion years ago.[3] Gaseous oxygen released from the oceans would have initially oxidized exposed land minerals such as iron. The prolonged period of gradual oxygenation of the atmosphere began after the saturation of iron binding sites and other oxygen sinks on land and in the ocean.

As oxygen levels increased during the Proterozoic Eon (2.5 to 0.54 billion years ago), the chemistry of the atmosphere was further altered by oxygen reacting with methane to produce carbon dioxide.[11] Methane is a much stronger greenhouse gas than CO_2, and the net effect was to diminish the greenhouse effect of the atmosphere. The first ice age (Huronian or Makganyene glaciation) began 2.4 billion years ago and lasted for a period of 300 million years. Initial oxygenation of the atmosphere may have produced a "snowball Earth" where snow and ice covered the entire surface of the planet.[11]

Over time, competing oxygen-intolerant anaerobic organisms were virtually killed off and proliferation of cyanobacteria in the oceans led to the "Great Oxygenation Event."[12] With the accumulation of atmospheric oxygen, ultraviolet radiation caused oxygen in the upper atmosphere to recombine as ozone (O_3). The ozone layer functions to absorb 97–99% of damaging ultraviolet solar radiation.[13] Prior to formation of the ozone layer, the mutagenic effects of UV radiation did not allow life to become established on land. The ozone layer formed 600 million years ago, and this time frame corresponds with the emergence of land-based life forms.

Our current atmosphere is characterized by high concentrations of oxygen produced by plant and bacterial life forms. By oxygenating the atmosphere, blue-green algae profoundly disrupted the climate of Earth and established conditions for the subsequent evolution of oxygen-dependent and land-based life.

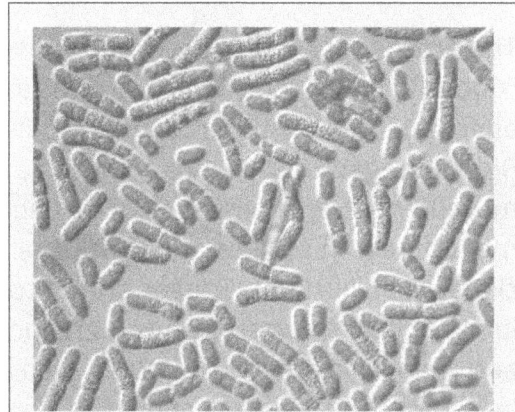

Figure 2. Cyanobacteria (Blue Green Algae). The most disruptive lifeform in the history of Earth. Responsible for oxygenating the atmosphere and facilitating the emergence of aerobic life, eradicating anaerobic competitors, plunging the Earth into a 300 million year ice age, and establishing the ozone layer that allows land-based life to exist.

1.3 Greenhouse and Icehouse Earths

The Earth has passed through the Huronian ice age (2.4–2.1 billion years ago), the Cryogenian ice age (850–630 million years ago), the Andean-Saharan ice age (460–430 million years ago) and the Karoo ice age (360–260 million years ago).[14] At present we are in the Pliocene-Quaternary glaciation that began 2.6 million years ago as the fifth ice age of the planet.[14] An ice age is defined by the existence of glaciers on the surface of the Earth. Extreme long-term radical changes in climate are roughly defined by transitions between a "Greenhouse Earth" and an "Icehouse Earth."[15] Icehouse conditions during the Cryogenian Period (second ice age) were so severe that ice and snow coated the entire surface of the planet, resulting in a "snowball Earth."[15]

Paleoclimatologists have developed models of long-term climate change based on an initial disruption of a climate equilibrium, leading to a transition out of a Greenhouse or an Icehouse Earth. The underlying mechanisms that could disrupt a climate equilibrium vary with the climatic era, and they remain open to debate.

The process of continental drift over the 3.5 billion years since the formation of Earth is poorly understood. However, 1 billion years ago, the total land mass of Earth was aggregated into a supercontinent called Rodina that was centered on the equator.[3] Rodina began to separate approximately 830 million years ago, with considerable movement of land masses since that time.[3] Noncyclic events associated with the movement of tectonic plates and shifting land masses may well have disrupted long-term climate equilibriums. Trigger events could include changes in volcanic activity, shifts in major ocean currents, the emergence of plant life on land acting as a new CO_2 sink, and changes to physical features such as the formation of mountain ranges affecting land-based CO_2 sequestration.[14] This hypothesis agrees with the lack of consistency in the time periods between ice ages.

The transition from a Greenhouse Earth into an ice age requires a trigger mechanism that initiates a decline in the concentration of greenhouse gases in the atmosphere and a subsequent drop in surface temperatures. As temperatures fall, a series of feedback loops continue to drive surface temperature toward a new equilibrium. An initial accumulation of snow and ice at the poles leads to an increased albedo effect (percentage of solar radiation reflected back to space) and a reduction in the efficiency of surface warming by sunlight. The accumulating ice acts as a CO_2 trap and, as ocean temperatures decline, more CO_2 is dissolved in ocean waters. Glaciation and the cooling oceans further draw down atmospheric CO_2 concentrations. These synergistic negative feedback mechanisms result in a continued drop in surface temperatures, an expansion of ice and snow cover, and eventually transition to an ice age.

The transition out of an ice age requires a trigger event leading to a release of greenhouse gases to the atmosphere. A reversed set of positive feedback loops progressively sustains a continued increase in surface temperatures. The melting of the ice releases entrapped CO_2, an increase in ocean temperatures releases dissolved CO_2, and a decrease in ice and snow cover reduces albedo and increases the efficiency of solar radiation heating the surface. Over time, the planet transitions to a "Greenhouse Earth."

The climate history of the planet can be described as prolonged periods of sustained high temperature, greenhouse climates, with brief ice age interludes. The Phanerozoic Eon covers the last 542 million year history of Earth and during this eon, the planet has existed as a "Greenhouse Earth" for 80% of the time while passing through two ice ages prior to entering the current ice age.[16] Direct data on atmospheric CO_2 does not extend beyond 800,000 years; however, indirect modelling methods for estimating the temperature of the planet and atmospheric CO_2 are consistent with a model whereby the concentration of CO_2 in the atmosphere is the driving force for long duration climate change, over the entirety of the Phanerozoic Eon.[17] The Phanerozoic Eon began with a massive diversification of life known as the Cambrian explosion. The major phyla of animals came into existence and more complex organisms such as trilobites evolved from simple single cell life forms.[18] The Cambrian Period lasted from 541 to 485 million years ago and ended with the transition from a Greenhouse Earth to the Andean-Saharan ice age.[18] The Earth emerged from the third ice age and transitioned back to a Greenhouse Earth climate that characterized much of the Devonian Period (419–369 million years ago).[19] During the Devonian, the planet was hot and extensive vegetation including massive forests of ferns and lycopods covered much of the land mass.[19] The expanding biomass of plant life led to increased oxygen production such that oxygen levels may have peaked at up to 35% of the atmosphere during the subsequent Carboniferous Period.[20] Gradually the vast biomass of plant life became an effective carbon sink and reduced atmospheric CO_2. A substantial portion of the carbon extracted from the atmosphere during the Devonian and Carboniferous Periods became buried beneath progressive layers of decaying vegetation. With time and pressure, plant sequestered carbon from these periods became coal formations within the Earth. The eventual drop in atmospheric CO_2 following the proliferation of plant biomass, triggered the transition out of a Greenhouse Earth and led to the Karoo Ice Age (fourth ice age).[21]

The planet transitioned out of the Karoo ice age during the Carboniferous and Permian geologic periods. The largest mass extinction in the history of Earth occurred 252 million years ago and marked the end of the Permian Period.[22] Ninety-six percent of all marine species and 70% of terrestrial species were eradicated during the Permian-Triassic (P-Tr) extinction event.[22] An impact event, massive volcanic eruptions, methane gasification from deep ocean deposits, a severe ocean anoxic event (deoxygenation) triggered by release of entrapped carbon dioxide or a combination of causes have been proposed to explain the P-Tr extinction.[22] Plant species generally survived, but total plant biomass was nearly eradicated and forests disappeared. There are no coal deposits dating to the early Triassic period following the P-Tr extinction.[22]

The Mesozoic Era (age of reptiles) began 252 million years ago after the P-Tr extinction and lasted for 186 million years, ending with another major extinction event.[23] The climate during the Mesozoic Era was that of a Greenhouse Planet and plant life thrived at high latitudes with polar forests coming into existence during the Cretaceous Period. As every 5-year-old child will attest, dinosaurs walked the Earth.

Sixty-six million years ago, the Mesozoic Era ended when an asteroid struck the Earth causing a vast dispersion of particles and water vapour to the atmosphere.[24] The occlusion of sunlight impaired photosynthesis such that 75% of all species died off, including all the non-avian species of dinosaurs and 100% of all terrestrial animals over 2.2 pounds.[24]

1.4 The Last 66 Million Years (Cenozoic Era)

The underlying database for the science of paleoclimatology becomes stronger the closer one gets to the present time. Oxygen exists in nature as stable isotopes with atomic weights of 16 and 18. When dissolved in water, the heavier ^{18}O requires higher temperatures to transition to the gaseous state. The ratio of $^{18}O/^{16}O$, then, is directly related to and is dependent upon the temperature of the water.[25] Cold water has a low ratio of $^{18}O/^{16}O$ and as the temperature of the water increases, the $^{18}O/^{16}O$ ratio increases.

Forarms is the informal designation applied to Foraminifera, an abundant shell-bearing marine micro-organism in existence today and with fossils dating back to the Cambrian Period.[26] Layers of sediment beneath the ocean floors are enriched in shell fragments of forarms, and the oil exploration industry uses forarm fossils as an indicator of potential petroleum deposits.[26] Deep sea drilling has provided a wealth of fossilized core samples containing shell fragments from Foraminifera. These shells are composed of calcium carbonate ($CaCO_3$), and the oxygen in the shell structure originated from dissolved oxygen sequestered from the deep ocean during the life of the tiny creatures. The oxygen isotope ratio of samples of shell fragments from benthic (seafloor dwelling) forarms provides climatologists with a "paleothermometer" for reading deep ocean temperatures over the last 66 million years.[26] Deep ocean temperatures can then be used to calculate global surface air temperatures with a reasonable degree of certainty.

At the beginning of the Cenozoic Era, global surface temperatures were roughly 10°C warmer than today.[27] Temperatures climbed from the beginning of the era and, 51 million years ago in the middle of the Eocene Epoch, average surface temperatures reached a peak of 28°C or 14°C above the current global average.[27] These conditions can be described as an extreme Greenhouse Earth with forests covering most of the land mass. In the absence of polar ice, sea levels were over 50 metres higher than today.[27] Fossil evidence dating back to the Eocene Epoch of the Cenozoic Era indicates that cypress and redwood trees grew on Ellesmere Island, that palm trees grew in Alaska, and that tropical rainforests existed in Europe and the Northern portions of North America.[28] Crocodiles and tropical snakes lived at high latitudes during the early to mid-Eocene.[28] There are no direct methods available for determining the concentration of greenhouse gases in the atmosphere as the planet warmed during the transition from the early to mid-Eocene Epoch. However, modelling based on indirect indicators supports an atmospheric CO_2 level of 2,000 ppm at the temperature maximum of the Eocene Epoch. In addition, methane production from swamplands likely added to total greenhouse gas production.[28]

Approximately 49 million years ago, a biological trigger led to a drawdown of atmospheric CO_2 and began the transition out of the severe Greenhouse Earth climate.[29] At that time, the land masses of the northern hemisphere were aggregated. North America and Asia were connected and small passageways of water existed between North America and Greenland, between Greenland and Europe, and between Europe and Asia. The Arctic Ocean was isolated from the main oceans of world with little to no mixing of waters. Fresh water flowing into the Arctic layered on top of the dense still seawater, resulting in a massive bloom of Azolla, a freshwater fern. The term Azolla refers to "super-plant" and under ideal conditions Azolla will rapidly bloom over the entire surface of a body of water. Ideal conditions existed at the surface of the Arctic Ocean 49 million years ago. A massive bloom of Azolla spread over the vast expanse of water with enough theoretical biomass turnover to drawdown atmospheric CO_2. During the Azolla event, carbon was continuously withdrawn from the atmosphere and deposited on the Arctic Ocean floor by successive waves of biomass turnover. The Arctic basin is characterized by immense, thick layers of fossilized Azolla and plankton that accumulated over a period of 800,000 years.[29] Modelling of biological carbon sequestration during the Azolla event is consistent with a drop in atmospheric CO_2 down to 650 ppm.[29] The vast plant biomass in place over the warm land mass and in the greater oceans would have contributed to a biological drawdown of atmospheric CO_2. By the end of the Eocene Epoch 34 million years ago, average global surface temperatures had dropped to 19°C.[28]

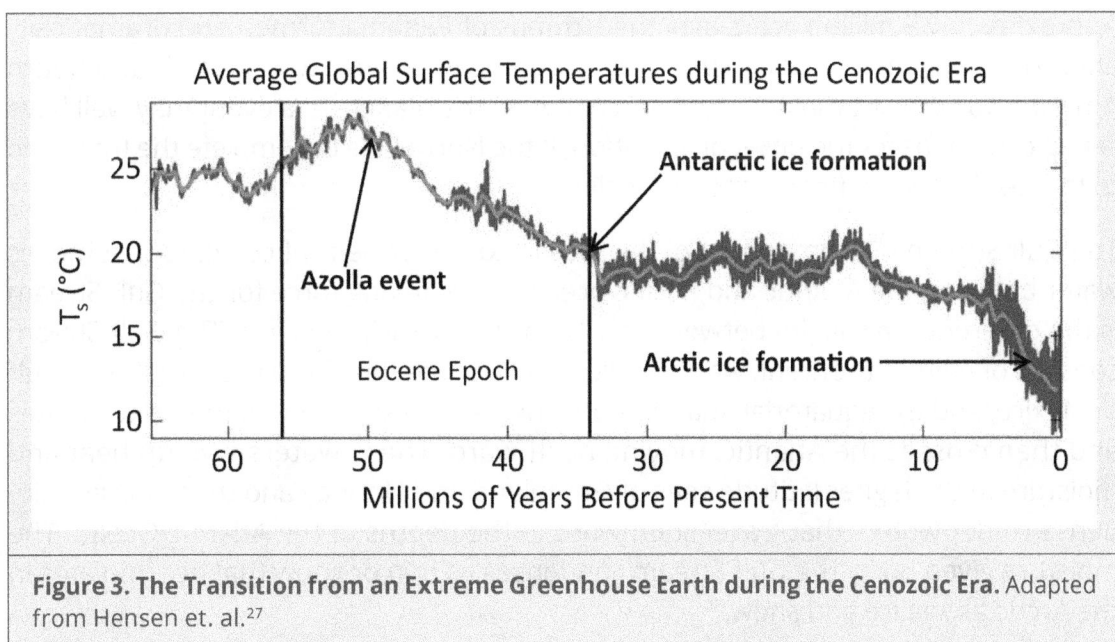

Figure 3. The Transition from an Extreme Greenhouse Earth during the Cenozoic Era. Adapted from Hensen et. al.[27]

The underlying causes of the current ice age are not fully understood; however, the patterns of continental drift over the past 50 million years leading to the current configuration of land masses would have altered ocean current patterns, which, in turn, may have acted as the primary triggers for glaciation beginning with the south pole and then, later, the north pole.

Approximately 40 million years ago, Antarctica began to separate from South America.[30] The Drake Passage opened and the powerful Antarctic Circumpolar Current came into existence. This current isolated the new continent from the warming influences of the greater ocean waters. Large scale cooling led to the onset of glaciation about 34 million years ago.[30] The advancing ice mass would eventually cover the entire continent.

The onset of glaciation at the North Pole is a more recent event with a more complex less certain origin. Five million years ago, North and South America were separated and tropical waters flowed between the Pacific and Atlantic Oceans through the Central American Seaway.[30] Global surface temperatures had cooled to 16°C, but there was no evidence of glaciation in the high Arctic.[27]

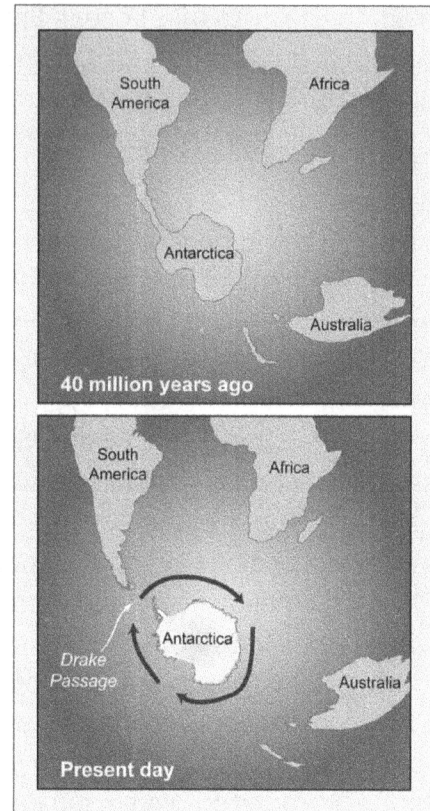

Figure 4. Separation of Antarctica and Formation of the Drake Passage. Antarctic Circumpolar Current isolates the new continent and glaciation begins.

Approximately 3 million years ago, the Isthmus of Panama formed and dramatically altered ocean currents.[30] Closure of the ocean gateway between North and South America was the last major geological change to the planet. This event may well have set up conditions for the onset of glaciation at the North Pole to complete the transition to the ice age that defines our current climate.

The Gulf Stream is a component of the "ocean conveyor belt" of currents that moves water between the Atlantic and Pacific oceans. The driving force for the Gulf Stream is the difference in salinity between the Atlantic and Pacific oceans. The Gulf Stream consists of high salt content, warm shallow waters that flow from the Caribbean, Gulf of Mexico, and the equatorial Atlantic along the eastern seaboard of the United States and then crosses the Atlantic, moving northward. These waters give up heat and moisture in the higher latitude seas of Greenland and Norway and then sink as salty dense colder waters that travel southward in the depths of the Atlantic Ocean. The moisture given up by the Gulf Stream condenses as rain or snow that accumulates in the Arctic as sea ice and snow.[30]

Prior to the formation of the Isthmus of Panama, ocean water from the Pacific flowed through the Central American Seaway diluting the salt content of the oceans in the Tropical Atlantic. In comparison to today, the driving force of the Gulf Stream was considerably weaker, and the stream was largely ineffective in releasing moisture to the air in the high latitudes of the Atlantic.

The closure of the Central American Seaway powered the Gulf Stream and delivered moisture to the arctic. This event coincided with a gradual cooling of surface temperatures such that the Arctic was primed for ice and snow formation. Moist air delivery to the Arctic via the Gulf Stream may well have caused an

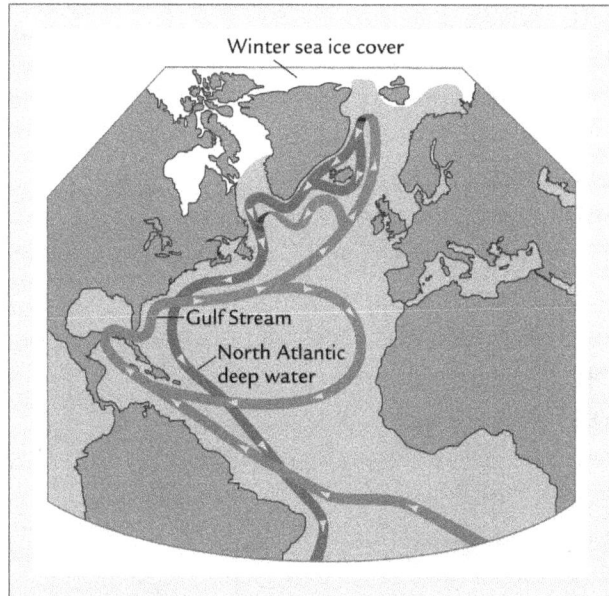

Figure 5. Atlantic Segment of the Ocean Conveyor Belt. Formation of the Isthmus of Panama 3 million years ago prevented mixing of Pacific and Atlantic waters and powered the Gulf Stream to deliver moist warm air to the Arctic.

initial formation of sea ice at the North Pole. The expanding snow and ice layer reduced surface heating by the sun, and negative feedback loops for transition to an ice age came into play. As of 2 million years ago, average global surface temperatures had dropped to 12°C, and the planet had completed the transition to the current ice age.[27]

1.5 The Last 2.6 Million Years (Quaternary Period)

Latitude 75°S and longitude 123°E places one at Dome C, deep inland on the continent of Antarctica, atop the Antarctic Plateau at an altitude of 3,233 metres above sea level.[31] Dome C is a flat elevation or "summit" of the Antarctic Ice Sheet located 1,670 kilometres from the geographic South Pole and is one of the coldest, most desolate and inhospitable places on Earth. The elevation is such that the oxygen content of the air is one-third less than that of sea level.[32] Dome C is located on the world's largest desert with low humidity and very little annual precipitation. Summer temperatures typically do not rise above -25°C and during winter, can go below -80°C.[32] During winter, the sun sets for the last time in the beginning of May and does not rise until late August. No native plants or animals live this far inland on Antarctica. Other than a slow steady accumulation of ice, Dome C has not changed over millions of years and, in some ways, can be considered as the epicenter of the current ice age.

In 2005, construction of a permanent French-Italian research station was completed at Dome C.[33] Concordia Station was built to extend the summer research activities at the site that had begun in 1992.[33] The severities of winter conditions are such that Concordia Station is completely cut off with no possibility of evacuation or supply delivery for 9 months of the year. During winters, the station is manned by a team of 13–15, composed of scientists, technical support, and medical personnel.[33] Among the more important research activities at the station is ice core sampling and analysis as part of the European Project for Ice Coring in Antarctica (EPICA).[33]

Permanent ice sheets such as Dome C in Antarctica continually thicken with snow accumulation over years.

Figure 6. Concordia Research Station. Dome C Antarctica. The harshest, most desolate land-based environment on Earth. Photographed March 2009 by Stephen Hudson.

The snow compresses to a granular intermediate or firn. Air becomes fully entrapped as the firn layer densifies under increasing pressure. At further depth, the firn layer transitions to ice with entrapped air bubbles.[34] By drilling, extracting and dating ice core samples at various depths, climatologists can analyze the composition of air that became entrapped in ice hundreds of thousands of years ago.

Concordia Station is one of 37 ice core sampling sites in the world, 12 of which are located in Antarctica, 19 in Greenland, and 6 in other non-polar regions with permanent year-round ice formations.[34] Collection and analysis of ice core samples have provided scientists with a powerful data set of past atmospheric compositions and has been a major contributor to our current understanding of the drivers of climate change.

Ice core samples provide an accurate tracking of atmospheric CO_2 from the past 800,000 years. Atmospheric CO_2 follows a regular pattern of oscillation between highs of 280 parts per million (ppm) and lows of 180 ppm that repeats every 100,000 years.[35] Another set of ice core data taken by the Russian Vostok Antarctic research station demonstrates the same time-dependent regular oscillation in CO_2 levels over the past 420,000 years.[36] This pattern of regular changes in CO_2

Figure 7. Ice Core Sample with Entrapped Air Bubbles. Perfectly preserved samples of atmosphere dating back hundreds of thousands of years. Photographed by Chris Gilbert, British Antarctic Survey.

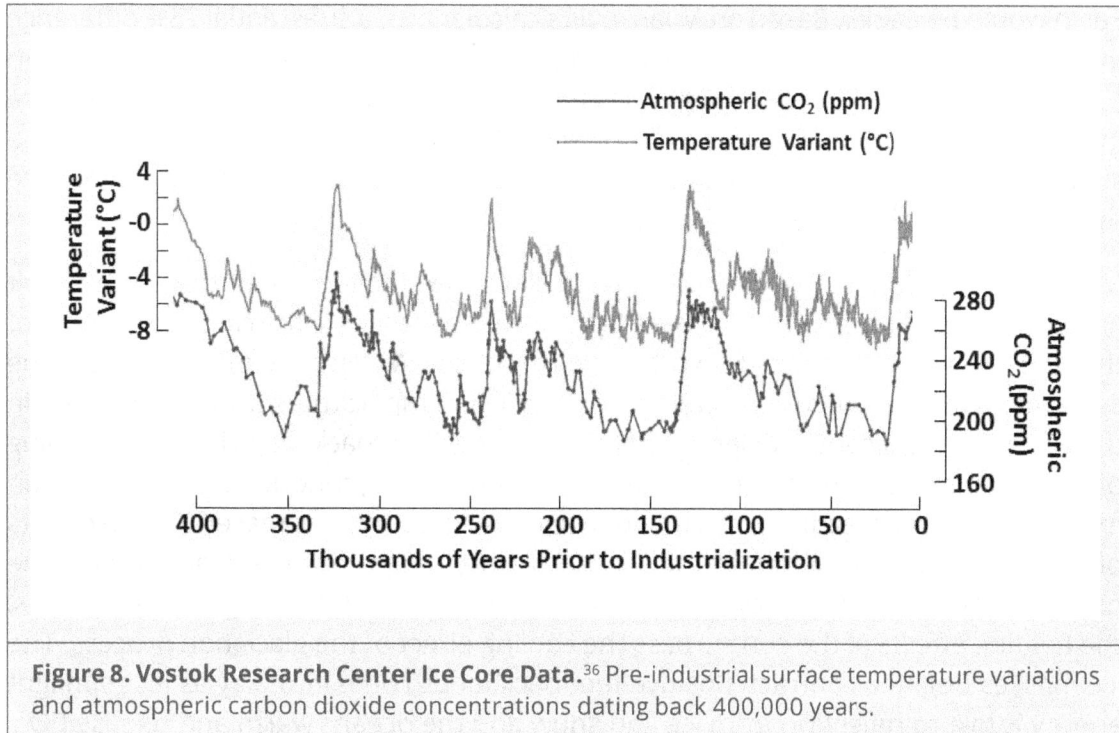

Figure 8. Vostok Research Center Ice Core Data.[36] Pre-industrial surface temperature variations and atmospheric carbon dioxide concentrations dating back 400,000 years.

correlates directly with average global surface temperatures and the occurrence of glacial periods. Within our current ice age, at approximately 100,000 year intervals, the planet passes through periods of extreme glaciation in the Northern Hemisphere that correlate with declining surface temperatures and declining atmospheric CO_2. The last glacial period ended approximately 12,000 years ago and, prior to industrialization, the planet was in a predictable interglacial period characterized by contraction of glaciers, average global surface temperatures of 14°C, and atmospheric CO_2 levels of 280 ppm.

The history of repeated 100,000-year cycles between glaciation periods within our current ice age implies a predictable cyclical driver of surface temperature fluctuations. Indeed, well understood systematic cyclical variations in the Earth's orbit will change the heat energy from sunlight reaching the Earth at mid-latitudes.

Milutin Milanković was a brilliant Serbian mathematician, astronomer, climatologist, civil engineer, and geophysicist best known for developing a theory of climate change based on variations in the orbital shape or eccentricity (100,000-year cycles), axial tilt (41,000-year cycles) and precession (23,000-year cycles) of the Earth's orbit[37]. In 1922, Milanković developed a set of mathematical models that allowed for the calculation of the amount of solar radiation contacting any given latitude on the planet at any time over the past 600,000 years.[37] Milanković believed that "orbital forcing" of sunlight at the mid-latitudes was the most important factor in defining surface temperatures. When the cyclical variations were such that orbital forcing of sunlight at the mid-latitudes was greatest, the Earth would be warmer, and when orbital forcing was weakest, the

Earth would be colder. Based on Milanković's calculations, a substantial 25% difference in maximum and minimum orbital forcing will exist.[38] The concept became known as Milankovitch cycles.[39] Although the reasons are not fully understood, evidence indicates that the 100,000-year cycle of eccentricity in the Earth's orbit around the sun is the more critical variable in defining the effects of orbital forcing on the surface temperature of the planet.

Milankovitch cycles, in theory, have existed since the stabilization of the Earth's orbit around the sun. The effect would have been greater when, as is the case today, most of the land mass was located in the Northern Hemisphere. Within the current ice age as the cycle moves toward weaker orbital forcing, surface temperatures drop and glaciers in the Northern Hemisphere begin to expand. Negative feedback loops based on cooling of the oceans and increased ocean sequestration of atmospheric CO_2, along with an expanding ice mass leading to further entrapment of CO_2 and increased albedo effects of reflecting sunlight, furthered the drop in surface temperature and accelerated the process of glaciation. As the cycle shifts toward stronger orbital forcing at mid-latitudes, the heating effects of the sun surpass the cooling effect of the glaciation process. The ice masses begin to contract, positive feedback loops come into play as less sunlight energy is lost to reflection from ice and snow and the oceans warm and release CO_2.

The model of 100,000-year Milankovitch cycles within our current ice age fits the data on periodicity of glaciation. Changes in atmospheric CO_2 follow, and will contribute to, the warming or cooling process as triggered by Milankovitch cycles. Other models based on greenhouse gas levels as the driver of glaciation cycles within our ice age would require primary predictable cyclical sources of CO_2 sequestration and release.

1.6 The Last 12,000 Years Prior to Industrialization (Holocene Epoch)

The Holocene Epoch is defined by the end of the last glacial period, approximately 11,700 years ago, and continuing beyond the present time.[40] The primary characteristics of the Holocene are a warmer interglacial climate within the current ice age and the more recent emerging global significance of human activities on the climate and ecosystems of the planet.

During the first 2,000 years of the Holocene, average global surface temperatures increased by 2–3°C as the climate completed the transition out of the last glacial age to the current interglacial equilibrium. Overall, the climate has been relatively stable over the past 10,000 years; however, modest fluctuations and trends in surface temperature changes occurred prior to industrialization. During this time period, the climate was influenced by a myriad of weaker non-cyclical and cyclical forces. These influences can include variations in solar flare activity, smaller variations in volcanic activity, changing ocean currents, variations in glacial discharge to the oceans, and changes in prevailing winds. In addition, orbital forcing is not as simplistic as a 100,000-year cycle in the eccentricity of the Earth's orbit around the sun. Axis tilt and precession (change in the orientation of the rotating axis) fluctuate through shorter cycles than eccentricity. Ice core data suggests these factors are weaker than eccentricity but will affect solar heat delivery to the key mid-latitudes. This complex interplay of weaker primary and secondary forces has influenced short-term modest changes in climate over the past 10,000 years prior to industrialization.

Any attempts to precisely define global climate changes over the entirety of the past 12,000 years are complicated by regional variations. The Antarctic, Greenland and Mount Kilimanjaro ice core samples, and sediment core samples from various locations, show differing profiles for temperatures variation, and there is no scientific consensus as to how to build a precise model of average global surface temperature changes during the Holocene. However, critical to interpreting the data is the magnitude of variation in temperature differences over time and between sampling sites. Using the average temperature at the middle of the 20th century as a baseline, regardless of sampling site or time, anomalies above and below this baseline have not exceeded 1°C over the past 10,000 years, prior to industrialization.[41]

The data indicates an overall stable global climate since the transition out of the last glacial period. A period of slightly warmer temperatures known as the Holocene Climatic Optimum existed from 8,000 to 5,000 years ago. In theory, the combination of axial tilt and distance from Earth maximized orbital forcing about 9,000 years ago. Albedo effects of the contracting glaciers may have countered the warming effects of the sun such that peak warming was delayed. From 5,000 years ago to the onset of industrialization, overall temperatures cooled slightly.[42] This process, termed "neoglaciation", is most likely due to a deviation away from

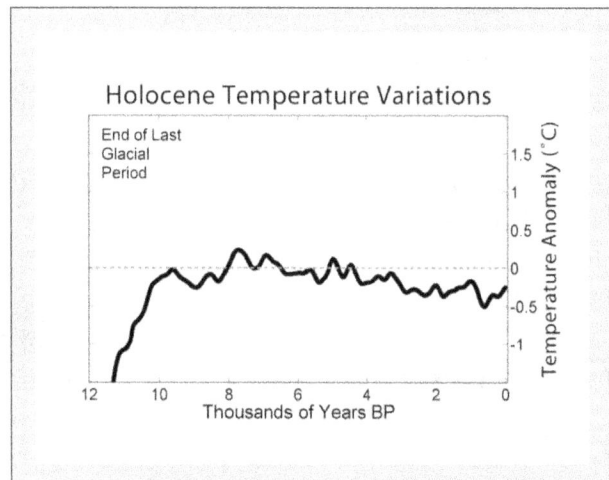

Figure 9. Average Surface Temperature Anomalies over the Past 12,000 Years Prior to Industrialization. Baseline temperature is the average global surface temperature from mid-20th century readings. Adapted from a Global Warming Art Project image originally created by Robert A. Rohde from publically available data.

optimal orbital forcing and thus less heating from the sun. The temperature differences between the peak warming of the Holocene Climatic Optimum to pre-industrial neoglaciation are greatest at high latitudes in the northern hemisphere. There is no evidence for cooling over the last 5,000 years in the southern Hemisphere.[42]

Atmospheric CO_2, as measured in ice core samples, have been relatively stable over the course of the Holocene. Levels increased from 240 to 265 ppm over the first 2,000 years followed by slight drift upward to 280 ppm over the last 5,000 years prior to industrialization.[43] This data is consistent with older ice core samples indicating that atmospheric CO_2 follows the transition out of a glacial period and plateaus at approximately 280 ppm during an interglacial period. The trends in atmospheric CO_2 concentrations over the past 10,000 years until just prior to industrialization do not correlate with the modest decline in surface temperatures over this time period. Hence, the slight increase in atmospheric CO_2 concentration was insufficient to overcome other drivers of average surface temperature change, over the past 5,000 years prior to industrialization.

Milankovitch cycles can be considered as an extremely long-term seasonal variation with a "summer" during maximum orbital forcing and a "winter" during minimal orbital forcing. Based entirely on the pattern predicted by Milankovitch cycles, and a pre-industrial atmospheric CO_2 concentration of 280 ppm, the climate should remain relatively stable over the next few thousand years, followed by the transition to the next glacial period 50,000 years from now.[43] However,

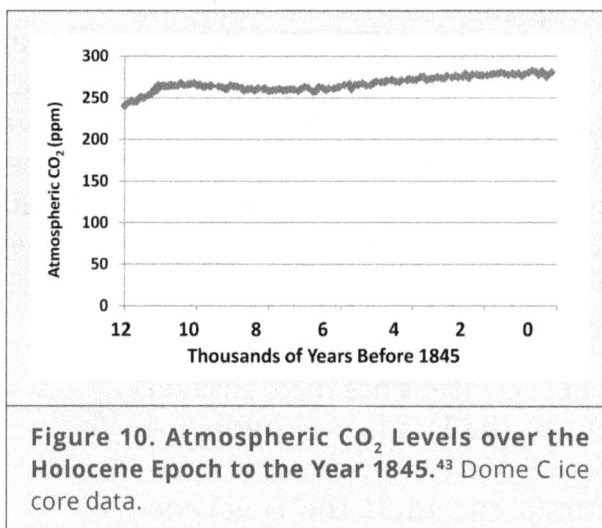

Figure 10. Atmospheric CO_2 Levels over the Holocene Epoch to the Year 1845.[43] Dome C ice core data.

the potential exists for the interglacial climate equilibrium to be disrupted by a potent non-cyclical event such as a substantial release of greenhouse gases to the atmosphere. Such an event could overwhelm the balance of orbital forcing and other weaker climate forces and act as a trigger to drive surface temperatures out of the current equilibrium.

Chapter 2: THE ANTHROPOCENE

To assign a more specific date to the onset of the 'Anthropocene' seems somewhat arbitrary, but we propose the latter part of the 18th century, although we are aware that alternative proposals can be made (some may even want to include the entire Holocene). However, we choose this date because, during the past two centuries, the global effects of human activities have become clearly noticeable. This is the period when data retrieved from glacial ice cores show the beginning of a growth in the atmospheric concentrations of several 'greenhouse gases', in particular CO_2 and CH_4. Such a starting date also coincides with James Watt's invention of the steam engine in 1784.[44]

Paul Crutzen—Atmospheric chemist and Nobel Prize winner for work on depletion of the ozone layer.

The term "anthropo" refers to man and the Anthropocene is a proposed geological epoch that would begin from the time that human activities affected the climate and ecosystems of Earth. "Anthropocene" is the title of a new academic journal launched in 2013 and has become a common term used in contemporary scientific literature. The Nobel Laureate and atmospheric chemist Paul Crutzen is among the notable advocates of a formal declaration of the Anthropocene Epoch. Recently, an expert working group on the Anthropocene within the International Union of Geological Sciences recommended that the IUGS recognize and declare a formal end to the Holocene and the beginning of the Anthropocene Epoch.[45]

In the 4.56-billion year history of Earth, up until the Anthropocene, bacteria and plants were the only living organisms that altered the planet to any significant degree. Blue green algae oxygenated the atmosphere leading to a 300-million year ice age, and facilitated the evolution of land-based life forms. A massive Azolla bloom over the Arctic Ocean 49 million years ago drew down atmospheric CO_2 and triggered the transition out of an extreme Greenhouse Earth. Since oxygenation of the atmosphere, the concentration of atmospheric carbon dioxide, and thus the surface temperature of the planet, has been influenced by variations in carbon fixation by the global biomass of plants.

The onset of the Anthropocene marks the emergence of an animal species with inherent capabilities to alter the atmosphere, and thus influence the climate and ecosystems of the planet. The difference is that, in theory, the actions of mankind are conscious and deliberate. From the time that James Watt advanced the design of the steam engine, mankind embarked on a path toward large scale industrialization and disciplined scientific discovery. The Industrial and Scientific Revolutions during the Anthropocene have led to a 9.2-fold increase in world population in combination with accelerated per capita energy use, implementation of intensive agricultural practices, and changes in forest biomass. The actions of man have altered and will continue to alter the climate and ecosystems of the planet. Science provides man with an understanding of how industrialization has changed the planet, and a powerful predictive tool for understanding the consequences of options for future industrialization.

2.1 The Industrial Revolution

In 1689, the Bill of Rights[46] was enacted in Britain, thereby establishing the prominence of parliament in the governance of the country. The emerging political enlightenment was part of an ongoing series of institutional and economic reforms that transitioned late 17th century Britain into a freer, more liberal society. By 1700, the secrecy, intransience and predatory practices of the guild system of commerce and manufacturing were in decline. In comparison to other nations at the time, Britain had developed a relatively advanced degree of technical expertise with a higher percentage of the total population working outside of subsistence level agriculture. A considerable workforce of artisans, craftsmen and engineers were engaged in activities such as mining, shipbuilding, and clock and instrument making. Under these conditions, the entrepreneurial and the inventive were naturally drawn to identifying opportunities, and then solving the problems that impeded progress in the commercialization of these opportunities. The combination of societal conditions that favoured free markets and entrepreneurship, along with a critical mass of expertise, fostered the Industrial Revolution in Britain.

By the end of the 17th century, wood had become scarce and costly for purposes such as home heating. Coal, in comparison, was abundant, provided more heat on combustion, and could be economically extracted from deposits in the Earth. Supply and demand created an industry for the mining, distribution and sales of coal. By the early 1700s, an estimated 2.25 million tonnes of coal were consumed each year in Britain.[47] Coal was used for domestic heating, cooking, and to heat water for washing and laundering. Industries that required heat in their manufacturing processes switched from wood and charcoal to the higher energy density of coal and were able to utilize new coal burning techniques to improve productivity. At the time, no other nation made such extensive use of coal, and the volumes of production from coal mines in Britain were well in excess of the remainder of the world.

2.1.1 Early Years of Coal Mining in Britain and the Invention of the First Steam Engine. Coal mining in the early 1700s was a dangerous, brutish process, powered primarily by a combination of human and animal muscle power. Water accumulation within pits and shafts was a constant, with flooding a frequent cause of colliery failures. Drainage was often not possible and hauling water to the surface was the only option to maintain mining operations. Horse engines or horse pumps consisting of a team of horses travelling in a circle and pulling a gear mechanism linked to a bucket chain were common fixtures of wet mining operations. The inefficiency of the water drawing process, limited the depth of operation of wet mines that were rich in coal deposits.[47]

Thomas Savery was a military engineer from Devon, England, with an interest in mechanics, who, in the 1690s, turned his attention to the problem of drawing water from coal mines. Savery invented and patented a primitive pump whereby steam produced in a coal-fired boiler was condensed in a vacuum vessel with piping down to the water that was to be pumped to the surface.[48] The pumping action of the engine was dependent upon atmospheric pressure, and this limitation and other design flaws were such that the Savery pump was never used in ongoing mining operations.

Thomas Newcomen of Dartmouth, Devon, advanced the original concept of Savery and others to invent the first practical steam engine in 1712.[49] Newcomen and Savery partnered and extended the original patent to include Newcomen's modifications in design. Newcomen replaced the vacuum vessel in Savery's original design with a cylinder and piston mechanism. Steam filled the cylinder forcing the machine side of the beam upward. At the top of the stroke, water was introduced to the cylinder to condense the steam and create a vacuum in the cylinder. The vacuum and counter-weight of the beam forced the piston down, and the oscillation of the beam was used to draw water up from the source. The engines were extremely inefficient in the transfer of steam energy to mechanical energy; however, when used at coal mines where

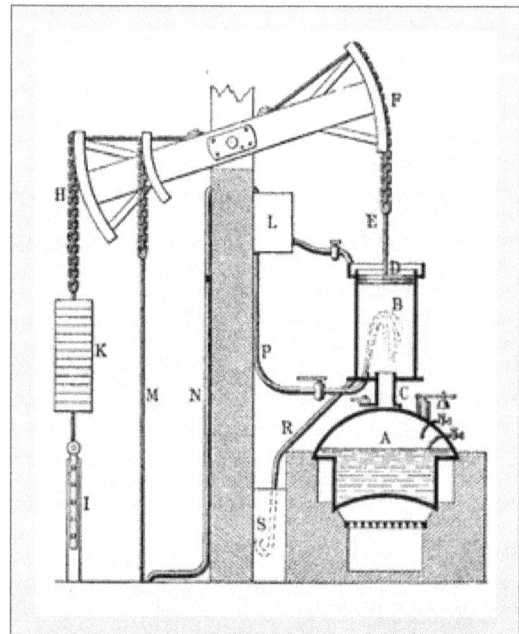

Figure 11. Newcomen Steam Engine. The first practical machine using thermal energy to drive a mechanical process.

fuel was plentiful and cheap, the engine proved to be a reliable and effective method of pumping water. Newcomen's engines became commonplace in mining operations in Britain and by the mid-1700s had been adopted in continental Europe. The Newcomen engine was the first use of coal combustion as an energy source to generate steam to power a mechanized industrial process. The engine had a long period of commercial use in colliers with some utilization in textile and other industries. During the 18th century, 1454 Newcomen engines were built and the engine remained in commercial use into the early 1800s.[50]

2.1.2 The Invention of the Watt Steam Engine.

James Watt patented his steam engine on the eve of the American Revolution, consummating a relationship between coal and the new Promethean spirit of the age, and humanity made its first tentative steps into an industrial way of life that would, over the next two centuries, forever, change the world.

Jeremy Rifkin (2002)[51]

James Watt was born in 1736 in the Scottish seaport of Greenock, a full 37 years after the initial issuance of Savery's "Fire Engine" patent.[52] Watt had little formal education beyond grammar school but, as a young man, exhibited considerable manual dexterity, along with an aptitude for mathematics and engineering. These skills eventually lead to employment at the University of Glasgow as an instrument maker. In 1764, Watt was asked to repair a lab-scale Newcomen engine. Watt experimented at length with the repaired engine and was able to demonstrate that the water used to condense the steam in the cylinder cooled the walls of the vessel such that most of the energy of the incoming steam was lost in reheating the cylinder.[53] Watt's core invention was to add a separate condensing cylinder to the basic Newcomen design. During the down cycle of the compression cylinder, a valve was opened and the steam was forced into the condensing cylinder. In theory, the latent energy of the steam would be efficiently converted to the mechanical energy of driving the piston. In 1765, Watt constructed a simple and successful lab scale demonstration of the condensing cylinder concept, and in 1769 a patent was issued covering the invention.[53] However, commercial production of a working engine based on Watt's improvements to the basic Newcomen design was a technical challenge requiring substantial time and resources. Eleven years passed from bench top demonstration to installation of the first full-scale Watt steam engine.

Matthew Boulton owned the successful Soho Manufactory and was an early adopter of the assembly line for the mass production of a range of goods including buttons, buckles and boxes.[53] Boulton understood the potential of Watt's invention and provided the financial backing and business acumen required to restart the project. Boulton successfully lobbied parliament to extent the lifespan of the original patent to the year 1800, and used his industrial contacts to source the best iron machinists available.[53] The precision boring techniques developed by John Wilkinson for the production of cannons and rifles were used to produce cylinders and pistons to the required specifications.[53] In 1776, two large engines were built. Wilkinson was given one engine, and the other was installed at the Tipton coal mine.[53] The engines were successful and were followed by additional machines installed at mine sites specifically to provide a fuel saving advantage relative to the conventional Newcomen engine.

Boulton saw potential for Watt's engine well beyond pumping water from mines, and encouraged Watt to further modify the mechanism of the engine such that the vertical motion of the piston was converted to rotational power to drive wheels for grinding, weaving and other applications best suited to rotational forces. In 1781,

Watt was issued a patent for the sun and planet gear assembly whereby the rod of the piston was attached to a fixed gear (planet gear) that drove a rotating gear (sun gear).[53]

The coal-fueled Watt steam engine was among the key inventions that led to a restructuring of society from the limitations and hardship of primitive muscle power to the efficiency and productivity of machine power.

Due to the limitation of the strength of materials used in boiler construction, the first commercial Watt steam engines were driven by low-pressure steam. In theory, high-pressure steam engines would be smaller, more powerful, and could be produced at a lower cost than low-pressure engines. However, the danger of exploding boilers placed a practical limitation on steam pressures. The development of safe high-pressure steam boilers was dependent upon advances in metallurgy.

Figure 12. James Watt. Inventor of the Watt steam engine that powered many of the disruptive transformations of the Industrial Revolution.

2.1.3 Iron and Steel: The Structural Materials of the Industrial Revolution. In the early decades of the Industrial Revolution, cast iron was readily produced in bulk at very high temperatures in a blast furnace. At sufficient temperatures, iron ore melts to produce liquid cast iron with a carbon content of 2–4%.[54] Molten iron could be directly cast into final products or cooled to pig iron for further processing. Cast iron, while useful for the production of items such as pots, pans, cannons and cannonballs, tends to be brittle and crack and cannot be shaped by heating and pounding with a hammer. Pig iron was either re-melted, and cast to final products at another site, or converted to wrought iron by oxidation of excess carbon in a refinery. The strength and hardness of the final product was dependent upon the carbon content. Wrought (worked) iron (also termed mild steel) has a very low carbon content of 0.02 to 0.08% while the much harder steel generally contains 0.2–1.5% carbon.[55] Wrought iron and steel were both tough and malleable enough to be worked into useful implements upon reheating and shaping.

At the time when the Watt steam engine came into production, wrought iron lacked the structural grade requirements to produce high-pressure boilers.

In 1783–84, the English iron master Henry Cort developed the puddling and rolling processes to cost-effectively produce structural grade wrought iron.[56] The puddling process required a skilled craftsman (puddler) to monitor and stir the molten pig iron. As the carbon content of the molten iron decreased, semi-solid balls of iron would form, and the puddler would gather these into a single mass that would be removed. The hot wrought iron would then be run through rolling mills to form flat iron sheets or rails. The rolling process was 15-fold more efficient than hammering the iron into sheets.[57]

The puddling process required craftsmen of exceptional strength with the endurance to work under conditions of extreme heat and fumes. The process was never mechanized and conditions were such that puddlers rarely lived to age 40.[58] Over the course of a 12-hour shift, a puddler and helper could produce approximately 1500 kg of structural grade wrought iron.[58] The

Figure 13. Puddler and Helper Remove 150 Pound Ball of Semi Molten Wrought Iron from a Puddling Furnace.

labour intensive nature of the process became a bottleneck to production as demand increased with the spread of the Industrial Revolution. Despite the limitations, for the next 100 years, high strength wrought iron produced by the puddling-rolling process would provide the structural material to build machines, engines, ships, locomotives and railroads. Landmarks such as the Eiffel Tower and the Statue of Liberty were built from puddled wrought iron.[58]

In 1856, Henry Bessemer developed the concept that oxygen blown through molten pig iron should react with carbon and thereby reduce the carbon content of the ore without the need for the puddling process.[59] Bessemer designed the converter as a large receptacle with nozzles (tuyeres) at the base for injection of compressed air. As air passed through the molten pig iron in the vessel during the blow, oxidation reactions decarbonized the iron, and the oxides either blew off as a gas or collected in a layer of slag that separated from the molten iron. In the fully developed process, the carbon content of the iron was controlled by adding alloys of carbon, iron, and manganese after the blow process.

The Bessemer process was rapid, efficient, and mechanized, and could produce high strength steel set to

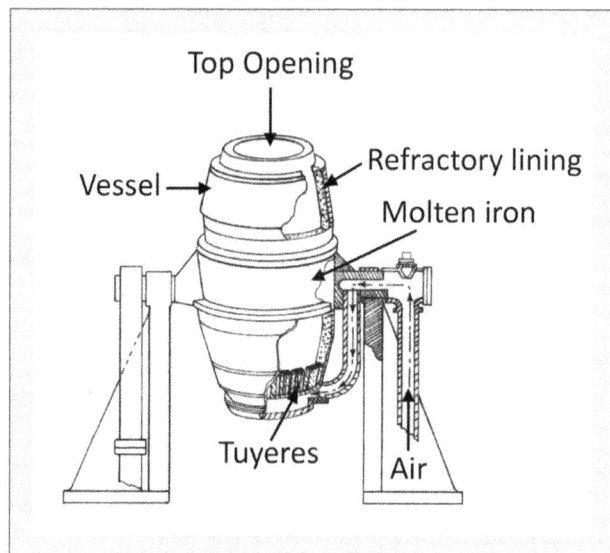

Figure 14. Bessemer Converter. Efficient, mechanized production of high strength steel from iron ore.

target carbon contents. Variations of the process were introduced with time, but all were based on the concept of oxygen decarbonization of molten iron. By 1884, the production of wrought iron rails had ceased. The quality of the steel manufactured in 1905 increased the life span of a typical rail from two to ten years, and the load bearing capacity of steel rails was such that the maximum rail car weight was increased from 8 to 70 tonnes.[60] In the US of 1867, the cost of structural grade wrought iron was $170/ton. By 1900, the cost of the superior steel rail was $14/ton.[60]

Perfection of the steam engine and the development of practical processes for producing high grade iron and steel were arguably the two most important advances of the Industrial Revolution. In addition, the era saw technical breakthroughs in textile manufacturing, machine tools, transportation (shipbuilding and locomotives), agriculture, hygiene, and the building of vast networks of railroads, canals and roads.

2.1.4 The Industrial Revolution as the Driving Force for Economic Transformation and the Development of Science.

Science owes more to the steam engine than the steam engine owes to science.

L. J. Henderson (1917)[61]

The prosperity and basics of life for the bulk of humanity changed little from the earliest times of recorded history until the 1800s. At any time prior to the onset of reaping the benefits of the Industrial Revolution, within any society, people (other than a small minority of the privileged) survived at a subsistence level. Concepts such as the middle class did not exist, and the vast majority of any given population struggled for survival, living poor, difficult lives, often cut short by disease, famine or violence. Life expectancy in 1800 was 30–35 years and similar to that of the people living in the Stone Age. In early 18th century Britain, 83 percent of the population lived in rural settings and survived at a subsistence level based on a combination of agriculture and domestic cottage industry work such as spinning and weaving of wool to cloth.[62] Pre-industrial societies were caught in the "Malthusian Trap" whereby any advance in technology simply allowed more people to live at the subsistence level.[63]

The early phases of the Industrial Revolution were characterized by the Dickensian horrors of exploitation of workers in factories. In Britain, working class conditions did not improve until a series of labour laws were enacted over the period from 1819 to 1903 to restrict and

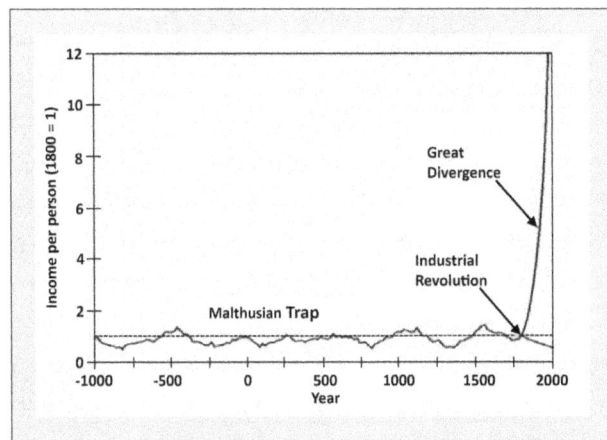

Figure 15. Adjusted Income per Person Over Time. Industrialized societies break from the Malthusian Trap.[63]

then abolish child labour, restrict work hours, and to improve safety and working conditions.[64] Starting around 1820, real income per person in industrialized countries broke from the Malthusian Trap and underwent sustained growth.[63] The advances in technology and the wealth created by the Industrial Revolution were sufficient to support an increase in population and allowed for material consumption and an increase in the standard of living of the working class to go well beyond the bare subsistence level of the previous generations. The economist Gregory Clark postulates that countries that were not included in the spread of the Industrial Revolution remained in the Malthusian Trap, and the resultant difference in income per person between industrialized and non-industrialized nations was termed the Great Divergence.[63]

By 1900, the disruptive impacts of the Industrial Revolution had largely reshaped western economies and societies at large. World population had grown from an estimated 790 million in 1750 to 1.65 billion by 1900.[65] In the industrialized world, economic indicators of productivity, growth and capital stock had emerged from plodding pre-industrial rates and were advancing toward modern levels.[63]

The inventions of Newcomen, Watt, Cort, Bessemer and others that became the platform for the Industrial Revolution were based on a combination of considerable direct experience in the subject at hand and an empirical approach to the inventive process. The science of thermodynamics was at an embryonic stage, yet Newcomen, and later Watt, understood the potential for converting steam energy to mechanical energy and invented practical machines to accomplish this conversion. Similar accounts describe the advances in metallurgy, production of chemicals and cement, gas production/purification from coal and subsequent large-scale distribution for gas lighting, and the inventions of machinery for factory production of textiles. The transformative inventions of the Industrial Revolution largely took place outside of the world of science and academic institutions. The wealth that was ultimately created by these inventions provided the support for the scientific revolution that followed.

Over the course of the 19th century and into the pre-war period of the 20th century, science gradually evolved to become a profession, with the formation of societies, and the exchange of results and ideas via meetings and journal publications. The first issue of the journal "Nature" was published in 1869[66] and the National Academy of Science was founded in the US in 1863.[67] Charles Darwin published the Origin of Species in 1859.[68] By 1900, the profession of science was practiced primarily at universities and research institutes, by professors and researchers with the highest level of academic training. The rigour and discipline of the scientific process of hypothesis testing through experimentation, data collection and analysis, and continued refinement of understanding was applied with an ever-increasing degree of sophistication to the fields of astronomy, medicine, biology, physics and chemistry. The tools of science such as costly telescopes and other instruments for observations and measurements of physical parameters were funded by the newly created wealth of the Industrial Revolution. By the late 1800s, scientists were beginning to consider fundamental questions as to the origin of the universe, the relationship between matter and energy, and the chemistry of life. Climatology would evolve as a branch of science dedicated to understanding the fundamental factors that define weather patterns and how changes in these factors have altered the Earth's climate in the past and can alter the climate in the future.

2.1.5 Coal Consumption and CO_2 Emissions from 1700 to 1905. Coal fueled the Industrial Revolution such that, by 1905, world coal production was estimated at 923 million tonnes annually, which represents a 370-fold increase over estimates for coal production in 1700.[69] By 1905, the US produced 38%, while all of Europe accounted for 58% of total world coal production.[70] Prior to 1905, coal was the only fossil fuel used to any significant volume. Combustion of 923 million tonnes of coal in 1905, produced 2,640 million tonnes of carbon dioxide gas that was emitted to the atmosphere.*

In 1896, the Swedish physicist/chemist Svante Arrhenius, while working as a professor at Stockholm University College, published a study entitled, "On the Influence of Carbonic Acid in the Air upon The Temperature of the Earth."[71] The paper was based on earlier observations of increased absorption of reflected sunlight from the moon when the moon was lower in the sky. Absorbance of infrared radiation by carbon dioxide gas was understood at the time. Arrhenius hypothesized that the presence of carbon dioxide in the atmosphere resulted in a partial absorption of infrared radiation, and that the degree of absorption would increase in proportion to the concentration of carbon dioxide in the atmosphere. When the moon was lower in the sky, the reflected radiation would pass through a greater depth of atmosphere, resulting in a higher degree of absorption. This same principle would apply to

Figure 16. Svante Arrhenius. In 1896, Arrhenius developed the hypothesis that atmospheric carbon dioxide could influence the surface temperature on Earth.

infrared radiation reflected from the surface of the Earth. Arrhenius hypothesized that the concentration of CO_2 in the atmosphere would largely define the proportion of infrared radiation coming from the surface of the Earth that either dissipates to space or is absorbed and re-emitted back to the surface. Based on the thermal absorptive capacity of carbon dioxide, Arrhenius calculated that a 50% decrease in atmospheric CO_2 would result in a period of glaciation, while a doubling of concentration would

---

* The carbon content of coal varies between 60–90% dependent upon quality. Combustion of coal is an exothermic (heat producing) reaction whereby 1 atom of carbon (atomic weight of 12) in the coal reacts with 1 molecule of oxygen (O_2 – molecular weight of 32) to produce 1 molecule of CO_2 (molecular weight of 44). On combustion with oxygen in the atmosphere, 1 tonne of pure solid carbon would produce 3.67 tonnes of CO_2 gas. The non-carbon components of coal result in an actual average production of 2.64 tonnes of CO_2 per tonne of solid coal combustion.

increase surface temperatures by 5–6°C. Arrhenius presented the hypothesis that changes in atmospheric CO_2 levels correlated with, and were the cause of, long-term fluctuations in the surface temperature of the Earth that occurred during transitions into and out of periods of glaciation.

Arvid Högbom, a colleague of Arrhenius, calculated the volumes of CO_2 emitted from the global combustion of coal. Arrhenius and Högbom hypothesized that man-made carbon dioxide emissions would eventually lead to global warming. However, given the rate of emissions in the late 1890s, Arrhenius believed that warming would take place over thousands of years. What Arrhenius did not foresee was a world population of 7 billion coupled with a 3.3-fold increase in greenhouse gas emissions per person by the year 2012.

Although the term greenhouse gas was not in use at the time, the basics of the greenhouse effect would have been well understood by Arrhenius and Högbom. The greenhouse concept for maintaining the surface temperature of the Earth can be summarized as follows:

1. Approximately 70% of incoming shortwave solar radiation passes through the atmosphere and warms the surface of the Earth. The other 30% is reflected back to space.

2. The heat from the Earth's surface is emitted as longwave (infrared) radiation that propagates through the atmosphere.

3. A portion of the long wave radiation is absorbed by greenhouse gases (primarily CO_2) that re-emit the radiation back to the surface, which retains more heat.

4. The temperature on the surface is dependent upon the balance between incoming short wave solar radiation and the concentration of greenhouse gases that absorb and re-emit infrared radiation back to the Earth's surface.

Arrhenius presented his paper as a working hypothesis that changes in atmospheric CO_2 levels lead to changes in the surface temperature of the Earth. At the time, there were no records of atmospheric CO_2 levels over the 5 billion year history of the planet, and there was no data compiled on yearly atmospheric CO_2 concentrations from the onset of the Industrial Revolution to the late 1890s when Arrhenius performed his calculations. It was not possible to apply the scientific rigour of model fitting to data to test the

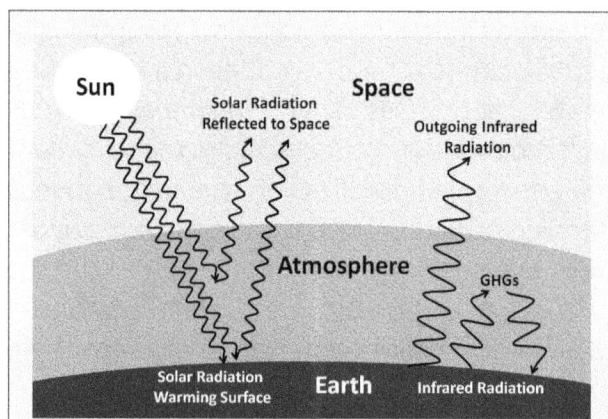

Figure 17. Greenhouse Effect on the Surface Temperatures of Earth. Greenhouse Gases (GHGs) absorb and re-emit infrared radiation back to the surface of Earth.

hypothesis. Unbeknownst to Arrhenius, atmospheric samples dating back 800,000 years and including the transition to industrialized society lay perfectly preserved in ice formations awaiting the attention of future generations of climatologists.

2.1.6 The Paris Exposition of 1900. There are no records of Professor Svante Arrhenius attending the Paris Exposition Universelle held over a 7-month period in 1900. If Professor Arrhenius had visited the Exposition, he would have travelled by train from Stockholm to Paris in the relative comfort of a passenger car hauled by a coal-fueled steam locomotive. The steam engine would have been a modernized version of James Watt's original invention powered by a high-pressure boiler. The train would have travelled over steel rails produced by the process of oxygen decarbonization of molten iron invented by Sir Henry Bessemer. The trip would have been lengthy and involved multiple transfers to travel as efficiently as possible on the 260,000 km of the total rail network of Europe. Given his expertise, Professor Arrhenius may have been one of the few passengers to give consideration to the black smoke belching from the stack of the locomotive. Likely, this would have been a fleeting thought when he recalled the overall quantity of global carbon dioxide emissions at the time, relative to the concentration of atmospheric CO_2 required to elevate surface temperatures.

Professor Arrhenius would have been 1 of an estimated 51 million visitors to the 530-acre site of the exposition located in central Paris.[72] The exposition was designed as a showcase for the technical, artistic and architectural achievements of the past century and to display the latest innovations at the forefront of the coming century. Announcement of the fair in 1882 was met with enthusiasm and support among the countries in Europe and abroad. The fair was vast in scope with 40 countries contributing pavilions and exhibitions. As described by the historian Margaret MacMillan, the Paris exposition of 1900 followed a prolonged period of relative peace and prosperity in Europe, and a spirit of cooperation and optimism prevailed among nations.[73]

Among the more popular exhibits of the exposition were the Palace of Electricity and the attached hall of mechanical and electrical engineering. The German company Siemens & Halske installed a house-sized 2,200 volt alternating current generator coupled with a Borsig steam engine.[74] The generator provided site-wide electricity to the lights and machines of the fair and was on display in the hall of machines. Five thousand multi-coloured incandescent light bulbs provided a spectacular display, lighting up the immense 70 metre high palace that became the focal point of the entire exposition.[72] Other demonstrations of the future potential of electricity included a moving sidewalk and an electric trolley car that transported visitors on the site.

The internal combustion engine and the novelty of the automobile provided a look into the future of land transportation. One hundred seventy-six gasoline powered, 40 electric and 21 steam powered automobiles were on display, along with the new engine invented by Rudolph Diesel.[75]

The timing of the Paris Exposition of 1900 coincided with the very early stages of centralized electric power generation with grid distribution, and the development and then, mass adoption of, petroleum-fueled internal combustion engines. Three years after the Paris exposition, Henry Ford launched the Ford Motor Company,[76] and in Kitty Hawk, North Carolina, Orville Wright successfully flew a heavier than air craft powered by a gasoline fueled internal combustion engine.[77]

By 1900, the innovations of the Industrial Revolution were on the cusp of evolving to a new age of widespread electric power generation and distribution, and a transition from the steam engine to the internal combustion engine. These and other advances would be coupled to a global population explosion. Coal fueled the Industrial Revolution and, to this day, combustion of fossil fuels in the form of coal, oil, and natural gas continues to be the primary energy source for industry, heating, transportation, and power production.

2.2 The History of Climate Change Science

Over the first half of the 20th century, the science of climate change can be accurately described as non-existent. Svante Arrhenius's paper on the radiative force and surface warming potential of atmospheric CO_2 was published in 1896, and for the next 55 years there was little follow-up or interest in the subject. The instruments available and the methodology for taking samples did not provide evidence of changing concentrations of CO_2, and there was some suggestion that water vapour in the atmosphere would absorb surface reflected radiation with little impact of greenhouse gases. In addition, the CO_2 absorptive capacity of the oceans was thought to be sufficient to efficiently remove excess CO_2 such that atmospheric accumulation would not occur. Mostly, however, Arrhenius's calculations and hypothesis were largely forgotten to the archives of science.

By the 1950s, detailed analysis established that the upper atmosphere was largely devoid of water vapour, and that the absorptive spectrum of water did not overlap with that of CO_2.[78] These findings suggested that water vapour was unlikely to impair any radiative force of atmospheric CO_2.

In 1955, Hans Suess, while working at the Scripps Institution of Oceanography, demonstrated that atmospheric carbon originating from fossil fuel emissions was not immediately dissolved in ocean water and tended to remain in the atmosphere.[78] Roger Revell, the Director of the Institution, came to the conclusion that the surface layer of the ocean had a limited capacity for rapid sequestration of atmospheric CO_2.[78] In 1957, Revelle and Suess co-authored a paper suggesting that fossil fuel combustion had the potential to lead to an accumulation of atmospheric CO_2, and that the resultant greenhouse effect could elevate surface temperatures on Earth.[79] At the time of publication in 1957, the hypothesis of elevated atmospheric CO_2 from fossil fuel combustion was largely theoretical with no strong supportive evidence. The previous year, Revelle had hired Charles David Keeling to establish a program of atmospheric CO_2 monitoring.

David Keeling was a young geochemist from Caltech when he accepted the position at Scripps.[80] Keeling had an interest in atmospheric chemistry and, while at Caltech, adapted an infrared device to accurately measure parts per million concentrations of CO_2 in air samples.[80] The objective of the new research program at Scripps was to set up monitoring stations at ideal locations on the planet and track short and long-term variations in atmospheric CO_2. In 1957, with funding from the US Weather Bureau and the National Bureau of Standards, Keeling established a base beyond the tree line, 2 miles above sea level on the Hawaiian island of Mauna Loa, and the next year began a systematic process of monitoring well mixed atmospheric CO_2.[78] By 1960, evidence of a progressive time-dependent increase in atmospheric carbon dioxide had begun to accumulate.[81]

In 2015, the American Chemical Society designated the Keeling Curve[82] as a National Historical Chemical Landmark.[82] Atmospheric CO_2 levels follow a regular annual cycle with a maximum in May, followed by declining levels over the spring and summer to a minimum in September. Seasonal variations in plant carbon sequestration activity in the Northern Hemisphere are the cause of the annual cycle in CO_2 levels with spring and summer plant growth leading to the September

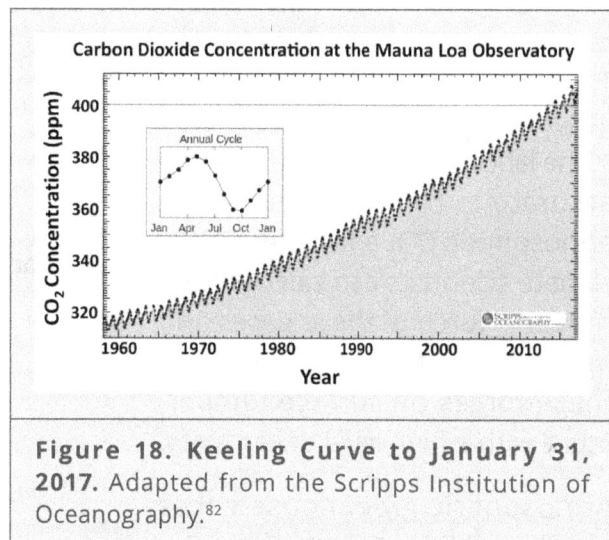

Figure 18. Keeling Curve to January 31, 2017. Adapted from the Scripps Institution of Oceanography.[82]

minimum and winter dormancy, resulting in the May maximum. Atmospheric CO_2 has increased steadily over the past 58 years. In May of 1958, the atmosphere contained 317 ppm CO_2, and in May of 2016, CO_2 averaged 408 ppm.[83] At present, Mauna Loa is one of over 100 stations on Earth monitoring atmospheric CO_2.[84]

The most recent reliable ice core samples come from the Law Dome in Antarctica and date to 1975.[85] The dating of these samples overlaps with direct atmospheric measures of CO_2, methane (CH_4) and nitrous oxide (N_2O). The concentrations of these greenhouse gases (GHGs) as measured directly in samples of air and corresponding date-matched ice core samples are nearly identical and validate the accuracy of measuring GHGs in pockets of air entrapped in older samples of ice. Law Dome ice core samples dated to 1845 have 285 ppm CO_2. This concentration is consistent with maximum levels of interglacial age atmospheric CO_2 measured in samples dating back over the past 800,000 years.[85] By 1900, CO_2 levels had increased to 297 ppm and by 1950 had reached 312 ppm.[85] Since 1960, the rate of accumulation of atmospheric CO_2 has accelerated;

60% of the total accumulation from the onset of the industrial era to the present has occurred in the past 40 years. As of 2011, atmospheric CO_2 had risen by 40%, CH_4 had increased 1.5-fold and N_2O was 20% higher, relative to pre-industrial levels[86]

The thermal absorptive capacity of gaseous carbon dioxide has been well understood since the time of Arrhenius. Greenhouse gases such as CO_2 absorb long wave infrared radiation reflected from the Earth's surface and thus reduce the amount of this radiation emanating to space. The radiation absorptive capacity of gaseous CO_2 has been determined in the laboratory. Given the extreme accuracy in the measurement of atmospheric CO_2 and other GHGs, climate scientists can calculate the radiative force of these gases and model the effects of changes in these forces on surface temperature with a high level of certainty.

Atmospheric greenhouse gases consist of CO_2, CH_4, N_2O, and a mixture of chlorofluorocarbons (CFCs or Freon), halogenated chlorofluorocarbons (HCFCs), and other halogen and chemically related gases. The radiative forcing of each component in the mixture of greenhouse gases is dependent upon the combination of concentration and potency for reflection of radiation back to the surface. Based on the difference between pre-industrial concentrations of GHGs and total atmospheric concentrations in the year 2011, CO_2 accounted for 64%, methane contributed 17%, and

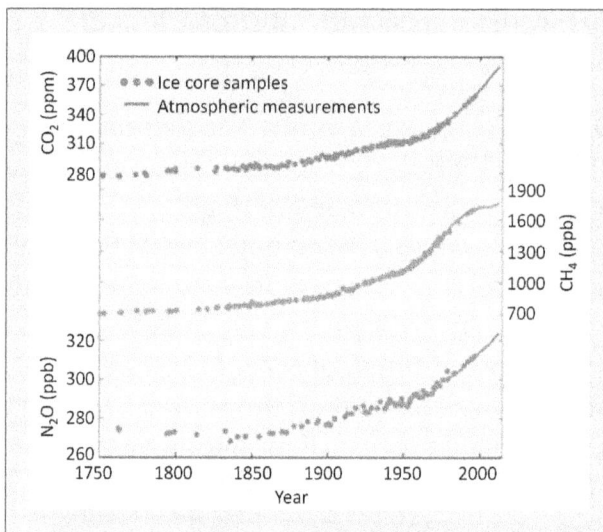

Figure 19. Globally Averaged Greenhouse Gas Concentrations since 1750. Data points are ice core samples. Lines are data from samples taken directly from the atmosphere. Adapted from IPCC Climate Change 2014: Synthesis Report.[86]

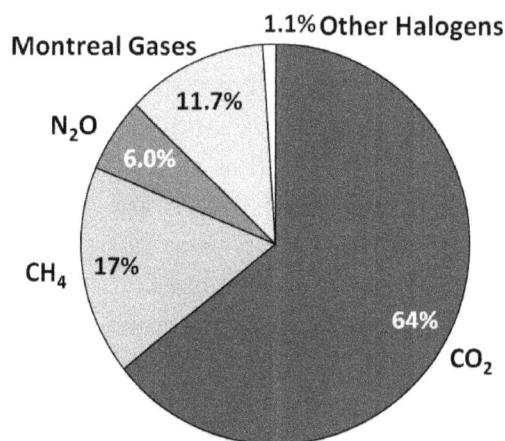

Figure 20. Year 2011 Relative Contribution of Greenhouse Gases to total Anthropogenic Change in Radiative Forcing.[87] Anthropogenic concentrations for carbon dioxide (**CO_2**), methane (**CH_4**), and nitrous oxide (**N_2O**) equal 2011 atmospheric concentrations less pre-industrial levels determined from ice core samples. **Montreal Gases** consist of chlorofluorocarbons (Freon) and other halogens and related compounds covered by the Montreal Protocol.

nitrous oxide accounted for another 6% of anthropogenic (man-made) radiative forcing.[87] The remaining 12.8% of added radiative forcing came from the mixture of CFCs, HCFCs and other halogen gases in the atmosphere.[87] Following the 1987 Montreal Protocol, the use of CFCs and other halogenated and related gases accounting for 91% of the total radiative forcing of these compounds in the atmosphere was phased out based on strong evidence of damage to the ozone layer.[88] These compounds remain in the atmosphere but are in decline and with time will become a lesser contributor to total GHG radiative forcing.[89]

Greenhouse gases are the major contributors of total anthropogenic emissions; however, aerosols released by human activity are known to impact the environment. Aerosols are defined as fine particles or tiny droplets in suspension and consist of sulphur dioxide, black carbon, primary organic aerosols, biomass burning aerosols, secondary organic aerosols, nitrates, and dust particles.[87] Black carbon aerosols from fossil fuel and biomass combustion absorb radiation and contribute to surface warming. Sulphur dioxide aerosols reflect incoming solar radiation and thus reduce surface warming. The net effect of aerosols is a modest negative radiative forcing and a slight reduction in solar heating of the planet's surface. Most aerosols are short-lived and unevenly dispersed in the atmosphere, and atmospheric concentrations of aerosols tend to vary by region. Regulations designed to curb acid rain and improve air quality have reduced aerosol emissions in many regions of the world.

Large volcanic eruptions can inject a pulse of sulphur dioxide into the atmosphere sufficient to cause a short period of lower surface temperatures. The 1883 Krakatoa volcanic eruption destroyed an entire island in what is now Indonesia. Emissions of aerosols from this event reduced average summer temperatures in the Northern Hemisphere by up to 1.2°C for the next few years.[90] Sulphur aerosols from the eruption fell to the surface as acid rain. Carbon dioxide is also released during volcanic eruptions; however, the degree of volcanic activity during the Anthropocene is such that volcanos account for only 1% of total GHG emissions.[91]

Records of fossil fuel use, cement production (in process chemical release of CO_2)* and flaring of methane (combustion to CO_2) between 1750 and 2011, indicate that these activities have released 1,375 gigatonnes of carbon dioxide (1 Gt = 1 billion metric tonnes) into the atmosphere.[92] Over the past 260 years, the net effect of anthropogenic changes to global forest areas and other land use activities have released another 660 Gt.[92]

---

* During the production of cement and in some other manufacturing processes, chemical reactions produce carbon dioxide that is released to the atmosphere. The term "in process" is used to describe these emissions.

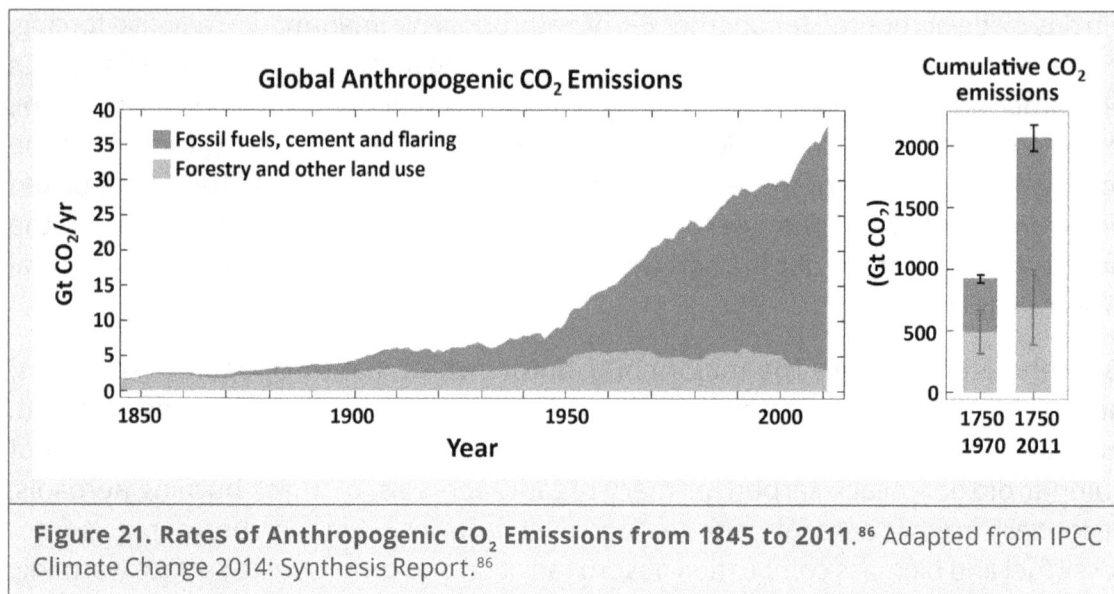

Figure 21. Rates of Anthropogenic CO_2 Emissions from 1845 to 2011.[86] Adapted from IPCC Climate Change 2014: Synthesis Report.[86]

In total, the activities of man have resulted in the release of over 2,000 Gt (2 trillion metric tonnes) of carbon dioxide. Some of this total has been withdrawn from the atmosphere by natural carbon sinks. Models of atmospheric CO_2 dissolving in oceans and withdrawn by land vegetation and other terrestrial sequestrations sites are consistent with the removal of about 60% of the total anthropogenic mass of CO_2 released into the atmosphere.[92] The net calculated theoretical accumulation of 880 Gt of atmospheric CO_2 corresponds directly to the 40% increase in measured concentrations of atmospheric CO_2 between 1750 and 2011. Further, there is no evidence of any imbalance between non-anthropogenic GHG release and sequestration over the past 260 years that could possibly account for such a massive accumulation in the environment. The 40% increase in atmospheric CO_2 levels relative to interglacial age concentrations stems from the activities of man. Data from 2002–2013 indicates that, during this period, 48% of anthropogenic CO_2 emissions accumulated in the atmosphere, which is consistent with the beginnings of a saturation of sequestration capacity and thus a declining buffering effect of ocean and land sinks.[93]

Table 1. Comparison of Measured Levels of Atmospheric CO_2 to Net Cumulative Emissions and from 1750 to 2011.			
	Atmospheric Measurements*		Net Anthropogenic CO_2 Added Since 1750 (Gt)**
Year	CO_2 (ppm)	CO_2 (Gt)	CO_2 Increase from 1750 (Gt)	
1750	280	2189		
2011	392	3065	876	880
*Ice core and direct measurements of concentrations converted to Gt of atmospheric CO_2. **IPCC Climate Change 2014 Synthesis Report.[86]			

In agreement with the projections of Savante Arrhenius and Arvid Högbom, accumulation of anthropogenic CO_2 in the atmosphere by 1900 was slightly above interglacial age levels and insufficient to cause surface warming.[86] By 1900, industrialized nations emitted 4.5 Gt/year of CO_2.[86] Ice core samples dated to 1900 contain 297 ppm CO_2, equalling a 11% increase over pre-industrial levels.[85]

Between 1900 and 1950, the rate of anthropogenic CO_2 emissions had doubled[86] and atmospheric CO_2 concentrations as measured in ice core samples had increased to 312 ppm.[85]

The onsets of the world population boom and post-war reconstruction beginning in 1947–48, coincide with an inflection point in the rate of anthropogenic emissions of GHGs. From 1948 to 2011, this rate increased 5-fold such that 60% of the total accumulation of anthropogenic CO_2 and other GHGs in the atmosphere has occurred in the last 40 years.[86]

Globally averaged combined land and ocean surface temperatures were stable between 1880 and 1930.[94] By 2012, surface temperature had increased by 0.85°C relative to the 1880–1930 average. Since 2012, global average land surface temperatures have spiked by another 0.2°C such that current temperatures are now over 1°C above pre-industrial temperatures. Changes in orbital (solar) forcing associated with Milankovitch cycles or any other natural forcing do not correlate with and could not have contributed to the increase in surface temperature over the past 90 years. There is no evidence of sufficient volcanic activity or any other non-anthropogenic source of atmospheric GHG accumulation, and normal variability in climate cannot account for the sustained warming trend. Between 1950 and 2010,

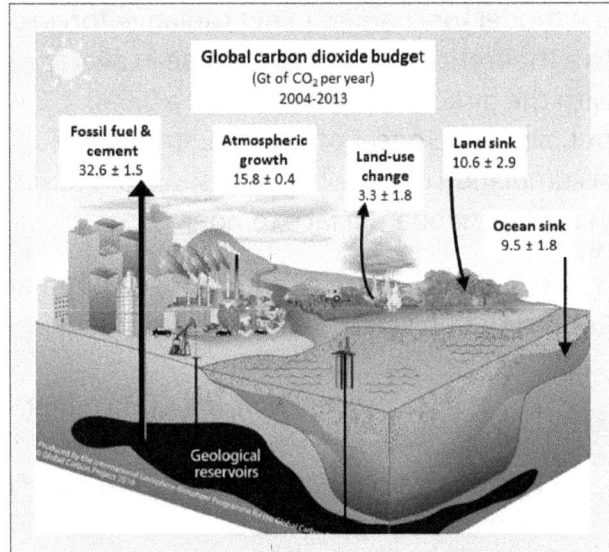

Figure 22. Overview of the Carbon Cycle. Emissions from industrial activities and land use change lead to an imbalance in the flows of carbon dioxide and continued accumulation in the atmosphere.[93]

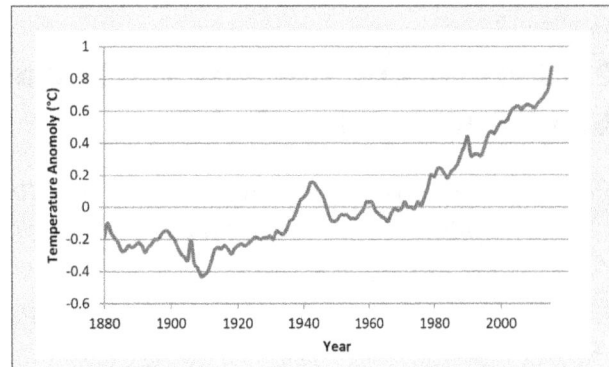

Figure 23. NASA Historical Average Surface Temperature Anomalies.[94] Baseline set to average surface temperatures between 1951 and 1980. Increase in surface temperature correlates with the radiative forcing of anthropogenic GHGs.

the model for changes in net radiative forcing of atmospheric components calculates to a theoretical 0.6–0.8°C increase in average surface temperatures, which correlates with the direct measurements of a 0.6–0.7°C warming over this same period.[86] Simply put, since the onset of the Industrial Revolution, the activities of man have led to an accumulation of greenhouse gases in the atmosphere causing at least a 0.85°C increase in average global surface temperatures.

By 1979, the evidence of anthropogenic climate change led the US National Research Council to publish the following statement:

> When it is assumed that the CO2 content of the atmosphere is doubled and statistical thermal equilibrium is achieved, the more realistic of the modeling efforts predict a global surface warming of between 2°C and 3.5°C, with greater increases at high latitudes.[95]

The N.R.C. in the US is the working body of the 3 national academies including the National Academy of Sciences founded in 1863 during the Lincoln administration for the express purpose of "providing objective advice to the nation on matters related to science and technology."

2.3 Agriculture, the Post World War Population Explosion, and Greenhouse Gas Emissions

> If the naysayers do manage to stop agricultural biotechnology, they might actually precipitate the famines and the crisis of global biodiversity they have been predicting for nearly 40 years.

Norman Borlaug — Father of the Green Revolution

By 1900, the Industrial Revolution had pushed world population to 1.6 billion people. At the time, limitations in the supply of the nitrogen containing mineral fertilizer restricted the potentiation for further food production and thus population growth. The atmosphere represented a potentially bountiful source of nitrogen; however, no practical method existed to convert gaseous atmospheric nitrogen to a form suitable for application as a fertilizer.

The Haber-Bosch process for the production of ammonia fertilizer ranks among the most important and impactful inventions affecting humanity and Earth. In 1909, Fritz Haber, a German chemist of Jewish decent, invented a bench top method based on a nickel catalyst to react atmospheric nitrogen with hydrogen gas to produce ammonia.[96] The reaction required extreme temperatures and pressures, and scale up of the bench top chemistry presented a considerable engineering challenge. Haber sold the rights to his invention to the German chemical company BASF and Carl Bosch developed a practical industrial scale process based on Haber's chemistry. In 1913, BASF began production of ammonia using what became known as the Haber-Bosch process.[96]

Ammonia fertilizer from the Haber-Bosch process is used in the production of food for 50% of the world's population. Of the current population of 7 billion people, an estimated 2.72 billion (39% of us) would not exist in the absence of enhanced agricultural output following fertilization of crop land using ammonia produced by the Haber-Bosch process.[97] The magnitude of the Haber-Bosch process is such that 450 million tonnes of fertilizer are produced annually accounting for 3–5% of annual global consumption of natural gas and 1–2% of world energy use.[96,98] The application of Haber-Bosch produced chemical fertilizers in combination with pesticides and herbicides has resulted in a 4-fold increase in world food production.[98] Fritz Haber* and Carl Bosch received the Nobel Prize in chemistry for inventing a process that, to this day, continues to feed the world.[98] The process has created a fascinating nitrogen cycle whereby 80% of the nitrogen in a human body fed by ammonia fertilized crops was, at one time, fixed from the atmosphere by the Haber-Bosch process.[98]

By 1950, the Haber-Bosch process had been implemented in developed nations. The ready availability of nitrogen fertilizer, along with the development of high yield seed varieties, herbicides, pesticides, and the modernization of agricultural practices dramatically increased food production capacity. Science-based advances in agriculture technology and practices were transferred to developing nations, most notably Mexico, Brazil, India, and Pakistan during the Green Revolution between the 1930s and late 1970s.[99]

In the 1960s, India adopted IR8, a new breed of high yield rice developed by the International Rice Research Institute.[100] When grown with fertilizers under optimum conditions, the yields of the so-called miracle rice were 10-fold greater than traditional rice.[100] India's Green Revolution allowed this densely populated continent to achieve self-sufficiency in food production and to eventually become the world's largest exporter of rice.[101]

In Brazil, vast areas of acidic non-productive soils were neutralized and transformed to productive agricultural lands by application of massive quantities of lime. This soil transformation project began in the 1960s and continued to the 1990s, and, eventually, Brazil became the world's second largest producer of soybeans.[102] Soybean production led to the development of a substantial animal feed industry and Brazil is second to the United States in total agricultural exports.[103]

---

* During the First World War, Germany was dependent on ammonia for the production of explosives. In the absence of the Haber-Bosch process, it is doubtful the war would have continued until 1919.[96] From the onset of war, Fritz Haber worked for the German government as the Head of the Chemistry Section of the Ministry of War and led the team that invented mustard gas. Haber left Germany in 1933 as part of the exodus of scientists and academics of Jewish heritage.[96]

In 1944, Norman Borlaug, accepted a position to head the Cooperative Wheat Research and Production Program in Mexico.[104] While working in Mexico, Borlaug developed high-yield, disease-resistant varieties of wheat. Borlaug became known as "The Father of the Green Revolution" and "The Man Who Saved a Billion Lives." Adoption of Borlaug's spring wheat varieties in Mexico eventually resulted in an 8-fold increase in per acre yields, and the new high yield varieties of wheat were subsequently adopted by India and Pakistan.[104] Borlaug became an effective promotor of the Green Revolution and is one of only 7 people to have been honoured with the Nobel Peace Prize, the Presidential Medal of Freedom, and the Congressional Gold Medal.[104]

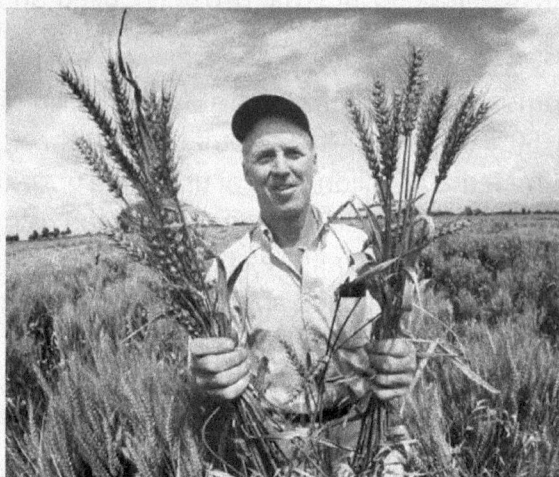

Figure 24. Norman Borlaug. The Father of the Green Revolution.

Since the 1950s, the world's population has increased by 4.8 billion.[105] The post-Second World War population explosion could not have happened without advances in agriculture. It is difficult to put an exact number on the impact of science-based advances in agriculture on the world population. Advances in medicine and sanitation have also been important contributors to the world population explosion. However, a study of population growth in India provides some indication of how the Green Revolution facilitated the population explosion in that country. Prior to 1960, agriculture in India had not evolved beyond traditional practices. The population at that time was 449 million[106] and the country was on the verge of mass starvation.[99] The Green Revolution started in India in the early 1960s and, as of 2015, the population has grown to 1.31 billion.[106] Based on these numbers, the Green Revolution fed an additional 861 million people (66% of the total current population) above the capacity for food production using traditional agricultural practices. Interestingly, the proportional increase in world population since 1950 is similar to the population increase in India from 1960 to 2015. It is reasonable to suggest that at least 60% of the current world population (4.4 billion

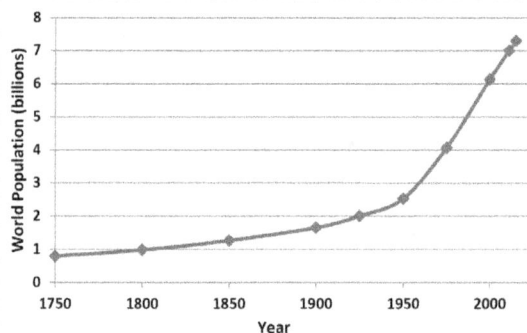

Figure 25. Global Population Growth since 1750.[105] Widespread adoption of Green Revolution agricultural practices facilitated the post war population explosion.

citizens of Earth) would not exist in the absence of development of chemical fertilizers, high yield disease resistant seed varieties, pesticides, herbicides, irrigation and other advances in agriculture, medicine, and sanitation.

Modern high intensity agriculture comes with considerable direct emissions of GHGs and indirect energy demands. The Intergovernmental Panel on Climate Change (IPCC) estimates that 56% of total anthropogenic emissions of methane and nitrous oxide come from agriculture.[107] Methane output from intestinal fermentation in farm animals and nitrous oxide plus methane outgassing from fertilizer application are the major contributors. On the basis of CO_2 equivalents, these emissions account for 10–12% of total anthropogenic GHGs in the atmosphere.[107] In addition, the energy demands of the agriculture sector to produce fertilizers, operate machinery and transport products is considerable.

By far, the most significant impact of modern agriculture on GHG emissions stems from an expanding world population. Overall, the global increase in annual rates of CO_2 and total GHG emissions since 1950 mirrors the growth in population. However, the impact of population growth varies between countries and is dependent on the per capita intensity of emissions. In 1950, GHG emissions from China and India were negligible. Between 1950 and 2013, the population of these 2 countries grew by 1.74 billion accounting for 37% of the global population explosion.[105] China's population growth coincided with industrialization such that the per capita rate of emissions exceeded the world average in 2013. The combination of a population boom and industrialization were such that China accounted for 28.6% of global CO_2 emissions in 2013.[108] India lacks the degree of industrialization typical of the US and Western Europe or China and thus has a relatively low per capita rate of emissions. The population explosion in India following the Green Revolution has had a modest effect on global emissions. The US is the second largest emitter with a per capita emissions rate among the highest on Earth. From 1950 to 2013, the population of the US increased by 157 million, and given the intensity of emissions from the US, this increase in population was a significant contributor to the massive growth in global emissions since 1950.

2.4 The Golden Age of Capitalism

The post-war population explosion coincided with a widespread economic boom among developed nations often referred to as the golden age of capitalism. At the end of World War II, much of the infrastructure of Europe and Japan lay in ruins, and the war effort had exhausted the treasuries of all major participating nations with the exception of the United States and Canada. During the 12-month period from the end of the war, the US shipped 35 trillion calories of food aid (equivalent to 1/6 of total food production) to Europe and Japan.[109] By 1947, the slow rate of post-war recovery in Europe, in combination with the threat of expansion of communism and the relative strength of the US economy, led the Truman administration to develop an ambitious aid plan designed to assist and direct reconstruction and economic recovery in Europe. The Marshall Plan operated from 1947 to 1951 and provided $12 billion in economic support to European nations.[110] The objectives of the plan were to accelerate reconstruction and to aid in establishing modern efficient industries operating without substantive trade barriers.

The rebuilding of the West German economy was integral to the overall vision of a new vigorous Europe with free market exchange of raw materials and manufactured goods between nations. Through the Marshall Plan, the US, as the dominant economic and military power emerging from the War, could influence the rebuilding of Europe, limit the spread of communism, and open up European export markets for its own expanding industries. The Soviet Union was included in the offer of assistance. Predictably, Stalin refused and set up barriers to exclude all satellite Eastern bloc nations from participation. In Japan, a similar program of US economic aid under the guise of Supreme Commander for the Allied Powers (SCAP) was established.[111] As was the case in Europe, US motivations in Japan were based on directing post-war economic recovery away from communism, toward a US style democracy and free market enterprise.

Without doubt, the Marshall Plan and other economic aid provided by the US to war torn countries accelerated reconstruction and helped to establish conditions that fostered the post-war economic boom. However, planning and execution of economic recovery came from within the individual countries and by 1951 these countries had transitioned to full autonomy. Post-war "economic miracles" occurred in Japan, France, Germany, Sweden, Italy, and Greece; however, economic booms were widespread and included the UK, the US, Canada, Australia, and the Soviet Union.

Post-war economic growth in the US increased the real income of all social classes. Automobile and home ownership and a higher standard of living characterized an emerging middle class. Sprawling, low density suburbs were erected to house the growing populace. By the 1950s, the middle class had the resources to spend on vacations, leisure activities and a range of new consumer goods. Infrastructure for transportation and the generation and distribution of electricity were constructed to support growing suburban populations and overall industrial expansion. Wartime innovations such as standardization of parts, mechanized materials handling, and the introduction of automation improved the productivity of industry.

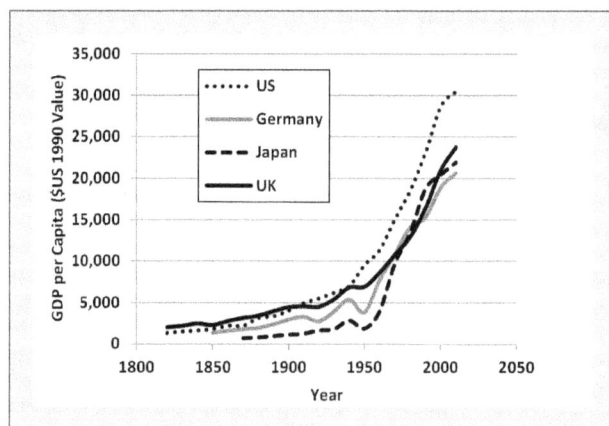

Figure 26. GDP per Capita Growth of Selected Advanced Economies.[112] Post-war economic booms were common among developed countries.

During the Anthropocene, inflation adjusted GDP per capita for the

US, the UK, Japan, and Germany followed the same post-WW II inflection point as per the world population explosion.[112] The post-war economic boom and burgeoning populations around the world created a steadily growing demand for energy. Economists define the post-war golden age of capitalism as ending in the early 1970s; however, on a worldwide basis, the demand for energy continued to increase on an annual basis that extends to present times.[113] Since 1950, a sustained population explosion and increased per capita GDP has driven fossil fuel combustion, which, along with the adoption of intensive

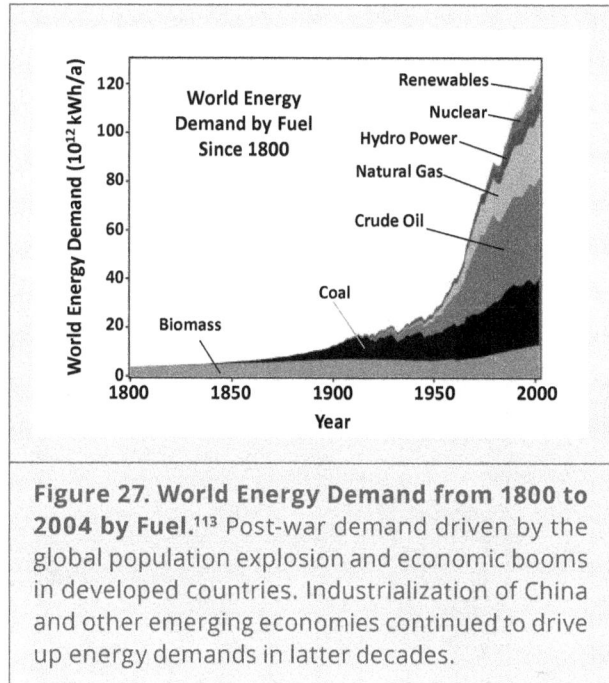

Figure 27. World Energy Demand from 1800 to 2004 by Fuel.[113] Post-war demand driven by the global population explosion and economic booms in developed countries. Industrialization of China and other emerging economies continued to drive up energy demands in latter decades.

agriculture and changes in forestry and land use, has resulted in a yearly increase in the rate of global GHG emissions.

A model whereby changes in population and GDP per capita are the only factors defining CO_2 emissions is overly simplistic. If the energy intensity of GDP changes, this will either increase or diminish the effect of a change in GDP on emissions. A growing economy that consumed less energy would be an example whereby a GDP increase over time would have less of an effect on GHG emissions. Similarly, a shift to low carbon non-fossil fuel sources of energy would be an example whereby a growing economy would have less of an effect on GHG emissions.

The IPCC has summarized studies of the contribution of all 4 factors (population, GDP per capita, energy intensity of GDP, and carbon intensity of energy) to changes in annual CO_2 emissions over 10-year intervals beginning in 1970. World population increase and increased GDP per capita were approximately equal drivers of increased GHG emissions over the period from 1970 to 2000.[86] The energy intensity of GDP has dropped over each successive decade between 1970 and 2010 indicating that economic growth over time is less coupled to energy consumption and thus the effect of a GDP per capita increase on CO_2 emissions is partially ameliorated. Changes in the carbon intensity of energy had a small effect on reducing emissions from 1970 to 2000 indicating a slight overall shift toward lower emissions per unit of energy. Despite a growing awareness and attempts to control emissions since 1990, the rate of carbon dioxide emissions was highest over the decade from 2000 to 2010. This was largely driven by a substantial increase in GDP per capita on a worldwide basis. The previous slight trend toward use of lower emissions energy sources reversed from 2000–2010 such that the carbon intensity of energy increased and became a net contributor to

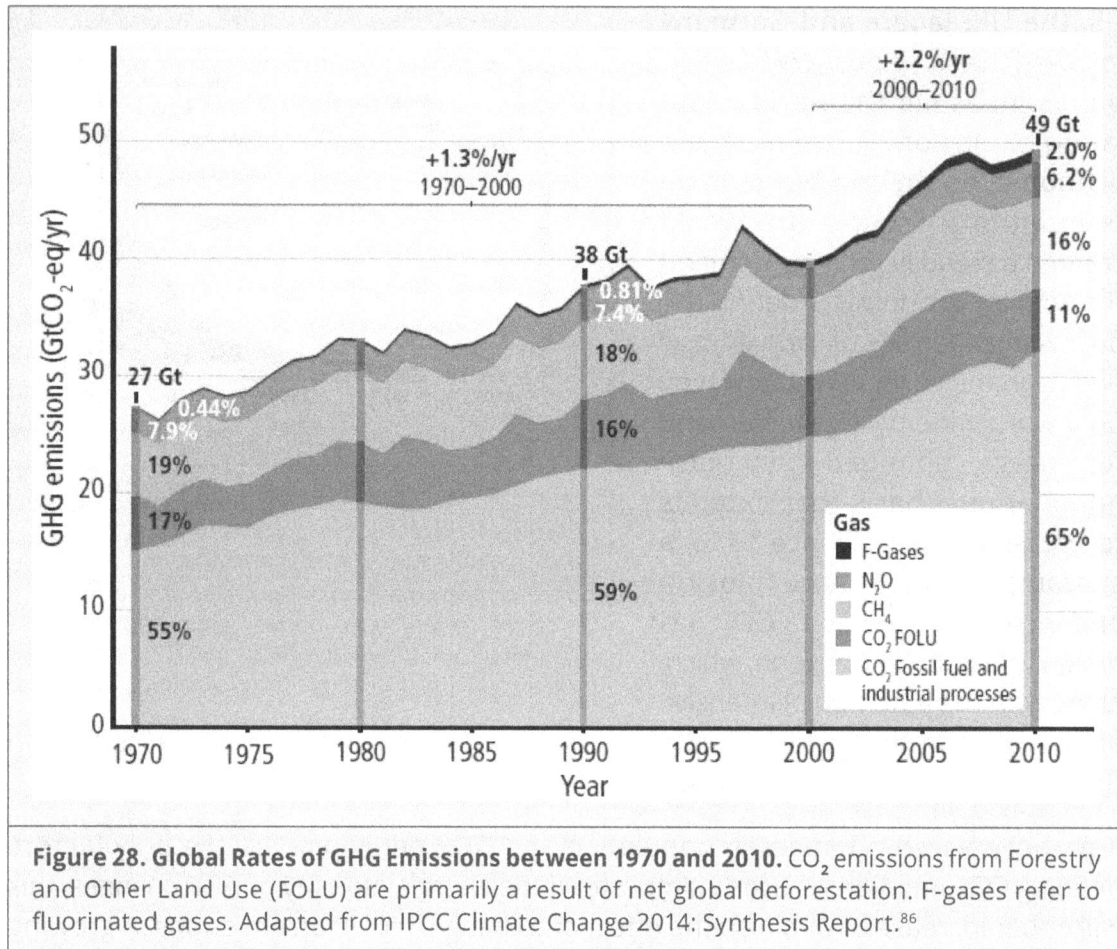

Figure 28. Global Rates of GHG Emissions between 1970 and 2010. CO$_2$ emissions from Forestry and Other Land Use (FOLU) are primarily a result of net global deforestation. F-gases refer to fluorinated gases. Adapted from IPCC Climate Change 2014: Synthesis Report.[86]

increased GHG emissions. Continued economic growth in China and other advanced and emerging countries coupled with the use of high emissions fossil fuels to meet growing energy demands were the driving factors behind the high rates of emissions from 2000 to 2010.

Data presented in the IPCC reports on total GHG emissions are based on the 100-year Global Warming Potential (GWP) of each gas. This calculation accounts for the radiative forcing potency of the gas and the chemical lifespan in the atmosphere. Emissions are then expressed as Gt of carbon dioxide equivalents such that the relative contribution of each gas to long-term surface warming is clearly evident.[86] As of 2010, carbon dioxide release from energy use and industrial processes accounted for 65% of total anthropogenic GHG emissions. Net CO$_2$ release from forestry practices and other land use (FOLU) contributed another 11% such that total CO$_2$ release accounted for 76% of global GHG emissions. Forestry emissions are primarily a result of the net loss of forested areas as the global rate of deforestation and forest degradation exceeds the combination of reforestation (re-establishing forested areas) and afforestation (establishing forests in new areas). Anthropogenic methane release contributed 16% of the total GHG emissions when expressed as CO$_2$ equivalents. Half of global methane release came from the

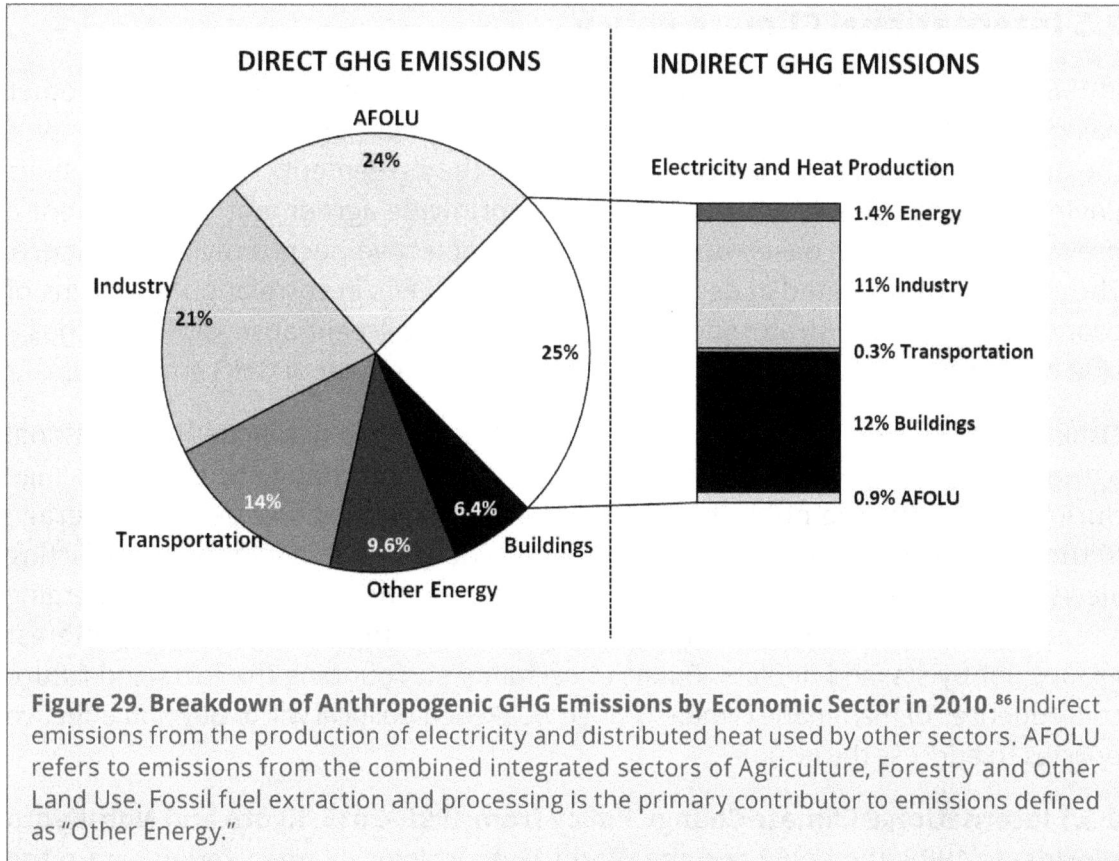

DIRECT GHG EMISSIONS

INDIRECT GHG EMISSIONS

AFOLU
24%

Industry
21%

25%

Transportation
14%

9.6%

6.4%

Other Energy

Buildings

Electricity and Heat Production

1.4% Energy

11% Industry

0.3% Transportation

12% Buildings

0.9% AFOLU

Figure 29. Breakdown of Anthropogenic GHG Emissions by Economic Sector in 2010.[86] Indirect emissions from the production of electricity and distributed heat used by other sectors. AFOLU refers to emissions from the combined integrated sectors of Agriculture, Forestry and Other Land Use. Fossil fuel extraction and processing is the primary contributor to emissions defined as "Other Energy."

agriculture sub-sector. Nitrous oxide, primarily from agriculture, contributed 6.2%, and the remaining 2% of emissions came from fluorinated hydrocarbons.

During the period from 1970 to 2010, the annual rate of total anthropogenic global emissions of GHGs increased by 81%.[86] CO_2 release from the combustion of fossil fuels accounted for 77% of the increase in annual GHG emissions over this 40-year period.[86]

The IPCC divides anthropogenic GHG emissions by economic sector.[86] In 2010, production of electricity accounted for 25% of global emissions with the bulk of electricity demand divided equally between the industry and buildings sectors. Agriculture, Forestry and Other Land Use (AFOLU) accounted for another 24% of emissions. Within the AFOLU sector, 11% of global GHG release came from forestry and other land use, and 13% originated from agriculture. Industry contributed 21% of direct emissions or 32% of total emissions, when indirect GHG release from the production of electricity used by industry, was included in the total. GHGs emitted by the buildings sector accounted for 6.4% of direct emissions or 18.4% of direct plus indirect emissions. The global transportation sector contributed 14% of total emissions in 2010 with considerable variation between countries dependent upon the state of economic development. Emissions from the extraction and processing of fossil fuel primary energy sources accounted for most of the remaining 9.6% of global emissions that are classified as other energy.

2.5 International Climate Policy

In 1987, the Montreal Protocol to phase out the use of ozone depleting CFCs and other halogens and chemically related compounds was signed.[88] By 2015, 196 nations plus the entirety of the European Union had ratified the agreement. The United Nations Environment Programme (UNEP) achieved worldwide agreement to implement a preventative program based on a solid body of objective peered reviewed science. The protocol has resulted in declining atmospheric CFCs in correlation with signs of ozone recovery in the stratosphere.[114] CFCs function as greenhouse gases and thus a side benefit of the Montreal Protocol has been an abatement of GHG emissions.

Other than the side benefit of the Montreal Protocol, no significant international agreements were developed to address the issue of anthropogenic climate change during the 1980s. The public had little understanding of the issue, and coverage in the popular media was often presented as a debate with a variety of conflicting views, many of which were not based on reputable science. The facts of an emerging scientific consensus were frequently challenged without foundation by special interest groups, lobbyists, and deniers. Public uncertainty surrounding the facts and future consequences of man-made climate change weakened political will to develop effective policies to address the issue.

2.5.1 International Climate Change Policy from 1990–2015: Kyoto and Non-Kyoto Worlds. In 1988, the UNEP and the World Meteorological Organization set up the Intergovernmental Panel on Climate Change (IPCC).[115] The IPCC does not conduct science but, through a series of expert working committees, assesses the vast body of scientific literature and other information on the subject of climate change and twice per decade produces objective, impartial and comprehensive reports. Each Assessment Report consists of three full reports covering the physical science basis of climate change (working group 1), the impacts, adaptations and vulnerabilities of climate change (working group 2), and the mitigation of climate change (working group 3).[116] In addition, a Synthesis Report summarizing the full reports is included in the overall assessment. The IPCC also publishes special reports and technical papers on selected subjects to supplement the Assessment Reports. The Fifth Assessment Report was released in 2014 with thousands of scientists and other experts contributing to the 5,000-page package. The IPCC is the internationally accepted authority on climate change, and in 2007 the organization was awarded the Nobel Peace Prize for "their efforts to build up and disseminate greater knowledge about man-made climate change and to lay the foundations for the measures that are needed to counteract such change."[117]

One of the core objectives of the IPCC is to provide policy makers with the required synthesis of factual information and conclusions on the causes and impacts of climate change, along with pathways of adaptations and mitigations of future climate change.

The first IPCC report was published in 1990 and provided the supporting documentation for the United Nations Framework Convention on Climate Change (UNFCCC) that was negotiated at the Earth Summit in Rio de Janerio in 1992 and enacted in 1994.[118] The agreement has worldwide support with ratification by 196 nations. The UNFCCC commits the nations of Earth to the goal of "stabilization of greenhouse gas concentrations in the atmosphere at a level that would prevent dangerous anthropogenic interference with the climate system." The agreement does not present specific targets or binding mechanisms to control future emissions. However, the framework acknowledges the reality of dangerous anthropogenic climate change and the need for the development of effective climate change policies. The UNFCCC commits the Conference of Parties (COP) to the agreement to meet on an annual basis to develop and implement protocols and instruments to achieve the overarching objective to limit future anthropogenic interference with the climate of Earth.

In 1997, the Kyoto Protocol was negotiated as an extension to the UNFCCC.[119] By this time, the IPCC had developed a set of projections with global surface temperature increasing from 1.1 to 6.4°C by the year 2100 dependent upon various options for future emissions.[86] Based on IPCC projections, the Kyoto Protocol attempted to set legally binding limits for GHG emissions.

The Kyoto Protocol came into effect in 2005 as a complex set of sub-agreements that apply to individual countries or groups of countries. The protocol was divided into a first reporting period (2008–2012) and a second commitment period from 2012 to 2020. Universal agreement was not achieved, and the world became divided into nations and regions that agreed to binding emissions targets for the first and second reporting periods (all states of the European Union, Australia, and several smaller advanced and emerging economies); nations that agreed to binding targets for the first commitment period only (Japan, New Zealand, and Russia); and the remainder of the world consisting of countries that either did not participate or failed to ratify the accord (the US).[119] Canada does not fit into any of these categories, in that Canada agreed to binding targets for the first commitment period, ratified the accord but then withdrew completely in 2012.[119] The level of ambition to control emissions varied between participant countries. The EU set targets to achieve an 8% reduction in emissions over the first reporting period and a subsequent target to reduce emissions by 20% over the second reporting period relative to emissions on record for 1990. In contrast, Australia committed to limit emissions to an 8% increase above the 1990 reference rate and Russia committed to not allow emissions over the first reporting period to exceed 1990 levels.

The follow-up Copenhagen Accord emerged from the 15th session of the Conference of Parties in 2009 and endorsed a continuation of the Kyoto Protocol.[120] Included in a lengthy list of statements, the accord recognized that the future surface temperature increase should be held below 2°C, and that deep cuts to global emissions should be implemented as soon as possible.[120] However, no specific emissions targets or binding mechanisms beyond an extension of the Kyoto Protocol were agreed upon and the meeting accomplished little.

The 16th COP meeting took place in Cancun, Mexico in 2010 and led to a follow-up series of pledges by individual countries to reduce emissions over the second reporting period of the Kyoto Protocol ending in 2020.[121,122] Cancun targets can be summarized as follows:

- **Australia.** Pledged a 25% reduction in emissions by the year 2020 relative to emissions on record for the year 2000 contingent upon the world concluding an ambitious agreement designed to stabilize atmospheric CO_2 equivalents at 450 ppm or lower. A second pledge of a 15% reduction was conditional to a less ambitious global agreement to curtail emissions. Finally, Australia provided an unconditional pledge of a 5% reduction in emissions.
- **Canada.** Set a target to achieve a 17% reduction in emissions by 2020 relative to 2005 levels. Final economy-wide emissions reduction to be aligned with that of the United States.
- **European Union.** The EU pledged to extend Kyoto targets and achieve a 30% reduction in emissions by 2020 relative to 1990. The EU formally stated that if surface warming is to be held to less than 2°C above pre-industrial temperatures, year 2050 global emissions must be cut by 50%, and advanced economies as a group should target a 25–40% reduction by 2020 and an 80–90% cut in emissions by 2050.
- **Japan.** Submitted a target of a 25% reduction in emissions by 2020 relative to a 1990 baseline.
- **Russia.** Pledged a 15–25% cut in emissions by 2020 relative to 1990.
- **United States.** Targeted an emissions reduction of 17% by 2020 using emissions on record for 2005 as the reference. The US also stated an internal goal to enact legislation that would lead to a 30% reduction in emissions by 2025, a 42% reduction by 2030, and an 83% emissions reduction by 2050.
- **China.** Pledged to reduce the emissions intensity of economic growth (emissions per unit of GDP) by 40–45% by the year 2020 relative to 2005.
- **India.** Used a similar formula to China and pledged to reduce the year 2020 emissions intensity of economic growth by 20%.

The Cancun pledges are a step forward from the Kyoto Protocol in that China, the United States and India were engaged. Further, the beginnings of a process of nationally defined contributions to global efforts to combat climate change emerged with the submission of Cancun pledges. This approach was to become the foundation of the 2015 Paris Agreement on Climate Change negotiated at COP21.

A subsequent meeting of the UNFCCC established the Durban platform where 195 nations agreed to negotiate a new international climate treaty by 2015.[123]

Annual rates of global GHG emissions increased by 40% between 1990 and 2012, and this fact can lead to the conclusion that the Kyoto Protocol was a failure.[124] However, declaring Kyoto to be an outright failure is overly simplistic and does not acknowledge the achievements of Kyoto participating countries that agreed to ambitious targets to curtail emissions. By 2012, the EU and other Kyoto participates had cut emissions by 13% and thus grossly exceeded the collective first reporting period target of a 5% reduction.[124] The combined member countries of the EU achieved nearly a 20% reduction in emissions relative to 1990.[124] The rate of emissions from the group of countries that originally agree to participate but either failed to ratify, withdrew or ratified without binding emissions targets, increased by 9% between 1990 and 2012. However, the global increase in GHG emissions between 1990 and 2012 came almost entirely from China and other countries that did not agree to participate in the Kyoto Protocol during the first reporting period. Emissions from these countries increased by 129% between 1990 and 2012 and China accounted for half of this increase in emissions.[124]

As of 2012, 22% of global GHG emissions came from China with the US contributing another 12%. A full 67% of global emissions came from the combined member states of the EU and the other top 8 emitting countries.[119]

By 1974, membership of the Organization for Economic Cooperation and Development (OECD) consisted of advanced economies in Europe plus Canada, the US, Iceland, Japan, Australia, and New Zealand.[125] The inflection point for increased GDP per capita

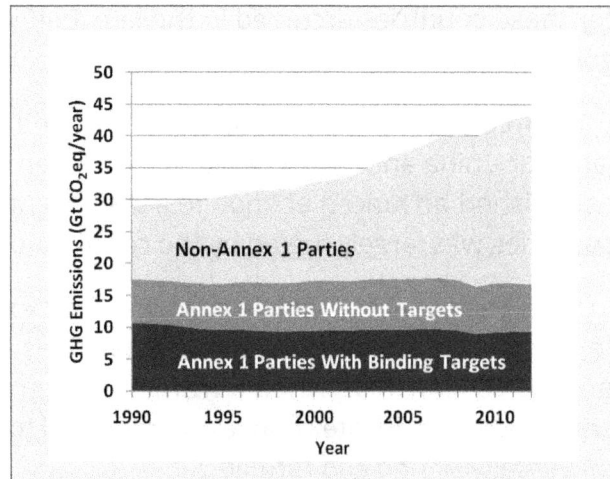

Figure 30. Global GHG Emissions from 1990 to the End of the First Kyoto Reporting Period in 2012.[124] Annex 1 Parties With Binding Targets consist of United Nations Framework Convention on Climate Change (UNFCCC) countries that fully participated in the Kyoto Protocol. Annex 1 Parties Without Targets originally committed to the Kyoto Protocol but either failed to ratify (the US), ratified and then withdrew (Canada) or ratified without legally binding targets for future emissions. Non-Annex 1 Parties did not participate over the first reporting period.

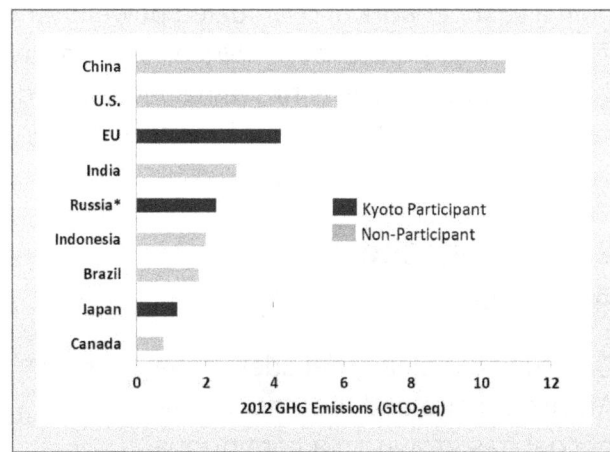

Figure 31. Year 2012 Global Leaders in GHG Emissions.[124] These countries accounted for 67% of global emissions. *Russia was a full participant in the Kyoto Protocol; however, the level of ambition to control emissions was relatively weak compared to the EU or Japan.

for these countries occurred in the late 1940s and early 1950s with the onset of the golden age of capitalism.

Economic development leading to increased GDP per capita occurred 40–50 years later for China and South Korea and other emerging economies. As was the case for established advanced economies, the recent economic growth in China and other countries was largely fueled by the combustion of fossil fuels.

In 1948, the science of climate change did not exist, and there was no counter-indication for extraction and combustion of fossil fuels to support economic growth. This was not the case in 1992 when a framework agreement acknowledging the danger of anthropogenic climate change and the need to reduce greenhouse gas emissions was universally signed and ratified.

The shortcomings of the Kyoto Protocol up to 2012 stem from the inability of the nations of the world to reach consensus as to an effective pathway to adequately control future emissions. The world divided into Kyoto participants with ambitious targets and other countries that either did not participate or participated with a low level of ambition to combat climate change. Nothing was accomplished to arrive at a fair and equitable solution that would facilitate economic growth in developing and emerging economies that was uncoupled from GHG emissions. China, India and most other countries in Asia, Africa, and South and Central America did not participate over the first reporting period to 2012. In China, coal use has increased by 3.5-fold since 1990 as the nation built a vast network of coal-fueled power stations. By 2014, coal combustion provided 77% of China's electricity.[126] Since 1990, for most of the global population, economic growth and productivity has remained coupled to energy consumption and a continued reliance on fossil fuels to meet energy demands.

2.6 Fueling the Anthropocene

Coal fueled the early decades of the Anthropocene and the annual use of coal has grown on a continuous basis since 1950. As the demand for coal increased, total fossil fuel supply expanded further with heavy demands for oil and natural gas. In comparison, hydro and nuclear energy are lesser contributors to the overall global Total Primary Energy Supply (TPES). However, these energy sources are signifi-cant contributors to the production

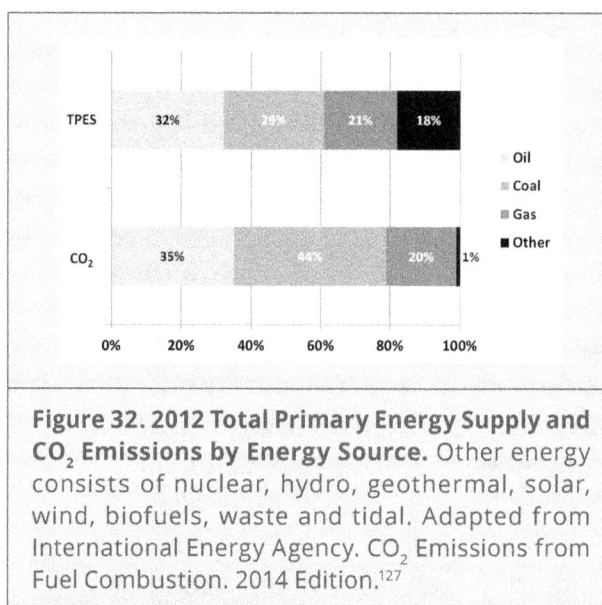

Figure 32. 2012 Total Primary Energy Supply and CO$_2$ Emissions by Energy Source. Other energy consists of nuclear, hydro, geothermal, solar, wind, biofuels, waste and tidal. Adapted from International Energy Agency. CO$_2$ Emissions from Fuel Combustion. 2014 Edition.[127]

of electricity. The use of biomass for domestic heating and cooking pre-dates the Anthropocene and continues in less developed regions of the world. Recently, sources of energy for the production of electricity have further diversified to include biomass and other other renewables in the form of wind, solar and geothermal.

In 2012, fossil fuels accounted for 82% of the global TPES. The combination of nuclear, hydro, biomass, and other renewables comprised the remaining 18%.[127] Carbon dioxide emissions from coal combustion on a unit of energy basis are almost double the emissions from natural gas.[128] While coal contributed 29% of TPES, combustion of coal accounted for 44% of total emissions from all sources of energy use.[127] Non-fossil fuel energy sources accounted for only 1% of total energy related CO_2 emissions.[127]

2.6.1 Production of Electricity.

Over the past 4 decades, the growth in global population, along with electrification and economic development in China and other countries, has led to an almost 4-fold increase in the production of electricity.[129] Between 1973 and 2014, coal-fired power plants produced about 40% of world's electricity supply. The use of oil as an energy source to generate electricity has declined over the past 40 years and oil has been largely replaced by a combination of natural gas and nuclear energy.[129] Hydropower continues to provide 15–20% of the global total supply of electricity. In recent years, renewable energy sources such as wind, solar, geothermal and biomass have become increasing significant and in 2014 contributed to over 6% of the total power production energy mix.[129] Given the continued heavy dependence on fossil fuels, and specifically coal, for the generation of electricity, this subsector of total energy supply is responsible for about 25% of global GHG emissions.

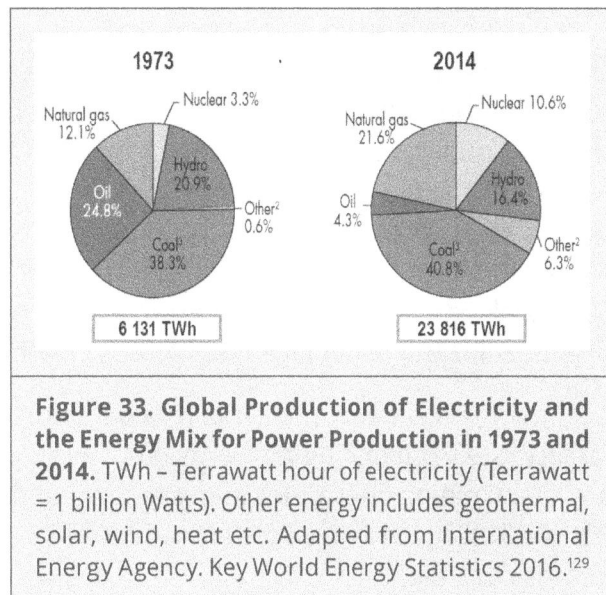

Figure 33. Global Production of Electricity and the Energy Mix for Power Production in 1973 and 2014. TWh – Terrawatt hour of electricity (Terrawatt = 1 billion Watts). Other energy includes geothermal, solar, wind, heat etc. Adapted from International Energy Agency. Key World Energy Statistics 2016.[129]

2.6.2 Coal.
Coal is the most abundant, lowest cost fossil fuel, and comes with the highest intensity of GHG emissions per unit of energy. There are approximately 900 billion tonnes of proven reserves of coal that would last for about 114 years at the current rate of use.[130] Power production accounts for 68% of worldwide coal demand with almost all of the remaining demand coming from thermally intensive industries such as steel manufacturing.[129]

In addition to carbon dioxide, the combustion of coal releases sulphur dioxide (SO_2), nitrogenous oxides (NO_x), and particulate matter (ash) containing mercury and other heavy metals. SO_2 forms aerosol particles in the atmosphere that return to the surface as acid rain. The environmental damage of acid rain is well documented, as are the damaging effects of particulate emissions on human health. In the US, pollution from fossil fuel use, and specifically coal combustion, led to the 1963 Clean Air Act. Over the years, the act has undergone a series of amendments to further strengthen emissions standards.[131] Implementation of measures such as the Clean Air Act has improved air quality in urban centers and other areas with large coal-fired industries and power plants. In 2010, the Clean Air Act prevented an estimated 160,000 early deaths from particulate emissions in the US.[132] However; stringent emissions standards do not eliminate the damaging effects of coal combustion. In the US, particulate emissions from coal plants cause an estimated 13,200 early deaths per year.[133] The total monetized cost of adverse health impacts from coal plants in the US is estimated at $100 billion per year.[133]

Emissions standards vary around the globe and the WHO estimates that air pollution results in the early death of 7 million people per year.[134] The WHO states that 88% of these air quality related early deaths occur in South-East Asia and the Western Pacific regions of the globe.[135] Outdoor air pollution is the cause of 3.7 million of these early deaths and coal combustion accounts for at least half of this mortality. The remaining 3.3 million early deaths per year are attributed to indoor air quality issues, primarily from cooking with coal and biomass.[134] Stroke, heart disease, lung cancer and respiratory diseases are the leading causes of early death from air pollution.

Coal mining has been and remains a dangerous occupation. In the US, since 1990, the fatality rate from mining accidents has averaged 37 deaths per year.[136] Safety standards are lax in mining operations in many parts of the world. In China, over the period from 2000–2013, coal mining accidents claimed 58,023 lives equal to a fatality rate of 4,144 deaths per year.[137]

Globally, the use of coal is the cause of 161 early deaths per TWh (terawatt hours) of energy content[138]. This mortality rate varies with environmental and safety standards. In the US, early deaths per unit of coal energy are 10% of the global average. In 2001, after reviewing the results of a 10-year study, the European Commission concluded that the costs of the damage to the environment and human health from fossil fuel combustion consumes an estimated 1–2% of GDP in the EU.[139] If the cost of these GHG-independent externalities were included, the price of electricity from coal-fired power plants would double.[139]

Table 2. Early Mortality per Terawatt hour of Electricity Produced by Various Energy Sources	
Energy Source	**Mortality[138]** (Early Deaths/TWhr)
Coal (global)	161
Coal (China)	280
Coal (U.S.)	15
Oil	36
Natural Gas	4
Biomass	24
Solar	0.44
Wind	0.15
Hydro (with Banqiao disaster)*	1.4
Nuclear (with Chernobyl disaster)	0.04
*1975 catastrophic failure of Banqiao Reservoir Dam in China resulting in 171,000 deaths.	

In addition to pollutants released to the atmosphere during combustion, the mining of coal and disposal of solid waste outputs can cause serious environmental damage. Vast quantities of fly ash from coal fueled power plants are committed to landfills and wet storage ponds, and many of these sites are classified as potentially hazardous to local communities.[140] Surface mining severely alters landscapes and requires large quantities of water for dust suppression. Drainage of contaminated mine water from poorly managed operations and acid rain resulting from high emissions coal plants can pollute local ground water and streams. Many of the deleterious effects of the coal industry on the environment have been mitigated with the installation of emissions control systems and improved mining practices. However, these improvements are not universally practiced and have no effect on GHG emissions from coal combustion.

In comparison to other energy sources, coal is by far the most damaging to human health and the environment, and has the highest intensity of GHG emissions per unit of energy. As will be discussed in subsequent chapters, emerging technologies to capture CO_2 emissions from coal combustion have the potential to partially abate emissions from coal-fired power plants and other industrial operations.

2.6.3 Oil. World demand for oil increased by 67% between 1973 and 2014.[129] Over this period, the use of oil for the production of electricity, and as a thermal fuel for home heating and industrial processes, has declined. This drop in demand was offset by increased consumption of gasoline, diesel and jet fuel products and, to a lesser extent, by increased use as a feedstock for the manufacture of plastics and other non-energy products. In 2014, the transportation sector accounted for 64% of oil use.[129]

The International Energy Agency (IEA) estimates that, as of 2011, there were 1.7 trillion barrels of proven conventional oil reserves.[141] Another 1.4 trillion barrels of recoverable unproven conventional oil are thought to exist, along with an estimated 2.8 trillion barrels of recoverable unconventional oil.

The concept of peak oil was initially conceived in the 1950s and is based on supply and demand relationships relative to a limitation in the known reserves of conventional oil. In the peak oil model, demand continues to grow, but reserves are finite. Production reaches a peak and

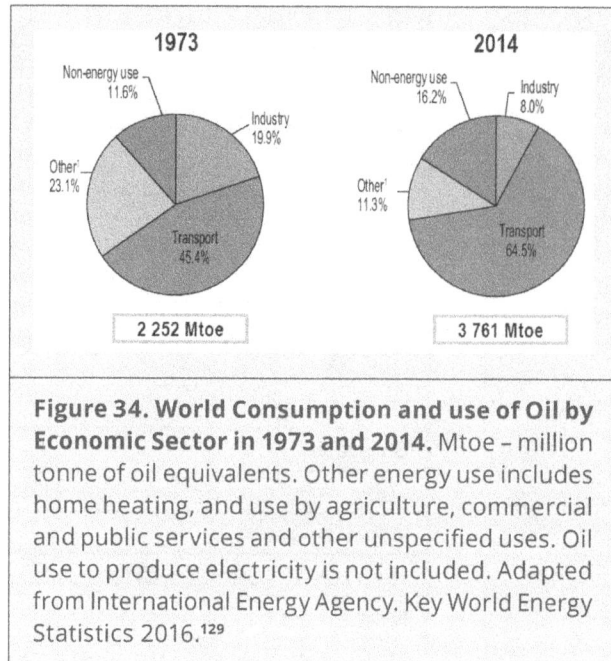

Figure 34. World Consumption and use of Oil by Economic Sector in 1973 and 2014. Mtoe – million tonne of oil equivalents. Other energy use includes home heating, and use by agriculture, commercial and public services and other unspecified uses. Oil use to produce electricity is not included. Adapted from International Energy Agency. Key World Energy Statistics 2016.[129]

transitions to a period of decline as prices escalate. However, as large deposits of conventional and then unconventional oil reserves were discovered, the projected dates of peak oil production were pushed further to the future. Unconventional potentially recoverable reserves greatly exceed conventional proven oil reserves. World shale oil reserves are estimated at 4.8 trillion barrels, and oil sands deposits, the largest of which are located in Canada, contain 0.8 trillion barrels of discovered oil.[141] Recovery and processing of unconventional oil is costly and often faced with environmental and regulatory challenges. A sizable portion of the world's total oil deposits are impractical to extract given current oil prices and available technology. That said, the volumes of oil deposits are substantial, and arriving at a peak oil scenario due to a limitation of reserves is unlikely for the foreseeable future.

Non-GHG emissions from combustion of gasoline, diesel, jet fuel and heating oil are similar to those of coal combustion but at a lower rate of emissions per unit of energy. Removal of lead from gasoline, the invention of the catalytic converter and implementation of emissions standards have had a beneficial effect on reducing the harmful effects of combustion of petroleum products. However, smog and outdoor air quality issues associated with the transportation sector remains a serious contributor to health problems. Annual early mortality from combustion of oil products on a worldwide basis is estimated at 36 deaths/TWhr of energy, which calculates to 400,000 early deaths/year.[138]

On April 20, 2010, the Deep Water Horizon offshore oil drilling rig was completing the final stage of drilling an exploratory well in the Gulf of Mexico when an eruption containing methane gas resulted in a catastrophic explosion and fire that destroyed the rig and opened a well head oil gusher.[142] By the time the well head was sealed on

July 15, a total of 4.9 million barrels of oil had been discharged into the waters of the Gulf causing extensive damage to marine and wildlife habitats and to the local fishing and tourism industries.[142] A court ruling declared that gross negligence and reckless conduct on the part of BP was the cause of the discharge and in July 2015, BP agreed to pay $18.7 billion USD in fines. Total costs to BP from the Deepwater Horizon oil spill are estimated at $42.2 billion.[142]

Risks of spill and subsequent environmental damage are part of the process of extraction and transport of oil and petroleum products. On July 6, 2013, a freight train carrying 74 cars of crude oil derailed in the town of Lac-Mégantic in Quebec, Canada.[143] The explosion and fire killed 47 people and destroyed half of the downtown.[143] On March 24, 1989 the Exxon Valdez oil tanker ran aground and spilled between 260,000 to 900,000 barrels of oil into the waters of Prince William Sound, Alaska, severely damaging an extensive wildlife habitat in a remote location.[144] The Exxon Valdez oil spill is considered to be among the worst man-made environmental disasters, yet by spill size the disaster does not rate in the top ten.

2.6.4 Natural Gas. World consumption of natural gas increased 3-fold between 1973 and 2014.[129] Natural gas is used to produce ammonia fertilizer, via the Haber-Bosch process, as a fuel for domestic heating and industrial purposes and, increasingly, to produce electricity. World-wide generation of electricity using natural gas increased 7-fold between 1973 and 2014 and was the primary driver of the global increase in natural gas production and use.[129]

Similar to the situation with oil, vast reserves of unconventional natural gas exist, along with conventional deposits. Since 2000, extraction of shale gas using hydraulic fracking processes has made the US the largest natural gas producer in the world.[145] Natural gas is transported by pipeline over land and can be transported over sea in liquefied form.

Natural gas is often touted as a clean fuel or a bridge fuel to a lower emissions energy world. Indeed, on a unit of energy production basis, CO_2 emissions from a natural gas power plant are half the emissions from a similar-sized coal-fired plant. Emissions of non-GHG pollutants are a minor issue for natural gas combustion, and the early death rate per unit of energy from natural gas combustion is 2% of the mortality associated with coal, and 11% of early death rate from combustion of petroleum products.[138] While natural gas is a markedly cleaner burning fuel than coal or oil, it is a carbon-based fossil fuel with a budget of GHG emissions associated with extraction, processing and combustion. Any emissions reduction policy based on replacing coal and oil with natural gas must account for the net effect on emissions.

2.6.5 Hydro-electric Power. In 1878, the world's first electric generator driven entirely by the flow of water was assembled to power a single arc lamp in an art gallery at Cragside, Northumberland, England.[146] The Schoellkopf Power Station No. 1 began operations at Niagara Falls in 1881 with electricity generated entirely by 100% renewable, zero-emission, gravity flow of water.[146] By 1920, hydro stations provided 40% of the electricity consumed in the US.[146]

In 2014, hydro generated 16% of world's electricity and was by far the most significant contributor among the basket of renewable energy sources.[129] Conventionally, hydro power is based on the construction of a large dam to create a substantial height for water accumulation above a set of turbines. Other forms of hydro include tidal power, wave power, and run-of-river power generation. Once the installation is complete, a well-designed hydro power station operates without significant emissions.

Figure 35. Three Gorges Dam on the Yangtze River in China. The world's largest power station.

Large hydro projects often involve massive reworking of natural water flows and reservoirs. These projects can disrupt natural ecosystems and displace local populations. Further, the magnitude of the construction project comes with a budget of GHG emissions. These factors often form the basis for opposition to the construction of large scale projects. However, in developing policies for new hydro installations, environmentalists and decision-makers must consider the full balance of ongoing zero-emissions renewable power generation, along with the impact of the construction project and water flow disruption on the environment. Well-designed hydro projects can account for and minimize environmental impacts, and can often be structured to facilitate irrigation of crop lands and mitigate the potential for flooding. Hydro power is extremely safe relative to fossil fuels with construction accidents as the only source of fatalities.

2.6.6 Nuclear Power. As of 2014, nuclear fission reactors provided 11% of the world's electricity.[129] Nuclear power is sometimes classified as renewable given the lack of GHG emissions from nuclear plants. Technically, conventional uranium stores are finite and thus classifying nuclear power as renewable is an inexact use of the term. However, even with increased deployment of present day nuclear power technology, total conventional plus unconventional, but economically recoverable uranium stores would last for 670 years, and reprocessing of spent fuels and weapons-grade nuclear material would result in an 8-fold extension of the supply.[147] New generation closed fuel loop nuclear power processes that are under development could allow for uranium supplies to last indefinitely. For practical purposes, nuclear fission can be considered as a renewable, zero-emissions energy resource.

Since 2007, the production of electricity by nuclear power has declined primarily due to a shut-down of plants in Japan and Germany.[147] A lack of public support has prevented growth of the nuclear industry. Accidents at Three Mile Island (1979), Chernobyl (1986) and Fukushima (2011) have raised public concern as to the safety of nuclear power plants. In particular, the Chernobyl disaster of 1986 gave rise to a widespread anti-nuclear movement in advanced economies. The risk/benefit balance of nuclear power is poorly understood by large segments of the public and specifically by those involved in the anti-nuclear movement.

> *As members of a select panel convened immediately after the accident, my colleagues and I established that the Chernobyl disaster tells us about the deficiencies of the Soviet political and administrative system rather than about problems with nuclear power.*

Hans Bethe — Nobel laureate for the discovery of fusion reactions in the sun.

On April 26, 1986 a grossly mismanaged electrical test was conducted at a poorly designed Russian built nuclear reactor at the Chernobyl power plant near the city of Pripyat in northern Ukraine. A detailed description of the underlying causes and outcome of the disaster is needed to fully to understand the risks associated with future development of the nuclear industry.

The design of the Chernobyl plant made use of diesel powered generators to provide emergency power to the back-up cooling water pumps that would be used in the event of a shutdown in power production. By 1986, the plant had been commissioned and operating for 2 years with a design flaw in the back-up power system. A significant time delay existed between a shutdown of the turbine and engagement of the back-up cooling water pumps.[148] Electrical engineers came up with a potential solution whereby, upon shutdown, the inertia of the still spinning turbine could power the water pumps until the generators kicked in. To test the concept, the reactor was to be powered down to no less than 700 MW.[149] Under stable conditions, the turbine was to be disengaged from the reactor followed by the recording of the residual momentum and steam output of the turbine spinning under inertia. Varying the power output of a reactor of the Russian RBMK design is accomplished through an automated system that inserts or withdraws control rods from the core. These rods contain "neutron poison" that absorbs

Figure 36. April 26, 1986. Chernobyl Reactor 4 hours after the Explosion. Photo taken by plant photographer Anatoly Rasskazov who received a dose of 300 Roentgens (fatal dose is 500 Roentgens) while photographing the disaster. Rasskazov died of cancer in 2010.

neutrons, thereby decreasing the number of neutron collisions with fission material (uranium-235). Inserting more rods poisons the reaction and decreases power output, while withdrawing rods has the opposite effect. The test was delayed following an unexpected spike in power demand and power down of the reactor did not start until 11:00 pm.[149] To make up for lost time, the reactor was powered down quickly causing a buildup of neutron-absorbing fission by-products in the core. These by-products poisoned the reaction, resulting in a further drop in power down to near complete shutdown levels of 30 MW.[149] The chief electrical engineer should have waited 24 hours for the fission by-products to dissipate; however, he ordered an immediate power up of the reactor. The improperly trained operators and the electrical engineers running the test either did not understand or were unconcerned that the reactor was running under dangerous unstable conditions. Various emergency indicators and alarms were ignored and automated safety systems that would have shut down the reactor were bypassed to allow the test to continue. The automated system for insertion and withdrawal of control rods was disabled to allow the operators to manually withdraw control rods to increase power under unstable operating conditions. At 1:23 am, the operators had the reactor running at 160–200 MW by manually withdrawing 205 of the 211 control rods to overcome the inhibitory effects of the fission by-products in the core. Only 6 rods remained inserted despite plant regulations that a minimum of 15 rods should be inserted at all times during operation.[149] To complete the test, the turbine was disconnected from the reactor, cutting off power to the cooling water pumps. The combination of having no cooling water, nearly all of the control rods withdrawn from the core, plus the unstable conditions in the reactor resulted in a rapid acceleration of the reaction. The reactor became increasingly unstable leading to a cascade of events that culminated in two catastrophic explosions.[149]

The 2006 World Health Organization report is the objective authoritative document describing the effects of the Chernobyl disaster on human health. Acute radiation sickness was confirmed in 134 emergency workers and of this total, 28 died from the illness in 1986 and another 19 died between 1987 and 2004.[150] One hundred sixteen thousand people were evacuated from the immediate area in 1986, followed by another 220,000 people from the surrounding area.[150] In total, 5 million people lived in contaminated areas, but the level of contamination was not sufficient to implement additional evacuation.[150] In 1986–87, 260,000 people were engaged in clean up and mitigation activities at the site and in the immediate 30 km zone.[150] The clean up effort continued on a larger scale further from the site of the disaster and eventually involved 600,000 people.[150] By 2005, a clear spike in thyroid cancer (6,000 cases) in young people living in the area was documented.[150] Thyroid cancer is among the least lethal forms of cancer with five year survival rates of 95%.[150] There were 9 deaths associated with the post Chernobyl spike in thyroid cancer.[150] Likely, consumption of milk contaminated with radioactive iodine was the cause of the elevated incidence of the disease. There is perhaps some indication of higher rates of leukaemia and cataracts among the group of high exposure site workers. However, beyond thyroid cancer in the younger population, there is no clear evidence of increased rates of cancer among the population exposed to radioactive fallout from the disaster.

There is no denying the magnitude of the Chernobyl disaster. The explosion ignited the graphite material in the core of the reactor compromising the containment vessel and causing an emission of radioactive particles in the plume of the fire.[151] The plant did not have secondary containment buildings and approximately 100,000 km² of land was contaminated with radioactive fallout.[151] The city of Pripyat remains abandoned as does the 30 km Chernobyl exclusion zone surrounding the disaster site.

The Chernobyl disaster was classified as the first level 7 or major accident in the International Nuclear and Radiological Event Scale.[152] On March 11, 2011 the second level 7 major accident occurred when a 13 metre tall tsunami struck the Fukushima Daiichi nuclear power plant in Japan.[153] The seawall was designed to protect the power plant from a maximum 10 metre tsunami and flooding caused a failure of the back-up diesel generators.[154] The 40-year old generation II reactors underwent a cooling systems failure, leading to partial core meltdowns of at least 2 of the 6 reactors and a number of hydrogen explosions.[154] Radiation was deliberately released from contained cooling water into the sea and was deliberately vented to reduce gas pressures from the containment vessels.[154] Total release of radioactive iodine was approximately 28% of the release from the Chernobyl disaster.[154] The WHO indicates there were 0 fatalities from exposure to radioactive materials from the Fukushima disaster, and that exposure to residents in evacuated areas was such that any health impacts are likely to be below detectable levels.[155] There is potential for a future spike in thyroid cancer among children exposed to radioactive iodine; however, given the screening program in place and the survival rate with early diagnosis, it is unlikely that any deaths will occur from cancer attributed to exposure to radioactive material from the Fukushima disaster.

By 2013, an estimated 2.7 million cubic metres of high level nuclear waste had been produced by the worldwide nuclear industry.[148] Approximately 90% of this material is stored under water in encased ponds at the plant sites where the material was produced.[156] Storage periods are generally 10–20 years, but waste can be stored in ponds for longer periods depending on space allocations. Ponds are often designed to hold the entire waste output for the lifespan of a reactor. Some countries such as France and Japan reprocess the waste to separate out uranium and plutonium and recycle this material back into fresh fuel and thus optimize utilization.[157] Other countries such as the US and Canada have programs based on direct disposal.[157] Eventually, waste material is removed from pond storage and placed in dry storage casks that are maintained on an interim basis in surface storage facilities.[157] Final storage solutions have yet to be implemented but will consist of deep underground placement in stable rock structures. Many countries have mandated that underground depositories will be designed to facilitate retrieval of the material for future use as fuel in generation IV reactors.

The issue of long-term final storage of high level waste, along with perceived issues of safety has held back expansion of the nuclear industry. The primary risk components of high level nuclear waste are long half-life stable radioactive elements such as technetium-99, iodine-129, neptunium-237 and plutonium-239, all of which require safe permanent separation from the biosphere.[157] The volumes are not excessive and deep underground safe storage can be designed. However, an alternative to storage is to utilize the waste from generation II and III reactors as fuel for generation IV fast reactors. These reactors would close the nuclear fuel cycle such that no waste material would ever leave the site. The European Sustainable Nuclear Industrial Initiative is funding the development of 3 promising generation IV fast reactor systems with construction of a prototype likely to begin in 2018.[158]

The safety record of the nuclear power industry is excellent. Including Chernobyl, the early death per unit of energy associated with nuclear power is the lowest of all energy sources and is 0.05% of the rate of early mortality from coal combustion.[138] The mortality rate associated with solar power is higher than that of nuclear due to accidents during rooftop installation of solar panels.[138] The 1986 Chernobyl disaster was a result of flaws in plant design and safety systems specific to the Russian nuclear program at the time. These failures and flaws are clearly understood and have been remedied such that another Chernobyl type disaster is highly improbable. Specifically, the remaining Russian RBMK reactors in service have been upgraded such that manual overrides of emergency safety systems during operation are no longer possible, and the plant cannot be operated under low power unstable conditions. Fukushima was the result of a natural catastrophe striking a 40-year old second generation nuclear plant. The magnitude of the tsunami severely damaged the plant and disabled the back-up water cooling systems, resulting in a partial core meltdown in 2 reactors. Release of radioactive material from containment was managed such that there were 0 short-term and likely 0 longer term fatalities from radiation exposure. In comparison, total confirmed casualties from the tsunami were 15,893 dead, 6,152 injured and 2,572 missing.[159] There are lessons to be learned for future plant design from the Fukushima disaster. The cost and other consequences of site contamination and mass precautionary evacuation were substantial. However, given the magnitude of the tsunami, the human health outcome of the Fukushima nuclear disaster provides proof of the safety of the industry.

In France, 75% of the nation's electricity is generated by a total of 58 nuclear reactors.[160] France has one of the lowest consumer costs for electricity among the 27 members of the EU.[161] Due in large part to the extensive use of nuclear power, as of 2012, the per capita rate of GHG emissions in France was 6.6 metric tonnes of CO_2 equivalents per person.[124] This rate of emissions was 30% of the per capita rate from the US and among the lowest for advanced economies.[124]

The tragedy of the Chernobyl and Fukushima disasters extends beyond the health issues and damage directly associated with these events. A factual assessment leads to the conclusion that well managed nuclear power is a safe, renewable, zero-emissions option for producing electricity. Instead, nuclear power is frequently and irrationally demonized as dangerous and damaging to the environment.

In the immediate future, newly constructed reactors will be generation III and III+ units incorporating design improvements that have evolved from experience in running generation II plants. Generation IV reactors are in the development stage and include fast reactors based on closed loop fuel cycles with highly efficient fuel use. Generation IV small unit modular systems have the potential to overcome the upfront capital costs that restrict installation in less densely populated regions or in countries with limited capital.

The future of nuclear power is uncertain. Germany has embarked on a program to phase out nuclear power by 2022, and the influence of the anti-nuclear movement in the Western world is such that any new installation faces strong opposition. Ideally, like any other option, nuclear power should be assessed objectively, based on benefits, risks, costs, sustainability, short and long-term impact on the environment, and the future potential for technology enhancement. In establishing positions on options for the production of electricity, environmentalists and policy makers must critically and objectively evaluate the balance of these factors. A single day of fatalities associated with the use of coal dwarfs the mortality record over the complete history of nuclear power. A rejection of the nuclear option may sustain carbon-intensive, coal-fueled power plants and thus contribute to continued GHG emissions, along with the costly damaging effects of coal extraction and combustion on human health and the environment.

2.6.7 Biomass. Biomass refers to any organic matter or material derived from non-fossilized recent organic matter. Evidence for controlled use of biomass for heating and cooking dates back to the time of Homo erectus 400,000 years ago.[162] Biofuels account for an estimated 10% of the total world primary energy use with most of this use in developing countries for the same domestic cooking and heating purposes as practiced since the dawn of man and the discovery of fire.[163]

Solid biomass fuels consist of wood, sawdust, wood pellets, wood chips, waste wood and low valued agriculture by-products (straw and hulls). Bioethanol and biodiesel are liquid biofuels derived from crop feedstocks and are used within the transportation sector.

2.6.7.1 Solid Biomass Fuels. On a per unit of energy basis, relative to coal, uncontrolled combustion of forest biomass releases a comparable amount of CO_2, lesser amounts of nitrous oxides and sulphur dioxide, with higher emissions of carbon monoxide, particulate matter and volatile organic compounds.[164] Biomass combustion for cooking and heating purposes in developing countries (primarily in Sub-Saharan Africa) is a major contributor to the 3.3 million early deaths per year associated with indoor air pollution.[134]

Industrial scale use of solid biomass consists of either direct combustion to produce electricity and process heat or indirect combustion following conversion to secondary products. The process of gasification converts biomass to synthetic gases under low oxygen conditions. These gases can then be combusted and used as secondary fuels. Often, when an economical source of solid biomass is locally available, this material will be co-fired with coal to produce electricity. Power generation from biomass has doubled since 2006 and currently accounts for 1.5% of the total world supply of electricity.[163] In the province of Ontario, Canada, coal-fired power plants have been completely converted to the combustion of wood biomass in a publicly-stated effort to switch to lower emissions, clean, sustainable power generation.[165] However, excluding CO_2 emissions, there is no supporting evidence that combustion of biomass is less harmful to human health and the environment than burning coal. In the US, with stringent emissions standards in place, early deaths per unit of energy for biomass harvesting and combustion are greater than that of coal.[138]

Conversion of coal power plants to combustion of solid biomass is often presented to the public as a cost-effective switch to clean, renewable power generation. However, implementation of large scale purpose harvesting of forests for production of bio-energy would be imposed on top of standard forestry resource management. With this scenario, harvesting, transportation and combustion of wood from forests removes net forest carbon sequestration capacity and releases a pulse of CO_2 into the atmosphere. A matching program of sustainable reforestation must be in place to re-establish the equivalent biomass of forest with the same carbon sequestration capacity.

Actual emissions from the harvesting and use of solid biomass as an energy source do not conform to a simplified carbon balance. GHG emissions from a given technology used to produce electricity must be based on the complete life cycle of CO_2 exchanges. This includes any atmospheric sequestration, along with the CO_2 emitted from all components of the complete chain of processes required to operate the technology. For power production from biomass, carbon dioxide is released at each step in the com-

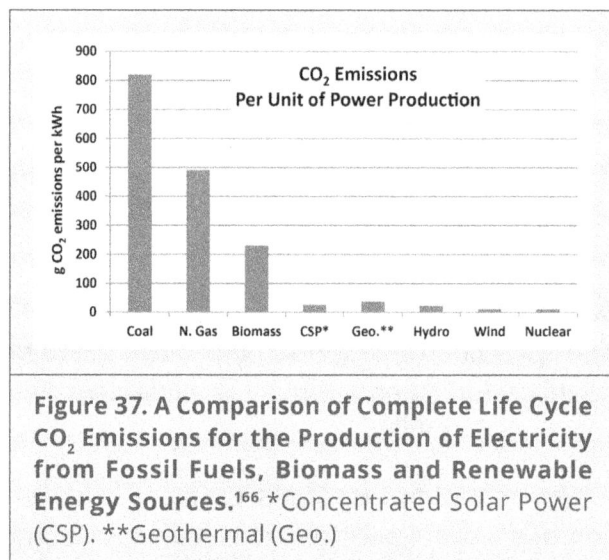

Figure 37. A Comparison of Complete Life Cycle CO_2 Emissions for the Production of Electricity from Fossil Fuels, Biomass and Renewable Energy Sources.[166] *Concentrated Solar Power (CSP). **Geothermal (Geo.)

plete sequence of harvesting, transport, processing and final combustion of the fuel. These total emissions exceed CO_2 sequestration from the atmosphere during growth of the biomass. Net CO_2 emissions from a biomass fired power station are approximately 30% of the GHG emissions from a comparable coal-fired facility and 50% of emissions from a natural gas power plant.[166] However, net GHG emissions from biomass are well in excess of true 0 emissions energy sources such as nuclear, hydro, solar, wind and geothermal. These calculations are based on an assumed balance between biomass removal and new growth to re-establish the same carbon sequestration capacity.

The European Commission has recognized the full dynamics of biomass combustion and has developed a set of sustainability criteria that govern the industry. Biomass fuels cannot be sourced from land that was converted from forest or other high carbon sequestration stock areas.[167] For existing facilities, industrial practices must confirm to standards whereby the complete life cycle of the biomass fuel, including harvesting, transport and processing, results in a minimum of 35% less GHG emissions relative to fossil fuels.[167] New biomass installations must operate with at least 50% lower GHG emissions than fossil fuel plants by 2017 and at least 60% lower by 2018.[167] Finally, a program of monitoring the sources of all biomass consumed in the EU is encouraged to ensure that imported biomass is sustainably produced.[167] World-wide adoption of governance policies such as the European sustainability criteria for use of solid waste biomass fuels can avoid scenarios whereby biomass replaces coal without meaningful cuts in net emissions.

Well managed, sustainable use of forest resources provides valued products such as lumber, pulp and paper. The basis of these products is structural carbon that originated from CO_2 sequestered from the atmosphere. The forestry, lumber and pulp and paper industries output considerable volumes of by-product biomass in the form of sawdust and waste wood that has little to no value. Historically, this material has either been left to decay in the forest, committed to landfill, stockpiled or combusted (open air or incinerated) without use of thermal energy. Large stockpiles of solid biomass waste will undergo anaerobic decay over time with release of methane as a potent GHG. The time span of the process is complex and dependent upon the moisture content of the waste, the climate at the location of the storage pile, and the fraction of degradable organic compound in the stockpile. Wood and straw waste are classified by the IPCC as slowly degrading waste with a storage pile half-life of up to 35 years.[168] Combustion of solid waste biomass for power production or other thermal processes makes use of the stored energy in the materials and eliminates stockpiling and the associated slow release of methane.

Conversion of coal power stations to combustion of biomass is superficially an attractive option based on convenience, cost savings and public perception. However, the net cost/benefit relationship of anthropogenic burning of solid biomass as a fuel source must be fully assessed with a rigourous analysis of fuel source, total inputs and total emissions including non-GHGs, and the short and long-term carbon balance. Science-based governance of biomass fuel use in combination with effective forest and land use management can potentially contribute to a reduction in carbon emissions.

2.6.7.2 Liquid Biofuels. Biomass can be converted to liquid fuels for use in the transportation sector. Europe, the US, and Brazil have substantial industries for biodiesel and bioethanol production from corn, wheat, sugarcane, and oilseeds crops. These industries arose from government initiatives that were designed to provide a measure of energy security based on assumptions of long-term elevated costs and apparent limitations in oil supplies. The feedstocks for generation 1 liquid biofuels are grown on land that could otherwise be used to produce food for human consumption or feed for farm animals. An expansion of the biofuels industry has led to the food/fuel land use debate that questions the long-term viability of growing typical crops for fuel. In addition, the cost of biofuel production exceeds that of petroleum products, and the viability of the biofuel industry is reliant on legislated inclusion rates in fuel mixtures or other forms of subsidy.

Bioethanol and biodiesel can be used directly to power internal combustion engines. These products are classified as "drop-in fuels" that do not require a reconfiguration of engine design. Bioethanol and biodiesel are often advertised as green, sustainable, low emissions alternatives to conventional gasoline and diesel fuels.

Unlike trees, crops used as feedstock for liquid biofuels can be planted and harvested on an annual basis and thus, in the absence of other considerations, provide an immediate carbon neutral cycling of fixation and release. However, growing, harvesting, transporting and processing these crops comes with a significant carbon budget that must be considered in calculating the overall life cycle CO_2 emissions from the use of these fuels.

By using a world average for atmospheric carbon sequestration per land area of forest, it is possible to model the area of forest land required to sequester and thus offset the CO_2 emissions from operating a typical light-duty vehicle fueled by conventional gasoline or bioethanol.[169] This modelling exercise assumes a closed loop for carbon sequestration during plant growth and release from biofuel combustion but takes into account all inputs and land use factors for fuel production. In the US operating a typical light-duty vehicle on gasoline requires 1.11 hectares of forest land to offset total emissions.[169] Operating the same vehicle on E85 (85% bioethanol from corn;

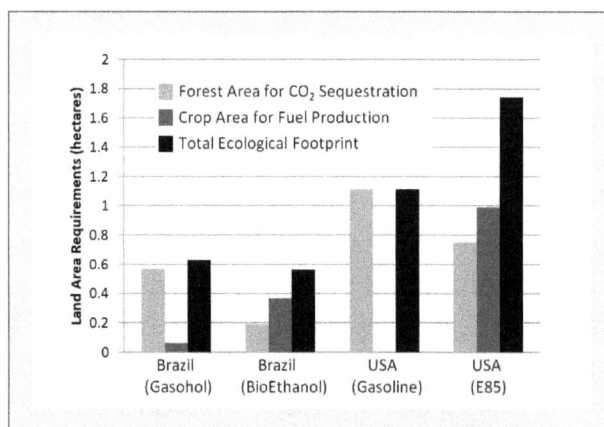

Figure 38. Impact of Fuel Options on the Ecological Footprint for Operating a Light Duty Vehicle in Brazil or the US.[168] Brazil Gasohol (27% sugarcane derived ethanol/gasoline blend). Brazil Bioethanol (100% sugarcane ethanol). US Gasoline (100% gasoline). US E85 (85% corn derived ethanol/gasoline blend). Total ecological footprint consists of the forested land required to sequester vehicle emissions plus the crop area required for biofuel production.

15% gasoline) requires 1.74 hectares for the combination of forest area to sequester emissions, plus land area for biofuel production.[169] Further, by the year 2036, fueling the US automobile fleet on E85, would require that the entire crop area, plus all of the range and pasture land of the country, becomes 100% committed to growing corn for fuel ethanol production.[169] Clearly, significant production of bioethanol from corn as a replacement for conventional gasoline is not viable and not warranted given the net impact on emissions and land use.

Brazil produces vast quantities of sugarcane, much of which is used for bioethanol production with end use as a component of gasohol. Gasohol is mix of bioethanol and gasoline that is mandated for use in road vehicles in Brazil. As of March 2015, gasohol must contain 27% bioethanol.[170]

Bioethanol production from sugarcane is a considerably more efficient process than bioethanol production from corn. Using the same modelling exercise as applied to bioethanol from corn in the US, fueling a typical light-duty vehicle in Brazil with 100% sugarcane-derived bioethanol results is a 28% reduction in the land area required for forest sequestration of emissions, plus biofuel production, relative to fueling the same vehicle with gasohol.[169] Complete conversion to fueling the entire fleet of automobiles in Brazil with 100% ethanol by 2036 would require that 10% of the current crop land area is dedicated to growing sugarcane for bioethanol production.[169]

Based on the experience of bioethanol production from corn in the US, it is tempting to dismiss liquid biofuels as a long-term option for fueling the Anthropocene. Indeed, with the exception of sugarcane, first generation biofuels from conventional crops offer little to no GHG abatement potential and come with significant land use implications. However, second generation biofuels produced from forestry and agricultural waste products or dedicated non-food energy crops are under development. These fuels must meet specific targets of complete life cycle inputs and emissions including land use criteria to overcome the limitations of first generation biofuels. Biomass to liquid (BtL) is a promising technology whereby biomass materials undergo thermal gasification followed by chemical conversion of the gases to liquid hydrocarbon fuels. Enzymatic breakdown of lignocellulosic (fibrous) plant material for production of ethanol is another potential process for conversion of second generation biomass feedstocks to liquid biofuels. Over the next 35 years, the International Energy Agency predicts a complete phase out of corn and oilseed derived biofuels with concurrent development of BtL biodiesel and lignocellulosic ethanol to compliment sugarcane derived bioethanol.[171] As will be discussed in subsequent chapters, sustainable second generation liquid biofuels may play an important future role in fueling niche applications within the transportation sector.

2.6.8 Wind. As of 2014, wind turbines generated 3% of the world's electricity and capacity is growing rapidly.[172] Wind is an extremely safe, zero-emission, fully renewable energy source. Studies indicate that wind turbines can readily provide up to 20% of the total supply to an electrical grid.[173] Beyond 20%, the economics of incorporating a variable energy source become increasingly complex. Advances in grid management, improved timing of energy use, and commercially viable methods for the storage of electricity have the potential to facilitate higher levels of penetration of wind and other variable energy sources into the full mix of options to produce electricity.

2.6.9 Solar. Solar power generated 0.85% of the world's electricity in 2014 and is growing at a rate of 40% annually.[172] Solar is a clean, safe, zero-emissions, renewable energy source, and the International Energy Agency (IEA) indicates that solar has the potential to produce 27% of global electricity supply by 2050.[174] As is the case for wind, solar is a variable uncontrolled energy source and, at high levels of penetration of the total energy mix, the production of electricity becomes complex and increasingly costly. Photovoltaic systems can be installed at any scale from large solar farms to small rooftop installations that supply household power and can potentially contribute, as a distributed source, to an electrical grid. Concentrated Solar Power (CSP) uses mirrors to focus sunlight into an intense beam that is used to heat water and generate electricity from steam turbines. As the technology has advanced, the upfront cost of installing solar systems has declined. In addition, alternate funding models have the potential to reduce or completely offset the capital costs of installation and could provide incentives for widespread adoption.

2.6.10 Geothermal and Heat Pumps. Shallow, accessible pools of heat beneath the Earth's surface can provide a clean, renewable, energy source for district heating and power generation. Interestingly, the bulk of this energy originates from decay of naturally occurring radioactive elements heating aquifers beneath the surface of Earth. As of 2014, geothermal energy generated 0.33% of the global electricity supply.[172] The total amount of geothermal energy is estimated at double the total world primary energy supply; however, the vast majority of this energy is unrecoverable.[175] Some countries such as the Philippines, El Salvador, Kenya and Iceland are located above readily accessible hotspots such that geothermal energy accounts for 14–29% of the total electricity produced by these countries.[176] The potential of geothermal energy is currently limited by the number of accessible hotspots. With available technology, growth of geothermal energy utilization may be limited to 1% of the world total primary energy supply. Advanced geothermal systems based on artificially creating aquifers by hydraulic fracture of bedrock formations or extreme deep drilling have the potential to extend the potential use of geothermal energy.

Heat pumps function to reverse the natural flow of heat from warmer to colder spaces by using vapour-compression systems that circulate a refrigerant through a heat source and a heat sink. Generally, heat pumps are reversible, functioning to either warm or cool a building dependent upon needs. Electricity is required to operate the system; however, net energy transfer is generally 3- to 4-fold in excess of power use. No emissions are associated with the actual transfer of heat between the heat source and the heat sink. Heat pumps can function as air source or ground source energy transfer systems. Ground source heat pumps use a network of underground pipes for the energy transfer and are more efficient that air source systems. The IPCC reports that buildings account for 18.4% of total GHG emissions from the combination of direct energy use and emissions from the production of electricity used by buildings.[86] Replacement of conventional heating and cooling with heat pumps can contribute to a reduction in emissions from the buildings sector.

2.7 Anthropogenic Climate Change to 2017

Emissions of anthropogenic greenhouse gases have accumulated in the atmosphere and increased total radiative forces by 2.3 Watts per square metre of surface area.[86] This additional energy directed to the surface of the Earth has increased average temperatures by 1°C, which has affected the land, cryosphere (ice and snow layers) and oceans. Higher levels of atmospheric CO_2 have increased the concentration of dissolved carbon dioxide in the oceans leading to ocean acidification and subsequent effects on the marine biosphere.

Elevated surface temperatures have begun to affect weather systems and the incidence and severity of extreme weather events. Overall, recent decades have experienced an increase in heavy precipitation events, warm spells, heat waves, and wild fires. The frequency of droughts has increased in the Mediterranean and in West Africa but decreased in Central North America and Northwest Australia.[177] There has been an increase in the frequency of strong tropical cyclones in the North Atlantic; however, the underlying causes of this trend are uncertain.

The rate of retreat of almost all glaciers on Earth has accelerated over the past 20 years. Thermal expansion of oceans, along with accelerated rates of melt water flows from the combination of glaciers, Greenland and Antarctic ice sheets and Arctic sea ice, have led to 3.2 mm/year rise in sea levels since 1993.[86] This rate of sea level rise is double that recorded over the period from 1901 to 2010.[86]

Ocean warming has driven a shift in the distribution of marine species poleward or deeper toward cooler waters. Oxygen minimum zones (dead zones) are expanding in the Pacific, Atlantic and Indian Oceans and are constraining fish habitats.[86]

Coral reefs are highly vulnerable to ocean warming and acidification. The current global coral bleaching event began in 2014 and is the longest and most-widespread in recorded history. As of February 2017, there are no indications of an end to this event. Satellite imaging from the National Oceanic and Atmospheric Administration indicates that the vast majority of the world's corals are under bleaching heat stress watch or warnings. There is a 60% chance that the projected level of heat stress damage will occur, regardless of climate change mitigation efforts.[178] Degradation and loss of coral reefs will severally damage vast complex reef dependent ecosystems, fish stocks and fisheries.

Climate change has had a negative impact on wheat and corn yields in many regions such that global aggregate yields of these critical crops are estimated to have declined by 1–2%.[179] The effect of climate change on agriculture varies by region with some positive effects noted at higher latitudes. Maize production and overall food security in southern Africa is highly vulnerable to climate change.

As of 2017, the effects on anthropogenic climate change have been largely confined to natural ecosystems with lesser effects on human activities and infrastructures. During the Anthropocene, as a species we have demonstrated a willingness to sustain and adapt to the damaging effects of mining, shipping and combustion of coal and, to a lesser extent, petroleum products in order to achieve the benefits and convenience that comes with the use of these fossils fuels. Given this history, an objective assessment of the current additional negative impacts of a 1°C elevation in average surface temperatures, along with capabilities for adaptation, may not warrant a global change in economic practices.

The impact of anthropogenic climate change cannot be evaluated solely on the basis of current observations. Accumulated atmospheric GHGs commit the planet to further climate change. The most aggressive mitigation pathways still require an orderly process of transition away from the fossil fuel-based energy infrastructure that provides much of the foundation of world economies. Further warming to a minimum of 1.5°C above pre-industrial temperatures is all but locked-in during the current century.

An assessment of the need for mitigation requires a detailed modelling of the effects of projected further surface temperature elevations under various scenarios of continued GHG accumulation in the atmosphere. As will be discussed in the next chapter, a continuation of current practices will result in extreme climate change. Sea level rise will place vast segments of low-lying, often heavily populated, coastal regions at considerable risk. Ocean warming and acidification will threaten the foundations of marine ecosystems and fish stocks. On land, food production and security will be severely compromised by excessive surface temperature elevation, and the frequency and severity of extreme weather events will increase.

Part 2:
WHERE WE ARE HEADED

Chapter 3: FUTURE CLIMATE SCENARIOS

Over the 4.5 billion year history of Earth, the climate has fluctuated between extremes of severe hothouse conditions with average surface temperatures up to 28°C and a snowball earth where snow and ice covered the entire surface. These transitions have reshaped oceans and land masses, eradicated habitats and caused the extinction of species while facilitating the emergence and proliferation of new species.

The age of mammals began 66 million years ago under hothouse Earth conditions driven by atmospheric CO_2 concentrations in excess of 2,000 ppm. Declining greenhouse gas levels following the Arctic Azolla event 49 million years ago led to a gradual cooling of the planet with a transition to the current ice age beginning 2.6 million years ago.

The 14°C difference between current temperatures and the hothouse earth of the past is used by some as the basis for arguments that the 1°C increase in temperature since the onset of the Anthropocene is of no consequence, and that the planet can readily withstand the relatively modest impacts of a man-made imbalance in the global carbon cycle.

Earth is in no danger of demise based on the activities of man. However, continued excess rates of anthropogenic GHG emissions places the civilizations built by man, along with existing habitats for other lifeforms at risk.

Arguments that the planet has naturally passed through extreme climate variation as the basis for a continuation of current practices for fueling the Anthropocene are absurd and irresponsible to future generations. Modern cities, industries, and food supply chains were built during the briefest of moments in the history of Earth. This moment occurred during an interglacial period within an ice age characterized by 280 ppm atmospheric CO_2 and associated stable average surface temperatures, weather patterns, ice masses, sea levels and ocean pH. There is some degree of tolerance for anthropogenic changes to these baseline climate conditions; however, we are approaching the limit beyond which climate change will negatively impact important sectors of global civilization and will damage the habitats and viability of other species.

3.1 Primary Effects on Surface Temperatures, the Cryosphere and the Oceans

3.1.1 Average Surface Temperatures. Reports published by the Intergovernmental Panel on Climate Change on the projected effects of climate change over the course of the current century focus on 4 scenarios of future anthropogenic GHG emissions.[86] These scenarios are termed Representative Concentration Pathways (RCPs) and differ in the magnitude and time dependent profile of future emissions. RPC2.6 assumes an aggressive mitigation pathway whereby GHG emissions peak between 2010 and 2020 and then substantially decline. In RPCs 4.5 and 6.0 arriving at peak emissions is delayed until 2040 and 2080 respectively followed by declining rates of emissions. RPC8.5 is based on a business as usual pathway whereby the rate of emissions remains correlated to population growth and increased GDP per capita such that emissions continue to increase through to the end of the century and beyond. These models are part of a continuum of possible future GHG emissions scenarios between extremes of immediate aggressive mitigation and no change in continued growth of emissions. Climatologists have further refined these models to include sub-scenarios based on various rates of decline from peak emissions.

The persistence or lifespan of a greenhouse gas in the atmosphere is defined by the rate of chemical conversion to other compounds. Atmospheric CO_2 is chemically stable and in the absence of an anthropogenic process to actively withdraw and sequester atmospheric CO_2, elevated levels will persist for an indefinite period extending beyond centuries. This fact contradicts confusing and erroneous representations that define a limited lifespan of atmospheric CO_2 based on the time from emission to ocean dissolution of a typical molecule to CO_2. This simplification ignores the full dynamics of the balance between dissolving and release of ocean CO_2, along with land-based cycles of sequestration and release that defines the persistence in levels of atmospheric CO_2.

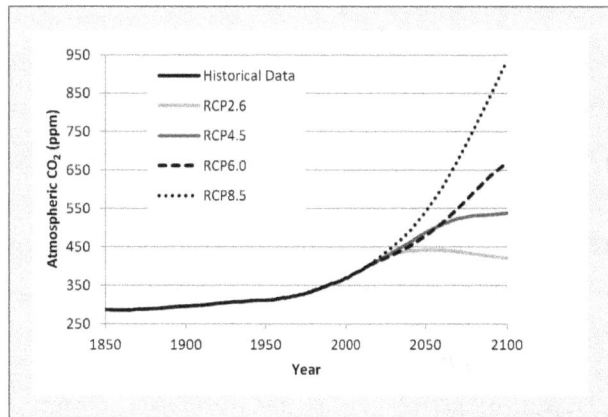

Figure 39. Historical Record and Estimates of Future Atmospheric CO_2 Levels Based on Representative Concentration Pathway (RCP) Models of Future Emissions.[180] RCP2.6 refers to aggressive mitigation of climate change with peak emissions prior to 2020. RCP4.5 and 6.0 refer to mitigation pathways with peak emissions in 2040 and 2080 respectively. RCP8.5 is based on a continuation of business as usual practices without climate change mitigations.

The potency of methane as a GHG in the atmosphere is 60-fold greater than CO_2.[87] However, methane is converted to CO_2 in the upper atmosphere with a lifetime of about

11 years.[180] A cut in current rates of methane emissions, as such, will result in lower end-of-century atmospheric concentrations of methane and total GHGs expressed as CO_2 equivalents. Nitrous oxide undergoes slow reactions in the atmosphere and thus will persist for long periods.[87]

Each emissions scenario provides an estimate of future atmospheric concentrations of carbon dioxide, methane, and nitrous oxide based on projected rates of emissions and chemical lifespans in the atmosphere. The net effective radiative forcing of the total concentration of GHGs as carbon dioxide equivalents is then calculated and used to predict future surface temperatures. In summarizing future emissions scenarios, most models will provide projections for atmospheric concentrations of CO_2, along with a range of surface warming outcomes that are based on the change in net radiative forcing for the sum of all GHGs.

Relative to pre-industrial temperatures, the RCP2.6 scenario of immediate and aggressive abatement of emissions results in a projected surface temperature increase of about 1.6 to 1.8°C by the end of this century.[181,94] By the year 2100, the actions of man have committed the planet to a minimum additional further increase of about 0.6 to 0.8°C above current average surface temperatures. The RCP8.5, business as usual scenario, will increase year 2100 average global surface temperatures by about 5°C.[181] IPCC scenarios of intermediate reductions in GHG emissions (RCP4.5 and RCP6.0) will result a year 2100 average surface temperature increase well in excess of 2°C, and likely in the range of 3–4°C for RCP6.0.[181]

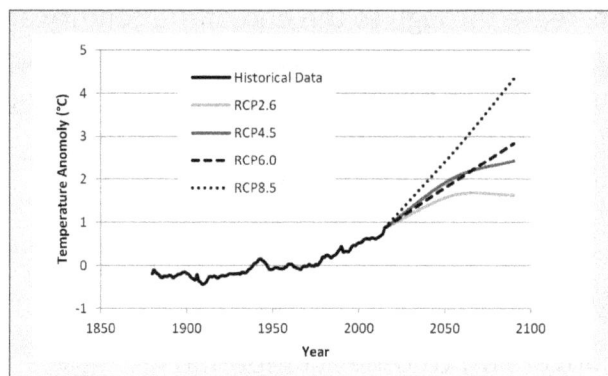

Figure 40. Average Surface Temperature Anomalies and Models for Future Surface Warming Based on RCP Emissions Scenarios.[94,181] Baseline set to average surface temperatures 1951–1980.

3.1.2 Cryosphere and Sea Levels.

> *The 'New Yorker' asked me to shoot a story on climate change in 2005, and I wound up going to Iceland to shoot a glacier. The real story wasn't the beautiful white top. It ended up being at the terminus of the glacier where it's dying.*
>
> James Balog — nature photographer, author of *Extreme Ice Now: Vanishing Glaciers and Changing Climate: A Progress Report.*[182]

Since 1979, the mass of arctic sea ice has declined by 3.5–4.1% per decade.[183] Arctic sea ice undergoes seasonal fluctuations with minimal perennial sea ice mass occurring in September after the summer melt. Perennial sea ice has declined by 11.5% per decade since 1979 as the annual period of sea ice melt has been extended by 5.7 days/

decade.[183] The RCP2.6 model of aggressive abatement of GHG emissions is consistent with a committed net loss of approximately 50% of the total September Arctic sea ice by the end of the century.[184] Other models predict progressively smaller Arctic sea ice mass with increased surface temperatures. Under the business as usual scenario of continued GHG emissions, September Arctic sea ice will have vanished by the year 2050.[183]

Almost all glaciers on Earth are in retreat and the rate of retreat has accelerated over the last 25 years. Over 80% of the total glacial ice loss has occurred in the Northern Hemisphere.[183] Studies on the Morteratsch Glacier in Switzerland provide an extensive data base beginning in 1878.[185] This glacier retreated at an annual rate of 17 metres per year over the period from 1878 to 1998.[185] Since 1998, the rate of retreat has nearly doubled to 30 m/year.[185]

Both the Greenland and the Antarctic ice sheets have declined over the past 20 years. Ice loss consists of a combination of increased annual surface melt and increased glacial discharge to the seas. The rate of ice loss from Antarctica has dramatically increased from 30 Gt/year from 1992–2001 to 147 Gt/year during the period from 2002–2011.[183]

The Western Antarctic Ice Sheet (WAIS) is classified as a marine-based ice sheet in that the bed of ice lies below sea level. This mass of ice accounts for 10% of the entire Antarctic Ice Sheet and is anchored at the

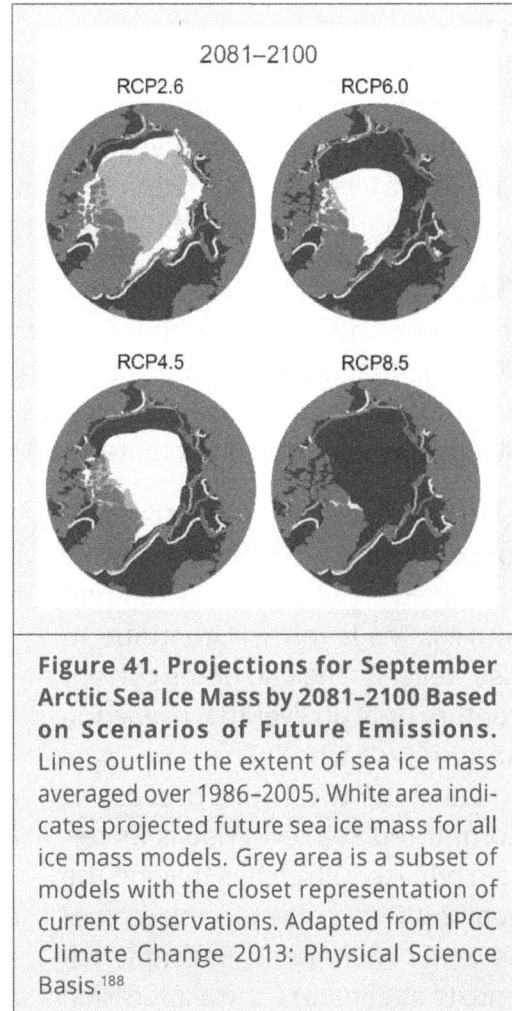

Figure 41. Projections for September Arctic Sea Ice Mass by 2081–2100 Based on Scenarios of Future Emissions. Lines outline the extent of sea ice mass averaged over 1986–2005. White area indicates projected future sea ice mass for all ice mass models. Grey area is a subset of models with the closet representation of current observations. Adapted from IPCC Climate Change 2013: Physical Science Basis.[188]

regions where the ice mass rests on underlying bedrock.[186] Glaciers within the WAIS are retreating at an accelerating rate and discharge from glacial retreat accounts for most of the ice loss. The Thwaites Glacier has been called the weak underbelly of the WAIS in that this glacier is anchored by an underwater ledge.[187] Retreat of the glacier within the perimeter of this ledge will cause the glacier to collapse into the ocean, resulting in seawater intrusion and the formation of melt channels that could destabilize the entire WAIS. Collapse of the WAIS would eventually raise sea levels by an estimated 3 metres.[187]

Recent modelling of the dynamics of the WAIS indicates that irreversible destabilization of the WAIS may well have begun.[187] In the absence of an anthropogenic trigger, eventual collapse of the ice sheet and subsequent sea level rise will likely occur over a prolonged period of thousands of years. The effects of sustained elevated surface temperatures on the time scale for collapse of WAIS are uncertain but with accelerated glacial retreat, a rapid catastrophic collapse within centuries is a distinct possibility.[188]

Approximately 90% of the additional energy retained by the planet following atmospheric GHG accumulation is stored in the oceans.[188] The warming of ocean surface waters accounts for most of this energy sink. Thermal expansion and glacial melt water discharge accounts for approximately 75% of observed anthropogenic sea level rise.[188] The combination of ice loss and thermal expansion of warming oceans is consistent with the observed 3.2 mm/year rise in mean global sea levels since 1993. The annual rate of sea level rise (SLR) from 1993 to 2010 is nearly double the annual rate over the 110-year period between 1901 and 2010.[188]

Extending beyond the current century, loss of Artic sea ice and the declining mass of the Greenland and Antarctic ice sheets will become significant contributors to future SLR. The magnitude of future SLR is dependent upon the level of atmospheric GHG accumulation and associated increases in effective radiative forcing. Committed future SLR by the year 2100 is estimated at 0.4 m for RCP2.6 and 0.63 m for RCP8.5.[189,*]

Glacial and ice sheet melting is a prolonged committed process following sustained elevated surface temperatures. Sea levels will continue to rise well past the end of the current century until an eventual new equilibrium is reached. The time scale to conclude the entire process of committed sea level rise is uncertain but likely stretches beyond the year 2300 and may take place over several millenniums. Recently, scientists at Climate Central modelled fully committed SLR for each of the 4 IPCC scenarios of future surface

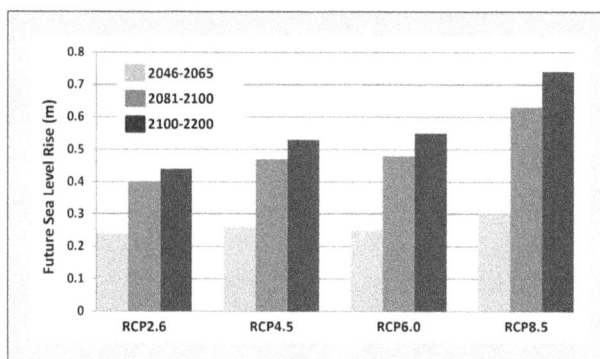

Figure 42. Committed Sea Level Rise by Future Emissions Scenarios. (Mean SLR estimates are relative to sea levels for the period 1986 to 2005). Eighty percent of the land area of the island nation of the Maldives (population 400,000) is less than 1 meter above sea level.[245,248]

warming.[189] Committed sea level rise was then mapped to determine current land mass and population centers that would be below projected sea levels after the melting process had reached equilibrium. Over 627 million people live in at-risk coastal locations less than 10 metres above sea level, and this figure includes two-thirds of the world's cities with populations greater than 5 million.[189] The final total land mass

---

* Actual estimates of sea level rise are based on probability models with a range of possible outcomes for each scenario of future emission. Typically this range is calculated as the upper and lower limits within a 95% confidence interval. For ease of presentation, models of sea level rise and other graphical representations of future outcomes of climate change have been simplified to show the highest probability outcome.

of coastal areas, and thus current global population that will be committed to a future existence below sea level, varies directly with the degree of surface warming over the course of the current century.

Table 3. Long-Term Sea Level Rise and Population Exposure*,[189]		
Surface Warming (°C)	Locked-In SLR (m)	2010 Population Below SLR (million)
1.5	2.9	137
2	4.7	280
3	6.4	432
4	8.9	627
*Long-term locked-in SLR extending beyond 2300		

Aggressive mitigation of GHG emissions cannot avoid a minimum 1.5°C increase in global surface temperatures and thus we have committed the planet to an eventual 2.9 m rise in sea level. Based on the current distribution of global population, land mass that is home to 137 million citizens of Earth will be below sea level. In China, 27.5 million people currently live on land that, in the absence of sea walls, will become submerged. The Maldives consist of 1,100 islands in the Indian Ocean and is the lowest-lying nation on Earth. Committed sea level rise will completely submerge these islands and eradicate the nation. Substantial portions of the cities of Miami and New York are committed to be below sea level. Under the best-case scenario of a 1.5°C increase in mean global surface temperatures, land that is currently home to 557,000 people in Miami and 1.1 million in New York City will be below sea level.

Predictions of extreme long-term sea level rise may come across as distant, and perhaps alarmist, given the uncertainty of the time span of the process. However, the dynamics of ice sheet melt is a non-linear process that includes positive albedo feedback mechanisms and an eventual irreversible transition to complete meltdown of the ice sheet. As the ice mass declines, the percentage of short wave solar radiation reflected from the surface declines and surface heating increases. This positive feed-back loop will contribute to further elevated surface temperatures and an eventual disappearance of the ice sheet. Irreversible complete meltdown of the Greenland ice sheet is predicted by a 2.5–3.1°C increase in average global surface temperatures.[188] The process would occur over thousands of years and raise sea levels by 7 m.[189] In the absence of terra-engineering projects to reduce surface temperatures by active withdrawal of atmospheric GHGs or blocking of incoming solar radiation, the Greenland ice mass will continually melt in a prolonged irreversible process.

Antarctica accounts for 8.3% of the total land mass on Earth, and approximately 10% of the total Antarctic ice sheet extends as ice shelves over the ocean.[189] Other than the West Antarctic Ice Sheet, the inland portion of the ice mass is relatively stable and climate change models generally do not consider a complete meltdown of the Antarctic ice sheet. However, if current rates of emissions were to be sustained over the next 100–200 years, atmospheric CO_2 concentrations would be well in excess of 1000 ppm, and the radiative forcing would be sufficient to trigger positive feedback loops that would commit the climate to an eventual complete transition out of the current ice age and into conditions of an ice-free hothouse Earth. Complete disappearance of the inland Antarctic ice sheet would raise sea levels by over 50 metres and reshape the land masses of the planet.[183]

During the current century, the risks of committed SLR are compounded by extreme weather events. SLR adds to the height of hurricane and cyclone induced coastal sea surges and thus increases the potential and potency of extreme weather events to inflict damage. SLR alone, and in combination with extreme weather events, results in a high risk of coastal flooding, erosion, water quality impairment and widespread damage to infrastructure, livelihoods and settlements. The IPCC divides the world into 9 regions based on geography. SLR is a significant contributor to future risks of damage from climate change in all regions, with the exception of the North and South Poles.[190] In Asia, vast populations living in low-lying coastal areas are exposed to risks from SLR and extreme weather events. In many situations, adaptations such as selective population relocation, land use change, and construction of physical sea restraints can be implemented to partially mitigate risk. The risks and associated costs of adaptations (where these are practical) increase with the magnitude of climate change driven SLR. Small island nations are particularly vulnerable to SLR and extreme weather. These nations are characterized by a high ratio of coastal to inland area, and the construction of sea walls to minimize risks may be impractical or well beyond available resources.

3.1.3 Ocean Warming and Acidification. The world's oceans have acted as a carbon sink with a net removal of approximately 28% of the total anthropogenic CO_2 released to the atmosphere since the onset of the Industrial Revolution.[191] Net withdrawal of atmospheric CO_2 is based on a combination of natural processes that withdraw, circulate, and slowly sequester carbon.

The solubility pump is the major driver of ocean sequestration of atmospheric gaseous CO_2. Atmospheric carbon dioxide dissolves in surface waters of the Gulf Stream and other waters and then equilibrates between dissolved carbonic acid and carbonate. Surface waters are eventually circulated to cooler depths in the ocean at higher latitudes and circulated around the globe via the ocean conveyor belt.[192] Outgassing of CO_2 from the ocean to the atmosphere occurs when cooler waters are upwelled and warmed at equatorial latitudes. The solubility pump favours acidification of the ocean and thus the ocean acts as a carbon sink. The net effect since the onset of industrialization is a 26% increase in ocean acidity (hydrogen ion concentration) as measured by a 0.1 unit drop in ocean pH.[86] The entire process is driven by elevated atmospheric concentrations of carbon dioxide.

Ocean floor sequestration and release of carbonate affects the net flows of the global carbon cycle. In addition, the ocean biological pump contributes to net removal of atmospheric CO_2.[193] Diatoms are hard shelled phytoplankton that fix dissolved CO_2 and produce calcium carbonate. Shell fragments become incorporated into the ocean floor and thus CO_2 is sequestered and removed from the carbon cycle. The White Cliffs of Dover are composed of calcium carbonate shells from plankton deposited in ocean formations during the Cretaceous Period when the oceans were 50 metres above contemporary sea levels.[194]

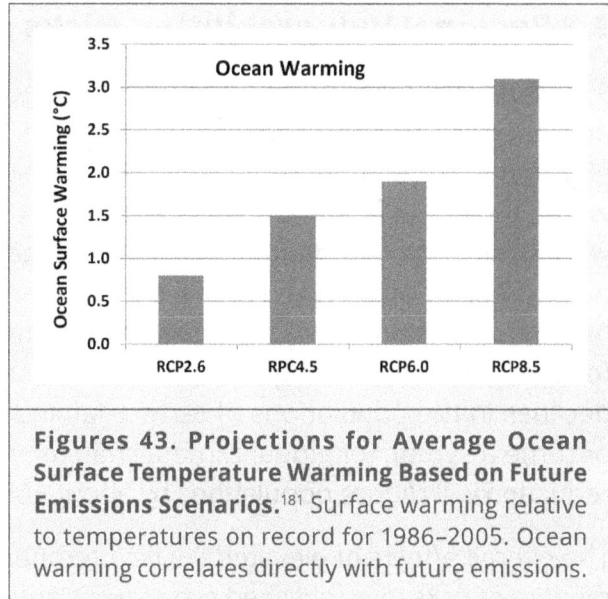

Figures 43. Projections for Average Ocean Surface Temperature Warming Based on Future Emissions Scenarios.[181] Surface warming relative to temperatures on record for 1986–2005. Ocean warming correlates directly with future emissions.

Carbon dioxide flows, between the oceans and the atmosphere, are based on the balance of solubility under changing conditions of CO_2 concentration, pH, and temperature. As the oceans continue to warm and acidify, CO_2 solubility declines, and the capacity of the oceans to act as a carbon sink decreases.

The degree of ocean warming varies with emissions scenarios and latitude. Warming of ocean waters at high latitudes in the northern hemisphere is predicted to be substantially greater than the warming of tropical or southern hemispheric ocean waters.

As the oceans warm, oxygen solubility declines, leading to an expansion of oxygen minimal zones (hypoxic or dead zones) at tropical latitudes.[195] These zones occur at a depth of 200 to 1000 metres and consist of waters with low levels of oxygen saturation and thus a limited capacity to support marine life. Expansion of oxygen minimal zones will constrain and lead to a redistribution of pelagic fish habitats. Pelagic fish are a mixture of larger predatory fish such as tuna and shark and smaller forage fish including herring, sardines, anchovy, and menhaden. Many of these species form the basis of industrial scale marine fisheries.

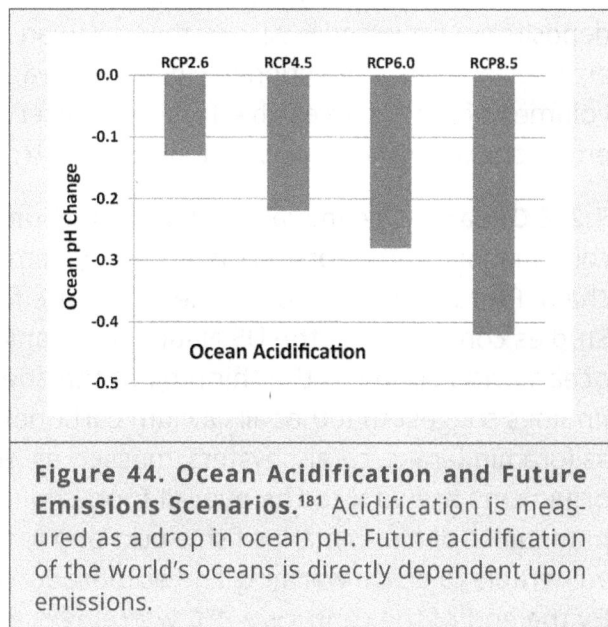

Figure 44. Ocean Acidification and Future Emissions Scenarios.[181] Acidification is measured as a drop in ocean pH. Future acidification of the world's oceans is directly dependent upon emissions.

3.2 Regional Vulnerabilities, Risks and Adaptations

3.2.1 Polar Regions. Declining sea ice has begun to impact sensitive Arctic ecosystems and habitats and no other region on Earth is responding as rapidly to sustained elevated surface temperatures. Committed surface warming has placed Arctic fresh water, terrestrial and marine ecosystems at high risk in the near future (2030–40) without options for adaptation.[196] Isolated indigenous communities are often based on a combination of mixed cash economies and subsistence level hunting, gathering, herding and fishing. In many cases, these communities will need to adapt or relocate following dramatic changes to permafrost, seasonal freeze-over and melt-out, and declines in the populations of terrestrial and marine mammals. The polar bear has become a symbol of global climate change and, by mid-century, committed habitat reduction will reduce populations by about 66% regardless of changes to emissions.[196]

The overall effects of elevated surface temperatures on polar regions are a complex mixture of risks to established ecosystems and human communities and opportunities for economic advances, based on the opening of seaways, longer shipping seasons and access to ports, and new opportunities for exploration, agriculture, commercial fisheries and tourism. Balancing the potential opportunities of opening up the Arctic with the vulnerability of existing natural ecosystems and human communities is a current and future challenge of adapting to anthropogenic climate change. The Russian Arctic contains approximately 90% of the country's natural gas reserves, and Greenland may well contain the world's largest remaining oil resources.[196] Substantial oil and gas deposits are projected in Alaska, the Canadian Arctic, and Norway. Future extraction of arctic oil and gas is questionable given the fragility of the environment and the known volumes of readily accessible fossil fuel reserves relative to the remaining budget of emissions within scenarios that avoid catastrophic climate change.

3.2.2 Oceans. Ocean warming, acidification, and reduced oxygen levels create a combination of stressors on marine ecosystems and key organisms in the oceanic food chain. Pteropods (shelled marine snails) are food for salmon, mackerel and herring. Studies completed by the US National Oceanic and Atmospheric Administration link ocean acidification to the thinning of the shells of West Coast pteropods.[197] These findings suggest that that all calcium carbonate-based shell forming organisms such as foraminiferans, corals, oysters, mussels as well as pteropods are at risk.[198] Calcifying organisms are vital to the overall food chain, and a decline in population has the potential to diminish marine fish stocks and to threaten fisheries. Coral reefs are highly vulnerable to ocean warming and acidification and have little capacity for adaptation. By the end of the century, a 4°C warming scenario places the entire global shellfish fisheries at high risk and will dramatically reduce the catch and species diversity of fisheries associated with coral reefs.[195,199] If surface warming is limited to 2°C, the risk to fisheries is reduced.[195]

The geographic distribution of a given marine species is dependent upon the physiology and, in particular, the thermal sensitivity of the species. Ocean warming will lead to a drop in the body size and the redistribution of tropical species to higher latitudes and/or to deeper, cooler waters. By mid-century, habitat shift will increase fisheries by 30–70% in higher latitudes while tropical fisheries will decline by 40–60%.[195]

Adaptations such as a reduction in quotas for commercial open ocean fisheries and an increase in production from aquaculture can be implemented to reduce risks to marine ecosystems and resources. However, tropical coastlines are highly sensitive to ocean warming, acidification, sea level rise, and extreme weather events. Coastal adaptations to reduce risks are often not possible. Coral reefs have little capacity for natural adaptation or to adaptation by human intervention. The entire ecosystem of coral reefs and associated fisheries is at high risk.

The potential for climate change damage to coastal ecosystems and fisheries is compounded by exposure of human infrastructure to sea level rise and extreme weather events. Scenarios of sustained elevated surface temperatures of 4°C will result in a high risk to socioeconomic security, food security, and livelihoods for vulnerable coastal populations.

3.2.3 Africa. Africa accounts for 20% of the total land mass on Earth and, with 1.1 billion people, 14% of world population.[200] Africa remains the poorest, least developed continent. In 2012, 42.7% of the population of Sub-Saharan Africa lived below the World Bank international poverty line and 23% were classified as undernourished. [201,202] Food and water insecurity, illiteracy, malnutrition, and a lack of infrastructure, in combination with tribal and military conflict, are such that African nations accounted for 36 of the 45 countries listed in the Low Human Development category of the 2015 United Nations Human Development Index.[203]

Conditions among the world's poorest citizens have improved over the past 15 years. Among the 973 million people in the developing nations of Sub-Saharan Africa, annual income per person increased over 3-fold between 2000 and 2014.[204] Over the same period, life expectancy went from 50 to 58 years, and the percentage of students that completed primary school increased from 55% to 69%.[206] Over the past 5 years, GDP per capita among developing Sub-Saharan African countries has increased at a steady rate of about 4.5% annually.[205]

Agriculture accounts for 32% of GDP among African nations and employs 65% of the total work force.[206] Much of the agriculture in Africa can be described as subsistence level family farming. In Sub-Saharan Africa, 98% of the crops are rain fed and productivity per acre is among the lowest in the world.[202] Agricultural productivity is held back by an overall lack of modern farming practices in combination with insufficient transportation and storage infrastructure, and underdeveloped markets and trade.

Potentially, the recent increase in GDP growth among Sub-Saharan African nations could provide the foundation for a much needed Green Revolution. However, the level of poverty and lack of infrastructure is such that these countries remain vulnerable to continued delays in the transition to intensive agricultural practices.

Overall, surface warming within the African continent is predicted to exceed increases in global average temperatures. In tropical regions, surface temperature elevations are projected to occur 10–20 years ahead of similar warming in other continents.[202]

Under the RCP2.6 scenario of aggressive global abatement of emissions, inland temperatures are projected to increase by 1.5–2°C relative to temperatures on record for the period from 1986 to 2005. This surface temperature elevation will be in place by 2050 and sustained indefinitely.[202]

Extreme inland surface warming is predicted under a business as usual scenario of continued excessive rates of global GHG emissions. Over the course of this century, surface temperatures are projected to increase by 4°C with pockets of warming of up to 6°C in southern and northern regions of the continent.[202]

Predictions of changes in precipitation patterns are less certain than temperature; however, under all RCP scenarios, a reduction in annual rainfall in Northern Africa and the southwestern region of South Africa is likely.[202]

Portions of West Africa may experience an increase in the number of extreme rainfall days by the end of the century. Overall, elevated surface temperatures correlate with increased frequency of extreme precipitation events from droughts to heavy rainfalls dependent upon location within the continent.

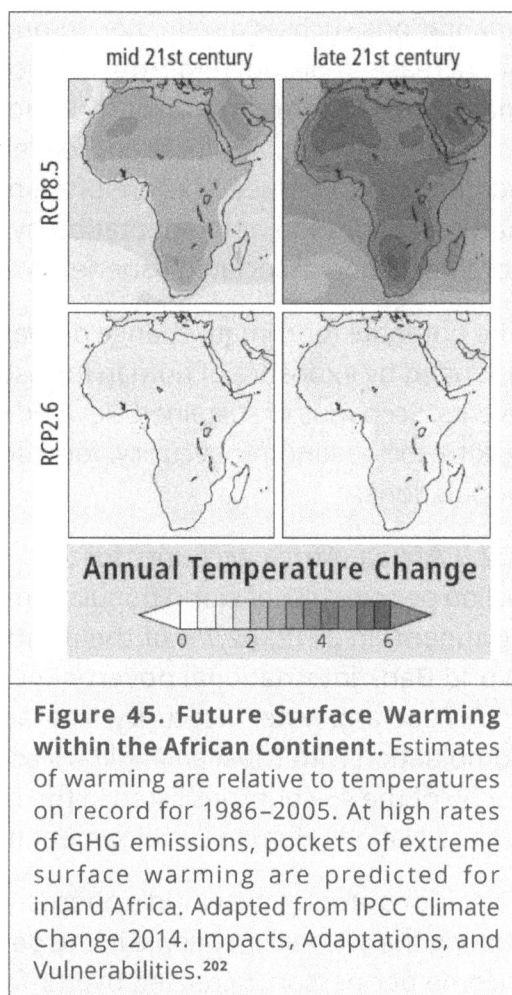

Figure 45. Future Surface Warming within the African Continent. Estimates of warming are relative to temperatures on record for 1986–2005. At high rates of GHG emissions, pockets of extreme surface warming are predicted for inland Africa. Adapted from IPCC Climate Change 2014. Impacts, Adaptations, and Vulnerabilities.[202]

Elevated surface temperatures and changes in precipitation patterns are very likely to have negative effects on crop production. Using detailed historical records of crops, temperature and precipitation at 35 stations in West Africa, researchers were able to model the effects of future climate change on projected yields of sorghum and millet. All models of surface warming result in declining crop yields with higher temperatures driving down yields by 20–40%.[202] Maize (corn) is the most abundant and important crop grown in Africa and is highly vulnerable to climate change. By mid-century, yields for maize production are estimated to decline by 22% across the aggregate of Sub-Saharan Africa.[202] There is some potential for adaptation by relocating maize production to regions of higher elevation. Cassava and other non-cereal crops are less susceptible to climate change, and a switch in crops away from maize may provide some degree of adaptation.[202] Modernization of agricultural practices will improve yields, and increased use of drought resistant varieties will provide capacity for adaptation at 2°C of surface warming. Capacity for adaptation declines with increasing surface temperatures.[202]

Diminished agricultural output is the major risk of climate change for highly vulnerable African nations. However, risks to livestock associated with heat stress and water security are substantial, with no capacity for adaptation under scenarios of a 4°C increase in surface temperatures.[202]

Fish account for approximately one-third of animal protein intake among Africans.[202] Ocean warming with projected declines in tropical fish stocks places the fisheries of West African coastal countries at considerable risk. The annual landed value of fish may decline by up to 21%, dependent upon the degree of ocean warming and the subsequent redistribution of wild fish away from low latitudes.[202] In addition, degradation of coral reefs with ocean warming and acidification threatens reef dependent ecosystems and fish stocks.

The poorer nations in Africa are presently at risk of under nutrition, along with food and water-borne diseases such as cholera and vector borne (insects and other carriers) diseases including malaria, Rift Valley fever, and leishmaniosis.[202] Poverty and associated issues of poor quality drinking water, food security, and availability of medical treatment are the primary drivers of these human health concerns. Future climate change has the potential to exacerbate these risks based on declining food production and other factors affecting human health and socio-economic stability.

The biome of the African continent varies from extensive arid and semi-arid deserts in the north to semi-tropical and tropical rainforests at equatorial latitudes to savannas, shrub lands and dry steppes in southern regions. Tropical ecosystems and savannas are home to over 1,100 species of mammals and provide habitats to some of the most abundant and diverse wildlife populations on Earth.[207] At present, land use change based on an expansion of agricultural lands, livestock grazing areas and harvesting of wood for fuel are changing the natural biome of the continent and compromising the natural ecosystems that support this wildlife.

The total forested area in the African continent has declined by 13.4% since 1990.[208] This rate of deforestation is 4.2-fold greater than the overall rate of world deforestation.[208] African forests account for an estimated 15% of the total carbon sequestered by world's forests. Deforestation in Africa contributes to anthropogenic emissions through the release of forest sequestered carbon.[209] As will be discussed, implementation of land use management practices has considerable potential to limit future deforestation in developing countries. However, surface warming and changes in precipitation patterns are associated with increased desertification and contraction of vegetative areas. Climate change could counter other efforts to limit deforestation and degradation of forested areas.

Sub-Saharan African countries are among the most climate vulnerable, least developed nations on Earth. Exaggerated surface warming and a higher frequency of extreme precipitation and other extreme weather events will disproportionally affect agricultural productivity and water quality. Existing agriculture practices and the current level of economic development within African countries provides little capacity for adaptation.

The African continent can be considered as the future ground zero following a catastrophic outcome of extreme climate change. Continued excessive emissions of greenhouse gases will expose a collective population over 1 billion to risks of malnutrition, famine, failure of clean water supply, disease and loss of life and livelihood. Continued poverty, political instability, climate migration, and human conflict, are likely secondary outcomes of a global failure to adequately combat climate change.

The United Nations 2015 World Population Prospects predicts a world population of 11.2 billion by the year 2100.[210] The UN predicts that 82% of the additional 3.9 billion people added to the planet over the next 83 years will reside in Africa and these citizens will account for 39% of total world population.[210]

Two-thirds of the remaining land on Earth with potential for conversion to agricultural use is located in Sub-Saharan Africa and South America.[211] To minimize loss of important carbon-sequestering forested areas, conversion of land to crop and livestock production to feed an expanding population must be based on the implementation of modern intensive agriculture practices.

The current annual per capita rate of GHG emissions from the aggregate of Sub-Saharan African countries is 3.4 tonnes of CO_2 equivalents per person, which is 50% of the world rate, and accounts for only 6.3% of total global emissions.[124] A full 35% of these emissions originate from deforestation and other changes in land use and not from combustion of fossil fuels.

The realities of future climate change are such that population growth, industrialization, and economic development among African nations cannot be coupled to the excessive use of fossil fuels and GHG emissions as was the case for the industrialization of China and established advanced economies. On the surface, placing emissions restraints on the economic development of countries with the lowest human development ratings imposes a double standard relative to the developed world.

Economic growth among African countries must follow sustainable pathways that are largely uncoupled from greenhouse gas emissions. Failure will be manifest as either another industrialized population in excess of a billion people driving up atmospheric GHG concentrations to what would become catastrophic levels or continued economic stagnation within the poorest countries on Earth.

3.2.4 European Region. The European region extends from the Arctic Ocean to the Mediterranean Sea, and from the Atlantic Ocean to the eastern and south-eastern boundary, defined as the watershed that divides of the Ural and Caucasus Mountains, the Ural River, the Caspian and Black Seas and the Turkish Straits. The IPCC includes Iceland and Turkey as part of the European region for assessing impacts, vulnerabilities, and adaptations to climate change.[212]

The European region, as defined by the IPCC, has a population of 818 million people living in 51 countries, and accounts for 7.3% of the total land area of the world.[212] Economic productivity varies considerably across the whole of Europe. In 2014, Switzerland had a GDP per capita of $85,590 USD, which is 8-fold greater than the world average.[213] In comparison, the GDP per capita of the Ukraine was $3,080.[213]

The total population of countries in the European Union is 510 million or 62% of the population of the entire region.[106] The 2015 United Nations Human Development Index ranked 26 of the 28 EU member states in the category of Very High Human Development.[203] In 2014, the EU accounted for 24% of global GPD with a GDP per capita of $36,423.

Overall, the European region can be described as densely populated, economically advanced, and culturally diverse. Population growth is near zero, and the population is expected to decline by the end of the century. Over the past 30 years, the combination of advances in agricultural productivity and low population growth has led to a decline in arable land use in agriculturally intensive Western European nations. In 1981, 61% of total land area in Germany, France, Italy, the UK, and Spain was defined as arable

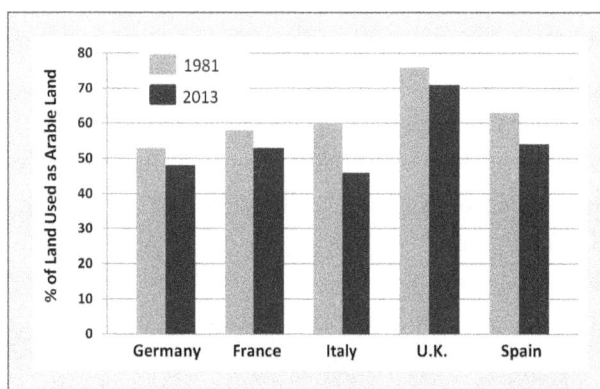

Figure 46. Change in Arable Land use in Western Europe since 1981.[214] With advances in agricultural productivity and very low population growth, land use for agriculture has declined and forested areas have increased.

land used for crops, pastures or fallow.[214] As of 2013, arable land use in these countries had declined to 53% of total land area.[214] Forested areas cover 34% of the total land mass of Europe and account for 25% of global forest sequestered carbon.[209] Forested area has increased by 2.1% since 1990 following advances in forest management and afforestation of land previously used for agriculture.[208] Afforestation in Europe is counter to the 3.1% rate of global deforestation over the same time period. Increased forest biomass in Europe has resulted in a net withdrawal of atmospheric carbon dioxide from the forestry and other land use sector and thus has contributed to reducing GHG emissions by the aggregate of countries in the region.

Future climate models for the European region indicate a pattern of temperature increase that follows overall global warming for each of the IPCC representative concentration pathways. Peaks of warming above the global average are projected for Southern Europe in summer and Northern Europe in winter.[212] The frequency of weather extremes including heat waves, droughts, extreme precipitation and severe storms will increase in direct proportion to the overall degree of global warming. The RCP8.5 business as usual scenario is very likely to result in a robust increase in the number and severity of annual heat waves across the entirety of the region with the exception of Scandinavia and northern parts of the United Kingdom.[212]

Future climate change will have profound effects on fresh water resources in southern Europe. The water content of the soils will decline based on the degree of surface warming such that ground water recharge and water table levels will be significantly reduced.[212] With declining precipitation and an increase in the incidents and severity of droughts in Southern Europe, demands for water to irrigate crops will increase and may outstrip supply options. Improved water management and changes to farming practices will likely be needed to provide a capacity for adaptation to climate change in Southern Europe.[212]

Sea level rise is projected to increase the severity of storm surges in coastal areas. Storm intensity will vary with location with a higher probability of severe events on the eastern coast of the North Sea and on the west coasts of the UK and Ireland.[212] In addition, the coastal areas of Germany, France, Belgium, Denmark, Spain, and Italy will be exposed to potential damage from sea level rise. Construction of physical defenses such as sea walls will reduce the risk and magnitude of damage from climate change-dependent storm surges. In the absence of protective adaptations, coastal flooding has the potential to affect 5.5 million people in Europe by the year 2080.[212] Direct costs of sea level rise are estimated at 17 billion Euros annually by the year 2100.[212] Coastal flooding in Turkey following a 1 m sea level rise would place 3 million people at risk, and adaptation would require $20 billion USD.[212]

Climate change-dependent flooding of river basins following extreme precipitation events are expected to increase for some river systems, and decrease for others within the European region. Specific river basins (including the Rhine and Danube basins) in Finland, Denmark, Ireland, France, and Germany are at risk of flooding.[212] By the year 2080, climate change-dependent river basin flooding could affect an additional 250,000–400,000 people in the EU and, when all factors are considered, could result in a 17-fold increase in economic losses due to flooding.[212]

The effect of future global warming on agricultural productivity varies considerably between Northern and Southern regions of Europe. Unlike the situation in Africa, there is considerable capacity for adaptation of agricultural practices to offset the effects of global warming. A future scenario of a 2.5°C elevation in surface temperatures is projected to have little effect on food production in Europe.[212] Capacity for adaptation is diminished at higher temperatures. A surface temperature increase of 5°C would reduce mean yields by an estimated 10% and would increase the frequency of unfavourable crop years throughout Europe.[212] Crop yield losses of 25% are expected for Southern Europe with 5°C of surface warming.[212] An expansion of climate suitable crop land in Northern Europe would increase yields in this region and it may well be possible to grow corn in Finland by the end of the century.[212] However, under conditions of severe climate change, the shift in crops to boreal forest zones would not compensate for the losses in production from the high productivity soil zones of Southern Europe.[212]

The effects of global warming on European fisheries and aquaculture are complex with some benefits to fish stocks based on relocation of pelagic fish to northern waters.[212] However, ocean warming is likely to have a negative effect on farmed salmon and wild salmon fisheries.[212]

Higher concentrations of atmospheric CO_2 and elevated surface temperatures are projected to enhance forest growth in Northern Europe.[212] However, in Southern Europe, wild fire risk will increase due to longer summers and a higher frequency of droughts and heat waves.[212]

Elevated temperatures and, specifically, the frequency and severity of heat waves will increase mortality among the elderly and those with predisposing illnesses.[212] Adaptive measures can be implemented to reduce risks; however, older buildings are often poorly equipped to provide adequate cooling in the event of extreme heat waves.[212]

Overall, the advanced state of development of European nations provides considerable capacity for future adaptation to climate change. Construction of protective barriers and other physical adaptations will reduce risks of flooding from climate change-dependent severe weather events. Food production is unlikely to be compromised given future scenarios of up to a 3°C increase in surface temperatures. Under conditions of extreme climate change, food production will substantially decline in Southern Europe, but this will be partially offset by an expansion of farm land in northern regions. However, the opening up of new arable land areas will require deforestation and thus a loss of carbon sequestering forest biomass.

Adaptation to climate change is required given the irreversible commitment to a minimal future surface warming of 1.5–2°C. However, under scenarios of a 4°C increase in surface temperatures, the cost of adaptations will increase, while the effectiveness of these adaptations will diminish. An over-reliance on future adaptation in the absence of serious cuts in emissions ignores the potency of extreme climate change to damage ecosystems, food supply chains, and human infrastructure. Further, nations in the developing world lack the infrastructure and resources required for substantial adaptation.

The EU has become a leader among advance economies in efforts to combat climate change. In 2012, per capita emissions for the EU28 were 8.2 metric tonnes of CO_2 equivalents per person, which was half of the per capita emissions rate of the US.[124] The EU is currently on track to markedly surpass the original Kyoto target of a 20% cut in emissions by the year 2020 relative to a 1990 baseline.[215] While the EU has made significant progress, current rates of emissions are still above the world per capita average and the EU is the third largest emitter of greenhouse gases after China and US.[124]

3.2.5 Asia. Asia is by far the largest continent on Earth accounting for 30% of the total land area of the planet.[200] The region is divided into 51 countries that are home to 4.3 billion or 60% of the world's population. Economic development varies considerably among Asian nations with Afghanistan, Yemen, Nepal, Pakistan and Papua New Guinea ranked in the Low Human Development category of the 2015 United Nations Human Development Index.[203] In contrast, 11 Asian nations are ranked among the 49 countries that comprise the Very High Human Development category.[203] The heavily populated nations of China and India are home to 61% of the total population of Asia.[106]

China is by far the largest economy; however, the GDP per capita in China is well below that of Japan and South Korea.[216] India has a population close to that of China but has yet to undergo widespread industrialization and has a GDP per capita that is only 4% of the figure for Japan.[213]

Table 4. Major Economies of Asia (2014 World Bank Data)			
	Population (millions)[214]	GDP ($billions USD)[216]	GDP per capita[213]
China	1,364	$10,354	$7,590
India	1,295	$2,048	$1,581
Indonesia	254	$888	$3,492
Japan	127	$4,601	$36,194
South Korea	50	$1,410	$27,970
Saudi Arabia	31	$746	$24,161

Since 1901, surface temperatures over the broad expanse of North Asia have increased by approximately 2°C.[217] This pattern of warming extends to portions of East Asia and Central Asia and markedly exceeds the increase in average global surface temperature since pre-industrial times.

By the year 2100, under the RCP2.6 scenario of aggressive mitigation of climate change, populated regions of Asia are projected to warm by 1.5–2°C relative to temperatures over the period of 1986–2005.[217] Under this scenario, surface warming in the extreme north is predicted to be slightly higher than average for the region.[217]

Surface temperature predictions for Asia are strikingly different under the business as usual RCP8.5 scenario of future GHG emissions. By the end of the century, extreme surface warming of over 6°C above the 1986–2005 mean is predicted for portions of North, Central, West, East and most of South Asia.[217] Temperature increases of 4°C are predicted for Southeast Asia and the southern portions of India under the RCP8.5 emissions scenario.[217]

Models of climate change-dependent shifts in precipitation patterns over the region are less certain; however, an overall increase in precipitation across North Asia is predicted.[217] The magnitude of changes in precipitation patterns are projected to increase under the business as usual RCP8.5 scenario of future emissions.[217] Severe climate change will be characterized by more frequent extreme, heavy precipitation events within the summer monsoons regions of India and Southeast Asia and an increase in the intensity of precipitation from tropical and extratropical cyclones that make landfall.[217]

Forests in the whole of Asia account for 11.4% of the total carbon sequestered and stored in the biomass and soils of the world's forested areas.[209] Tropical forests in Southeast Asia provide over half this carbon sequestration capacity. The land area of tropical forests in Asia has declined by 8.4% since 1990 due to an expansion of palm plantations and aggressive forest harvesting practices, some of which are illegal.[208,218] Forested areas in other regions of Asia have increased since 1990, and this increase in plant biomass partially offsets the losses from deforestation in tropical regions.[208]

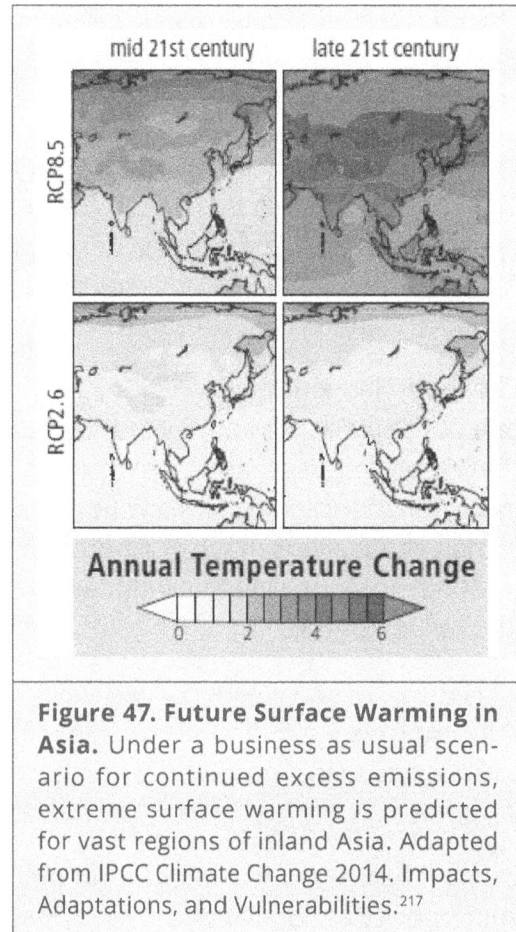

Figure 47. Future Surface Warming in Asia. Under a business as usual scenario for continued excess emissions, extreme surface warming is predicted for vast regions of inland Asia. Adapted from IPCC Climate Change 2014. Impacts, Adaptations, and Vulnerabilities.[217]

Future surface warming is predicted to facilitate an expansion of boreal forest land area in North Asia and a northern expansion of subtropical evergreen forests in East Asia.[217] The effects of future climate change on forests in South and Southeast Asia are uncertain. Improvements in forest management and land use practices have considerable potential to reduce deforestation in Southeast Asia, and these measures can be effective tools to maintain biodiversity and increase the overall carbon sequestration capacity of global forests.

Currently, areas of West and Central Asia and parts of China struggle with issues of water supply.[217] At higher surface temperatures, evaporative losses in combination with demands for irrigation and household use are likely to exacerbate water stress in these regions. In China, studies suggest insufficiencies in water supply for agriculture may begin sometime between 2020 and 2040.[217]

The net effects of climate change on agricultural production in Asia are complex and vary with location. Rice is the staple food of the region and heat stress is projected to reduce the overall yield of rice with western Japan, eastern China, the south Indochina peninsula, and the northern section of south Asia as the more vulnerable rice producing regions.[217]

In the Asian portion of Russia, the frequency of climate-dependent food production shortfalls in the major crop growing regions is predicted to increase from twice per decade to 5–6 crop years per decade under scenarios of extreme climate change.[217] A higher frequency of droughts will be the cause of crop failures and shortfalls in production and could threaten Russia's food supply system.[217]

The Indo-Gangetic plains region has been called the bread basket of South Asia and is one of the world's great wheat growing areas accounting for 14–15% of global production.[217] Under a scenario of extreme surface temperature elevation, predictions of a localized drier climate, along with severe heat stress and a diminished water flow from Himalayan glaciers, have the potential to reduce wheat yields by up to 50%.[217] Such a catastrophic decline in production would threaten the food supply of an estimated 200 million people.[217]

As is the case for climate change modelling in Europe, increasing surface temperatures may improve yields and productivity in some regions. Specifically, warming may lead to an increase in crop yields in northern areas that currently lack an adequate growing season.

Overall, future models of extreme surface temperature are predicted to increase the vulnerability of populations in many parts of Asia to food production and food security risks. Projected regional declines in rice and wheat yields, in combination with a higher frequency of crop failures following extreme weather events, likely outweighs the benefits of a northern expansion of the total crop production area and projections of increased productivity in some areas. Adaptations such as diversification of crops to match changes in growing conditions, and the development of heat stress tolerant crop varieties may partially ameliorate production losses. However, a future based on assumptions that agricultural practices can be adapted to maintain productivity with increasing severity of climate change in Asia is highly risky. Under a business as usual scenario of continued excess rates of global emissions, sustained food production is unlikely, given the projections for extreme surface temperature elevations and changing precipitation patterns.

Aquaculture and wild fisheries are an important component of the overall food supply chain and a significant contributor to the economy and livelihoods in many regions of Asia. Asia accounts for an estimated 85% of the global production of fish from the combination of wild capture and aquaculture.[217] Overall, body weights of marine fish in Asian waters are predicted to decline by 14–24% by mid-century due to habitat change under the business as usual scenario of continued high rates of global emissions.[217] Climate change will result in a redistribution of fish stocks away from the tropics to northern waters. Indonesia and other low income tropical Asian countries with a significant reliance on wild fish capture are vulnerable to warming oceans and subsequent declines in fisheries production.

Thirty-four percent of the world's coral reefs are located in Southeast Asia and most of these structures are found on the coasts of the Philippines and Indonesia.[219] Some of the coral reefs in Southeast Asia provide the habitat to the highest level of biodiversity among all global marine ecosystems. The value of the coral reef fisheries in Southeast Asia is estimated at $2.4 billion USD annually and is an important con- tributor to the local economy.[220] The damaging effects of human activities such as overfishing and illegal fishing using explosives (blast fishing) have led to the degradation of 85% of the coral reefs in Indonesia, and 70% of

Status of Coral Reefs in SE Asia

Figure 48. Current Status of Coral Reefs in Southeast Asia.[221] Destroyed (90% loss and unlikely to recover). Critical Stage (50–90% loss). Threatened (20–50% loss). Good Condition (not under threat).

the coral reefs in the Philippines.[219] As of 2008, in the whole of Southeast Asia, 60% of the total reefs were classified as destroyed or are at a critical stage.[221] Climate change-dependent decalcification of shell constructing organisms following ocean warming and acidification exacerbates other activities of man that are destroying the region's coral reefs.

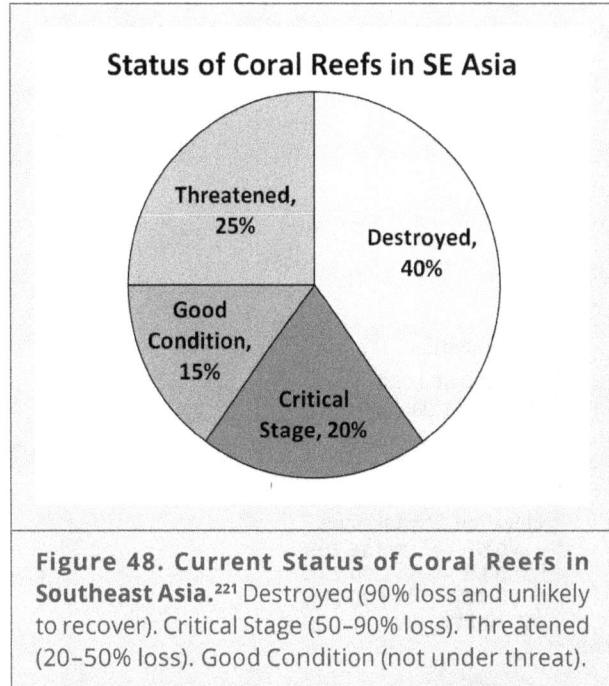

Asia accounts for an estimated 90% of the world population exposed to tropical cyclones.[217] Of the 130 port cities defined as at-risk of storm surge flooding, 10 cities account for half of the total population, and all but Miami are located in Asia.[222] Climate change-dependent sea level rise will increase the storm surge potency of cyclones such that the flood potential of what was a 1 in 100-year storm may well be equal to the flood potential of a future 1 in 10-year storm event.[217] Given the coastal location of Asian cities, a business as usual emissions scenario and resultant year 2100 sea level rise will place 362 million people at risk of flooding from 1 in 10-year tropical storms.[217]

Future heat-related mortality will increase in relation to the degree of surface warming with higher death rates in the poorer regions of Asia with limited capacity for adapta- tion. The elderly, young children, the malnourished, and people with cardiovascular or respiratory diseases will be pre-disposed to heat stress.[217]

Overall, Asia is second to Africa and Small Island States for vulnerability to the damaging effect of future climate change. Above average increases in surface temperatures are predicted for much of the region. Surface warming and extreme weather events will directly impact human health and productivity.[217]

Asian Cities At Risk of Sea Level Rise[222]

City Population
○ 100,000-500,000
○ 500,000-1,000,000
⬤ More than 1,000,000

% Urban Population in Low Elevation Coastal Zones
0%
0-5%
5-10%
10-15%
15-20%
20-25%
20-25%

Figure 49. Exposure of Population Centers in Asia to Risks of Sea Level Rise. Adapted from Fuchs, R.J. 2010. Cities at Risk: Asia's Coastal cities in an age of Climate Change. Asia Pacific Issues, No 96. East-West Center, Honolulu.[222]

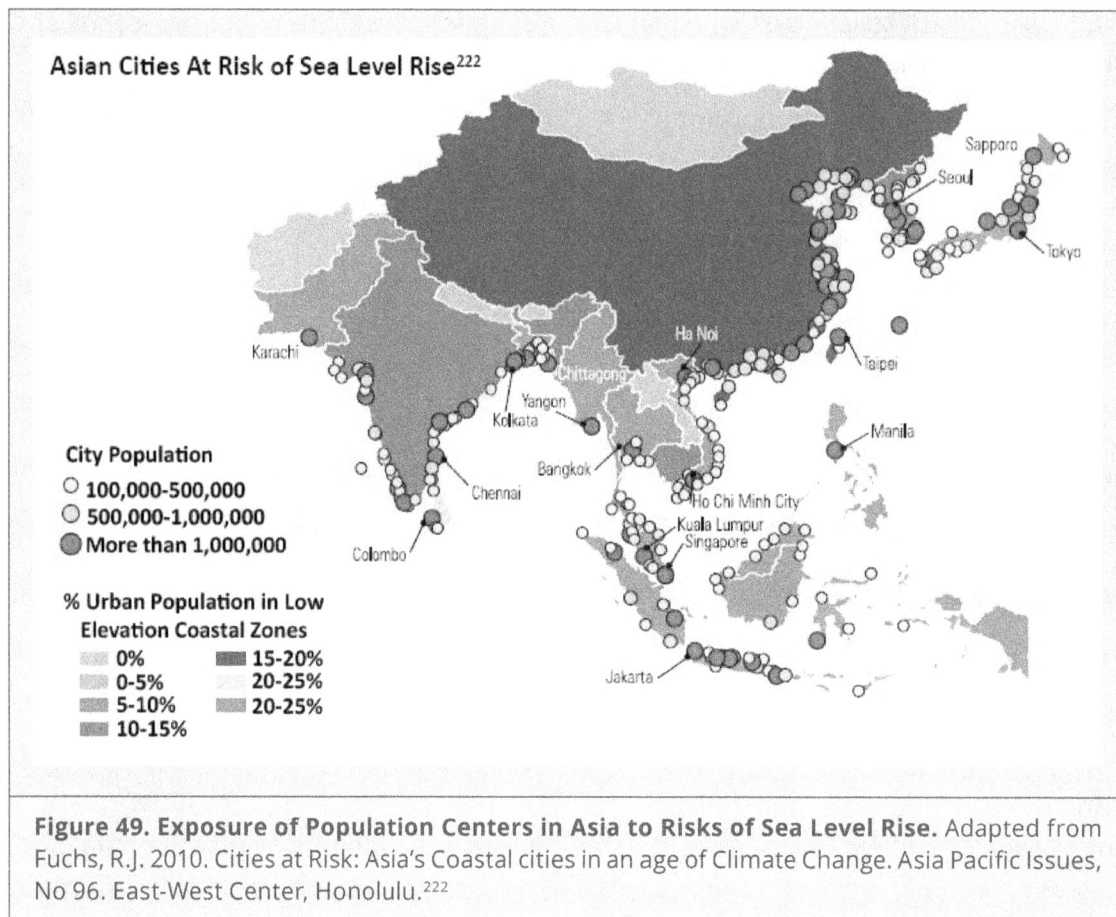

In many regions within Asia, water security, food production, and food security, are at considerable risk. Climate change-dependent sea level rise, extreme precipitation events, and cyclone storm surges place vast populations living in low-lying coastal and river basin areas at risk of flooding. Among developing countries, climate change-dependent flooding, poverty, and food and water security issues are expected to place populations under additional risks of diarrheal and vector-borne diseases.[217]

Construction of sea walls, dikes, reservoirs and diversion channels will be a future requirement to adapt to the threat of climate change-dependent flooding and storm surges.[217] Strategic planning to improve the sewage, drainage and water supply networks of low-lying cities can provide some capacity for adaptation. As an example, upgrading the dated drainage system in Mumbai would reduce the potential for large scale flood damage by an estimated 70%.[217]

Engineered urban adaptation to the risks of climate change-dependent flooding requires considerable capital that is often beyond the resources of developing nations. The poorer countries of Asia with limited capacity for adaptation will likely remain vulnerable to the potentially devastating effects of climate change-dependent extreme weather events.

Modernization of agricultural practices and the use of high yield varieties of rice, wheat, maize and other grains has occurred over most of Asia and allowed for self-sufficiency in food supply. The success of the Green Revolution in Asia provides some capacity for adaptation of agriculture to changes in growing conditions. Under the RCP2.6 future scenario of aggressive abatement of emissions, adaptation of agricultural practices should allow for sustained food production. However, if the current excessive rate of GHG emissions continues indefinitely, adaptations are unlikely to be effective and vast populations in Asian countries will be vulnerable to reduced yields and crop failures.

Other than Japan, Asian countries were not part of the post-WW II economic miracle. The economic boom in South Korea began in the 1960s[223] and rapid growth of the economy in China did not take hold until the 1990s.[213] Industrialization in Japan, South Korea and China followed the world pattern of growth in GDP coupled to increased emissions from fossil fuel combustion. Widespread industrialization has yet to occur in India and much of Southeast Asia and the early stage of economic development in these countries is reflected in the relatively low rates of per capita emissions.

In 2012, the sector of forestry and other land use accounted for 42% of the total emissions from Southeast Asian countries. This rate of emissions equalled 3.2% of the global total release of greenhouse gases. The rate of destruction of tropical forests and peatlands in Southeast Asia offsets any gains in forest biomass in other regions of Asia.

The combined populations of India and Southeast Asian countries account for 26% of the world population and, when forestry and land use is considered, GHG emissions from these countries amount to 13.5% of the world's total. As is the case for developing nations in Africa, to avoid global climate change catastrophe, economic development for the 1.9 billion people of India and Southeast Asia must follow low carbon pathways and deviate from the historic pattern of GDP growth that is coupled with GHG emissions.

3.2.6 Australasia. The IPCC defines Australasia as the lands and territories of Australia and New Zealand. Australia ranks second to Norway in the 2015 Human Development Index of the United Nations, and New Zealand ranks 9th on the index.[203] On a per capita basis, Australia has one of the world's most productive economies and is among the group of countries that underwent a post-WW II economic boom beginning in the late 1940s.[213] Australia and New Zealand are rich in natural resources and these economies are largely export driven. Both nations are net exporters of food products, and agriculture accounts for 11% of the total value of exports from Australia, and 56% of the export value of goods produced by New Zealand.[224] The total population of the 2 countries is 29 million[106] and, in 2012, GHG emissions accounted for 1.6% of the world total.[124]

Warming over the land mass of Australasia since 1901 is comparable to the global average of a 1°C increase in average surface temperatures.[224] Under the RCP2.6 scenario of aggressive mitigation of climate change, predictions of a future minimum surface warming for Australasia are consistent with global averages and should stabilize at 1.5–2°C above pre-industrial temperatures.[224] However, above average surface temperatures are predicted for the interior of the Australian continent under the business as usual RCP8.5 scenario of future emissions.[224] Predictions of a 6°C increase in surface temperatures for central Australia by the end of the century relative to 1986–2005 averages are consistent with hotspots of warming for central regions of all continents under a future scenario of continued, excessive emissions of GHGs.

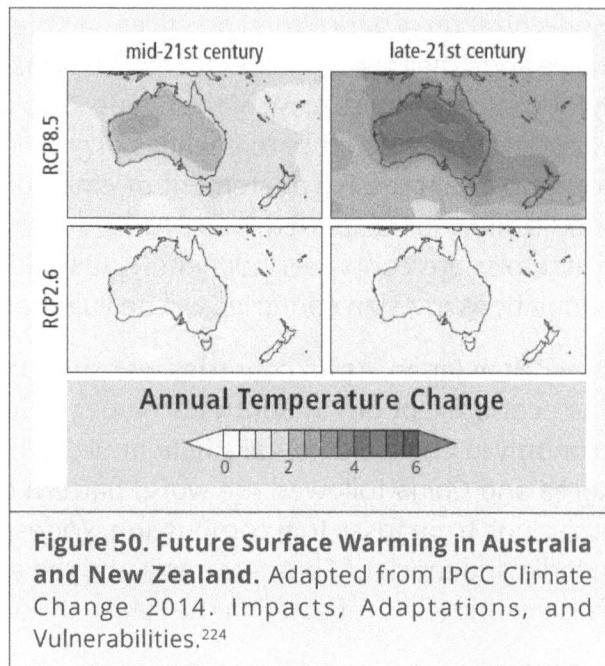

Figure 50. Future Surface Warming in Australia and New Zealand. Adapted from IPCC Climate Change 2014. Impacts, Adaptations, and Vulnerabilities.[224]

Average annual rainfall is projected to decrease in Southwestern Australia based on the RCP8.5 business as usual scenario.[224] Future climate change-dependent shifts in precipitation patterns for the remainder of the continent are uncertain, and range from predictions of increased to decreased precipitation in the agriculturally important southeastern region.[224] Given existing water supply issues in much of Australia, future climate models of the impact of both increased and decreased precipitation patterns have been developed. The effects of climate change on the frequency and severity of cyclones and other major storm events in Australasia are uncertain.[224]

Australasia has the lowest precipitation rate for inhabited continents. The lack of moisture, irregularity of rainfall patterns, and the importance of agriculture to the economies of Australia and New Zealand have led to the construction of extensive water flow infrastructure, along with the development of water use policies, to manage the resource and maintain agricultural productivity. Australia has constructed over 500 large dams and has the world's highest per capita water storage capacity.[225] In 2004–2005, irrigated land accounted for 1% of the total arable land, yet produced 23% of the gross value of agricultural commodities.[226] The Murry and Darling River basins drain the highly productive southeast region of Australia. This region receives little rainfall and water flows are managed to achieve a balance between water extraction for irrigation and the environmental sustainability of the rivers. Extensive, multifaceted urban water conservation and recycling programs have reduced per capita water consumption in Melbourne by 40% and in Brisbane by 50% since the early 2000s.[224]

The advanced state of water storage infrastructure, water allocation programs, and the efficiency of water use provide Australasia with considerable capacity for adaptation to future fluctuations in water supply. However, a business as usual emissions scenario leading to surface warming of 4–6°C will increase evaporative losses and likely increase the frequency of droughts and other extreme weather events. A future of extreme climate change could disrupt water supply beyond the capacity for adaptation.

Over 60% of Australian beef products are exported, and Australia is second to Brazil for beef exports.[224] The total export value of lamb, poultry, pork, wool, and dairy products is comparable to the value of beef.[224] Livestock production systems in Australia rely heavily on pastureland that is vulnerable to changes in precipitation patterns and elevated surface temperatures. Heat stress can affect animal productivity, and a future scenario of 3°C in surface warming across Australasia is predicted to reduce the value of beef, sheep and wool products by 4% with similar declines in dairy production in most regions.[224]

Adaptations of farming practices such as selection of appropriate cultivars and shifts in sowing times has the potential to maintain or even increase crop production under conditions of moderate climate change in Australasia.[224] However, crops are vulnerable to shortages in water supply and thus increased frequency of droughts and regional decreases in seasonal precipitation could negatively affect wheat, rice and sugarcane production.

The risk of wildfires is expected to correlate with the increase in surface temperature in most of Southern Australia.[224] Models of reduced rainfall show an increase in the risk of wildfires with the highest vulnerability in Southeastern Australia.[224]

Flood risks following extreme rainfall events are predicted to increase in many regions of Australasia, with a higher vulnerability in northern regions.[224] Sea level rise and storm surge potential are likely to increase the vulnerability to flooding near river mouths.[224] Flood risk management such as the introduction of specific building codes, managed relocation, and systems of floodways and floodplains can reduce the potential for injury and death, and structural damage in flood prone areas.[224]

The RCP8.5 business as usual scenario for future emissions predicts a sea level rise of 0.53–0.97 m by 2100.[224] In Australia, a sea level rise of 1.1 m would affect $226 billion USD in physical assets.[224] Climate change adaptations have been incorporated into government recommendations for planning of coastal settlements and infrastructure. Capacity to withstand a sea level rise of 0.5 m by the year 2090, along with long-term provisions to withstand up to a 0.8 m rise in sea level are recommended for coastal structures.[224]

As is the case for coral reefs around the globe, Australia's Great Barrier Reef is highly vulnerable to ocean warming and acidification with little natural capacity for adaptation to changing ocean conditions. The complete complex of reefs stretches along 2,300 kilometres of coastline, is the largest single structure built by lifeforms on Earth, and is home to over 1,500 species of fish.[227] The Great Barrier Reef was designated a world heritage site in 1981.[224] Ocean warming and acidification resulting from atmospheric CO_2 levels of 450–500 ppm are predicted to increase coral bleaching and the incidence of disease and mortality within the reef.[224] Future atmospheric CO_2 concentrations in excess of 450 ppm are predicted for all scenarios of future emissions with the possible exception of the RCP2.6 scenario of aggressive mitigation. Extreme ocean warming and acidification, following a business as usual scenario of sustained emissions, will cause irreversible catastrophic damage to one of the great marine ecosystems on Earth.

Australia and New Zealand have considerable infrastructure and polices in place to reduce existing risks associated with irregularities in precipitation patterns, water shortages, river basin flooding and heat waves. These factors and the advanced state of development in the region allows for future adaptations to manage risks associated with moderate climate change. However, the magnitude of future climate change following a business as usual scenario of continued excess accumulation of GHGs in the atmosphere will likely overcome the capacity for adaptations and substantially increase the risks of declining yields and failures in agriculture production, damage to coastal infrastructure, and direct effects of heat waves on human health.

GHG emissions originating from Australasia are a minor contributor to the global total, and one could put forth a position that efforts to reduce emissions from a region with a population of only 27.5 million are misplaced. This argument ignores collective global responsibility. Emissions must be considered on a per capita basis for all nations. Australia met its Kyoto accord targets for the 2008–2012 reporting period; however, these targets were weak to the point of non-existence and did little to curtail emissions.[215] In reality, at 27.4 tonnes of CO_2 equivalents per person, Australia's per capita rate of emissions is among the highest on Earth and 3.3-fold greater than counterpart per capita emissions from advanced economies within the European Union.[124]

3.2.7 North America. The IPCC defines North America as Canada, the US, and Mexico. Caribbean and other island nations that are typically considered as part of North America are placed in the small islands category for assessment of vulnerabilities, impacts and adaptations to climate change. Canada's northern arctic is considered as part of the Polar Regions in the IPCC designation.

As of 2016, the total population of Canada, the US, and Mexico was 489 million or 6.6% of the world total.[228] The combined economies of these three countries accounts for 26% of global GDP, with the United States, as the world's largest economy, accounting for 85% of the total GDP of North America.[216] The United Nations ranks the US in 8th, Canada in 9th (same ranking as New Zealand), and Mexico in 74th place among all countries in the 2015 Human Development Index.[203]

In the US, 44% of the total land area is used for agriculture, while in Mexico, arable land accounts for 55% of the area of the country.[214] In addition, extensive land areas in the populated southern regions of Canada are devoted to agriculture. North America is a net exporter of food with efficient production of a wide range of grains, oilseeds, fruits, nuts, vegetables and animal products coming from the diverse growing areas of the continent. The US is the world's largest producer of corn with nearly half of the total crop used in animal feeds. Currently, the production of bioethanol fuel consumes about 30% of the US corn crop.[229] Corn and soybean are primarily grown in the US corn belt extending east to west from Ohio to Nebraska and north to the Canadian border. North America has extensive livestock industries. The US is the world's largest producer of beef, poultry, and dairy products, and is second to China for pork production.[230] Agricultural goods are traded between the three countries of the region and exported around the world with Japan, China, and Europe as the primary export markets.[231]

Water stress is common to arid areas in the Midwest and the Southwestern regions of the US. The interconnected water system of California is the largest in the world and serves to irrigate 5.7 million acres of farmland and supplies water to over 30 million people.[232] Natural ecosystems in the Southwestern US are under threat of decline from a lack of moisture due to water extraction to meet urban demands, along with the irrigation requirements of agriculture.[233]

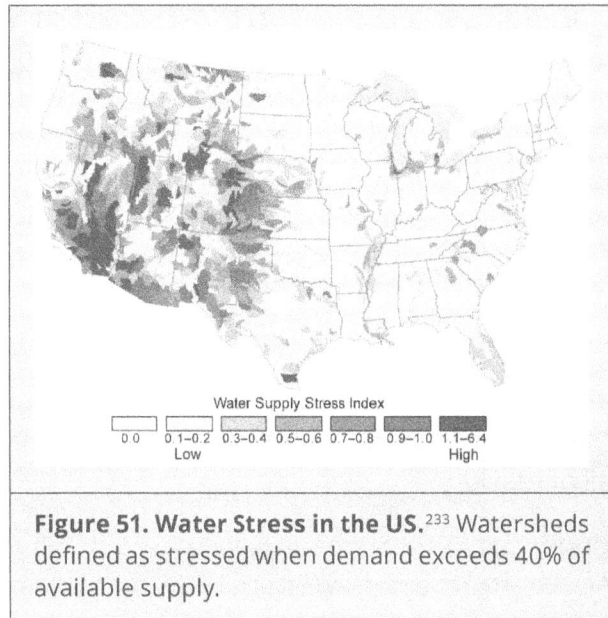

Figure 51. Water Stress in the US.[233] Watersheds defined as stressed when demand exceeds 40% of available supply.

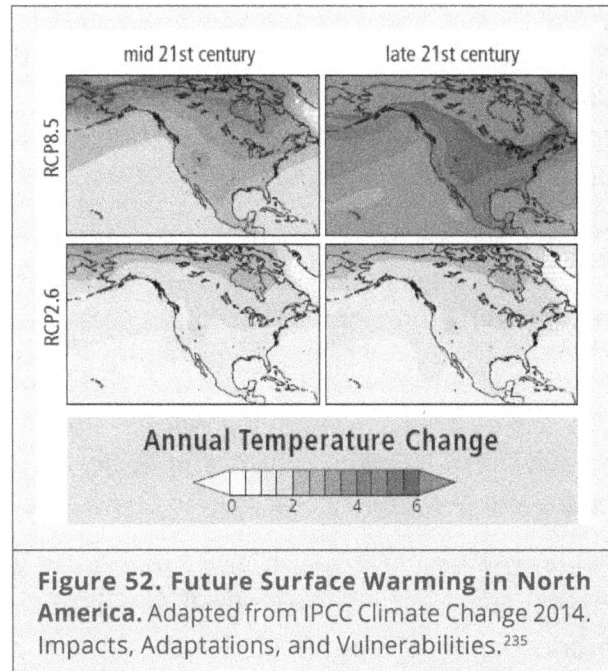

Figure 52. Future Surface Warming in North America. Adapted from IPCC Climate Change 2014. Impacts, Adaptations, and Vulnerabilities.[235]

The arid and semi-arid regions of Central and North Mexico account for 80% of the nation's GDP and are home to 75% of the population.[234] Insufficiencies of water supply, inefficiencies of use, increased demand, and inadequate irrigation infrastructure are such that agricultural productivity and urban centers are vulnerable to water shortages.[233]

Since 1901, surface temperature changes in North America have followed the global average with perhaps less warming in the Southeastern US and Mexico, and slightly higher temperatures in Central Canada.[235] By the end of the century, under the RCP2.6 scenario of an aggressive mitigation of climate change, warming over North America will stabilize between 1.5–2°C above pre-industrial temperatures with slightly warmer conditions predicted for Canada.[235]

Based on the RCP8.5 business as usual scenario of continued high rates of emissions, extreme inland surface temperatures, well in excess of global average surface warming, are projected for North America.[235] By the end of the century, surface temperature elevations of 5–6°C, relative to 1986–2005 temperatures, are predicted for much of the US and Mexico. Even higher temperatures are predicted for parts of Canada and all of Alaska .[235]

Under the RCP8.5 scenario, increased precipitation is projected for most of Canada, Alaska, and the northwestern US.[235] A decrease in precipitation is predicted for Mexico and the Southern and Southwestern regions of the US.[235]

Predictions of extreme heat, dryness and reduced precipitation under a business as usual future of continued excessive emissions will exacerbate current water stress in the Midwest and Southwestern US and most of Mexico, and will place these regions at risk of water shortfalls.[235]

Diversion of water supply to heavily populated regions with intensive crop irrigation needs may trigger regional disputes over water authority and substantial water transfer costs between regions.[235]

A future of extreme dry, hot summers and increased frequency of droughts will increase the risk of intense damaging wildfires for many regions in North America.[235] The Pacific coast, Northern plains and Rocky Mountains are highly vulnerable to climate change-dependent increases in the frequency and intensity of wildfires.[235]

Under the RCP2.6 scenario of aggressive action to curtail emissions, future climate change is unlikely to affect overall agricultural productivity in North America.[235] Animal production will not be affected under these conditions. Adaptations such as climate-dependent relocation of growing areas and selection of new, thermal-tolerant varieties should maintain crop yields under modest conditions of climate change.[235]

Overall, agricultural productivity in North America is projected to decline under future conditions of extreme climate change.[235] Central North America is among the most heat stress at-risk regions on Earth.[235] In the highly productive corn belt and southern agricultural regions of the US, surface warming will exceed the thermal threshold of grains, livestock and dairy.[235] In these regions, corn, soybean and cotton yields are projected to decline by 30–82%, dependent upon the particular crop and the severity of thermal stress.[235] Heat stress, in combination with reduced forage quality, will negatively affect cattle weight gains and milk production.[235] In Mexico, heat stress under a future scenario of extreme climate change may reach the danger level for farm animals.[235] Coffee production in Mexico, and the livelihood of an estimated 500,000 farmers, is highly vulnerable to extremes of climate change.[235] Output from the irrigated growing regions of California is unlikely to be affected by climate change; however, these projections are based on an assumption of continued adequacies of water supply.[235] Agricultural productivity in Canada and the Northern US will increase with longer and warmer growing seasons and increased precipitation.[235] Climate-dependent relocation of crop growing areas and other changes in agricultural practices are likely outcomes of extreme surface temperature elevations. However, these adaptations cannot compensate for declining production from water and heat stressed regions.

Extremes of surface warming and increased frequency of heat waves will have direct effects on human health. The advanced state of development in the US and Canada will facilitate adaptations such as increased use of air conditioners and heat wave warnings, and thus reduce the potential for heat stress to impact human health. Mexico has less capacity for adaptation and is at higher risk of heat stress related mortality and morbidity.

By the year 2070, coastal storm surges in combination with a 0.5 metre sea level rise will result in a 10-fold increase in asset exposure and a 3-fold increase in population exposure to storm and flood damage for the cities of Miami, New York, and New Orleans.[235] By the end of the century, under a business as usual emissions scenario, the number of extreme precipitation days in New York City is projected to double and, in absence of adaptations, the cost of river flood damage to the city of Boston is estimated at $57 billion.[235]

Overall, the US and Canada have considerable capacity for adaptation to a modest degree of climate change. Coastal and urban infrastructure can be built to reduce the risk of flooding and storm damage for at-risk cities. This capacity for adaptation is likely to be insufficient to adequately limit exposure to the potency of sea level rise and storm surges under a future scenario of extreme climate change.

The US is the world's largest producer of agricultural commodities, and food production is well in excess of domestic needs. Moderate declines in total agricultural output are unlikely to cause domestic food shortages. Agriculture accounts for only 1.2% of GDP in the US and 1.8% of GDP in Canada, and a drop in production will have a marginal impact on the economies of these countries.[236] Further, under conditions of moderate surface temperature increases, adaptation of agricultural practices and increased agriculture productivity in Canada and the Northern US can compensate for projected declines in output from the US Midwest. However, under conditions of extreme climate change, the projected magnitude of heat and water stress damage to highly productive agricultural regions of the US will result in a net loss of capacity to produce food in the whole of North America.

In Mexico, under a business as usual emissions scenario, predictions of extreme heat and decreased precipitation coupled with the current limitations of water supply management places a population of 128 million at risk of food and water shortages during the current century. When compared to the US, Mexico lacks a similar capacity for adaptation to extreme weather events and thus is at greater risk of deaths, injury, and infrastructure damage.

Forested areas in North America account for 13% of the total carbon sequestered by the world's plant biomass.[209] Since 1990, rates of afforestation and reforestation in North America have exceeded the rate of deforestation such that the quantity of forest sequestered carbon has increased by 6.3%.[209] This rate of expansion of forested areas is comparable to that of Europe. As is the case for Europe, the forestry and land use sector in North America functions as a carbon sink and, to a limited extent, partially offsets emissions from industry and other economic sectors. Globally, the buildup of forest biomass in Europe and North America does not compensate for the rates of deforestation and forest degradation in Africa, Asia, and South America.

North America is a major producer and consumer of fossil fuels. In 2014, the US led all other countries in the production and use of oil and natural gas, and produced 12% of the world's coal.[237] Energy demands are such that the United States is a net oil importer. Domestic production of natural gas and coal in North America approximates consumption.[237]

In 2012, Canada, the US, and Mexico were home to 7% of the global population but were responsible for 16% of world emissions.[124] The production of fossil fuels is an emissions intensive process and a significant contributor to the overall rate of release of greenhouse gases from oil and gas producing countries. Per capita rates of emissions in Canada and the US are among the highest in the world and twice the per capita emissions rate of the European Union.[124]

3.2.8 Central and South America. The natural biomes of Central and South America consist of a full spectrum of climates and ecoregions including vast tropical rainforests; sub-tropical and temperate forests, wetlands, grasslands and savannas; deserts and semi-deserts; montane; and mangroves, scrubland and tundra. The Amazon basin covers 40% of the total area of South America[238] and contains over half of the world's rainforests.[239] The plant

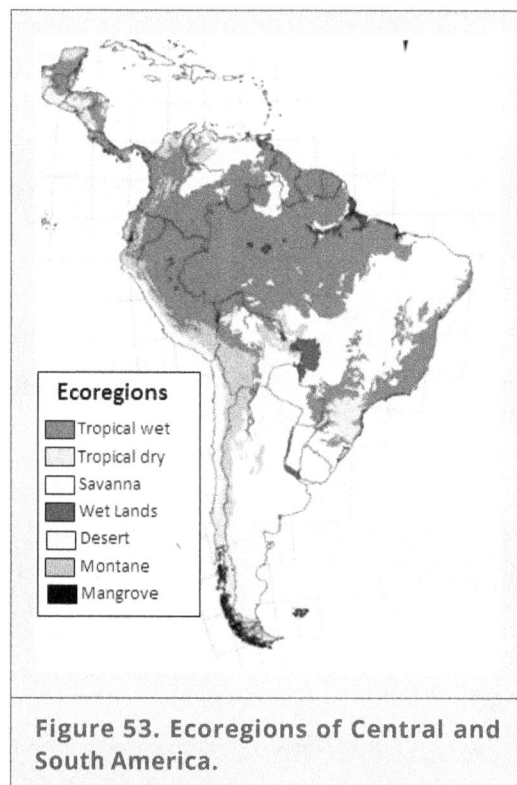

Figure 53. Ecoregions of Central and South America.

biomass of the Amazon accounts for 21–28% of the world total of forest sequestered carbon.[240,209] The Amazon provides habitats to an immense variety of plant and animal life. The biodiversity is such that 1 in 5 species of fish and birds live in the Amazon, along with 427 species of mammals, 428 species of amphibians, 378 species of reptiles, 40,000 species of plants and 2.5 million species of insects.[240]

Among the nations of Central and South America, the United Nation's 2015 Human Development Index ranks Argentina (40th place) and Chile (42nd place) in the Very High Human Development category.[203] The remaining nations are ranked in the High and Medium Human Development categories, and there are no countries listed within the Low Human Development scale of the index. As of 2014, the total population of the region was 415 million, with approximately 50% of this total living in Brazil. Brazil has by far the largest economy accounting for half of the total GDP of the region.[216] The GDP per capita of Brazil and the larger nations of South America tend to approximate the world average of $11,000; however, the GDP per capita of the Central American nations of Guatemala, Honduras, and Nicaragua are well below this level of economic activity.[213]

Sixty-three percent of the total land area of South America is within the borders of Brazil and Argentina.[200] Since the 1960s, these nations have adopted and advanced Green Revolution agricultural practices to become economic powerhouses in the sector of world food production. Brazil is a world leader in plant breeding to adapt non-native crops and grasses to grow in tropical climates. Over a 40-year period, vast quantities of lime were applied to the inland Cerrado region of Brazil, south of the Amazon basin. In so doing, Brazil transformed non-productive acidic savannah soils into arable land suitable for growing soybeans and other crops. Brazil and Argentina rank second and third behind the US in soybean production, and these countries are the world's third and fourth largest producers of corn.[241] Domestically produced soybean and corn provide the feed inputs for massive animal industries. Brazil and Argentina are the second and fourth largest producers of beef cattle and Brazil ranks among the top five nations for the production of milk, pigs, and poultry.[241] Overall, Brazil is second to the US for total value of exported agricultural products.[241]

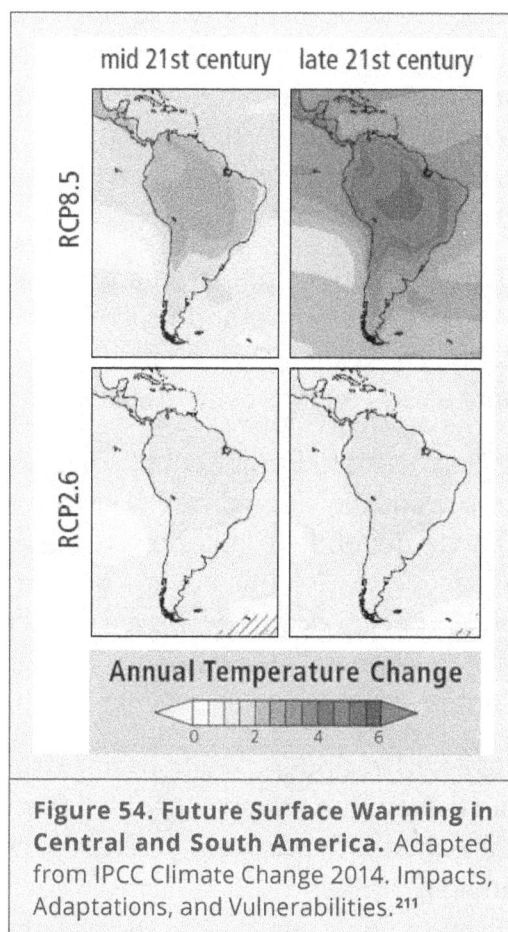

Figure 54. Future Surface Warming in Central and South America. Adapted from IPCC Climate Change 2014. Impacts, Adaptations, and Vulnerabilities.[211]

Brazil is by far the world's largest producer of sugarcane.[241] Bioethanol produced from sugarcane is blended with gasoline at mandated inclusion rates of 27% and thus is an important contributor to fueling Brazil's fleet of vehicles.[242]

Under the RCP8.5, business as usual scenario, surface temperature changes across Central and South America are projected to follow the global trend of inland temperature hotspots. By the end of the century, temperatures in central South America will have increased by 6°C when compared to averages for the region recorded from 1986 to 2005.[211] Temperature increases in coastal regions will approximate global averages for each future scenario of GHG emissions. With aggressive action to mitigate climate change, surface temperatures are projected to follow global averages of a committed 1.5–2°C increase relative to pre-industrial times.[211]

Precipitation models based on a scenario of continued excessive rates of emissions predict a 10% reduction in rainfall for Central America, a 15% decrease for the tropical region of South America east of the Andes, and an increase of 15–20% for southeast regions of South America.[211] The effects of future scenarios of aggressive to moderate abatement of emissions on rainfall patterns in Central and South America are uncertain.[211]

The projected impacts of moderate climate change on agricultural productivity vary considerably across the whole of the Central and South America. The highly productive crop growing regions south of the Amazon may see increased yields of soybean, corn, and sugarcane, with the potential for southern expansion of crop areas based on warmer, longer growing seasons and sustained rainfall.[211] This potential for increased agricultural output may require the development of new varieties adapted to warmer growing conditions.

Under conditions of extreme surface warming, corn production is likely to decline over much of Central and South America. Heat stress may negatively affect soybeans beyond the capacity for adaptation leading to reduced yields in the highly productive agricultural regions of central Brazil.[211]

In Northeastern Brazil, Central America, Columbia, the Andes regions, and Chile, overall crop production is predicted to decline with increasing surface temperatures.[211] Climate change-dependent limitations to water supply in some of these regions may further impact yields. Corn, beans, and rice are the main crops produced by the poorer nations of Central America and yields are projected to decline in relation to the severity of future climate change.[211] Ninety percent of agricultural produce is consumed domestically in Central America and thus climate change threatens the food security of this region.[211]

Heat stress, under conditions of extreme climate change, is likely to reduce the productivity of the important animal farming industries of Brazil and other countries in the region with predicted declines in the production of dairy and beef cattle, chickens, and pigs.[211]

Hydropower accounts for 66% of the total electricity produced in Central and South America.[211] In Central America, under extremes of climate change, the projected decrease in precipitation in combination with increased evaporation has the potential to reduce water inflow to key river systems by 20% and thereby reduce the capacity to produce hydropower by 33–53%.[211] Semi-arid regions of central Chile and North East Brazil are vulnerable to declining water in-flows that would threaten the water supply for irrigation and water flow rates to hydropower stations. Based on a future of extreme climate change, the rate of melting of the Andes glaciers will accelerate, followed by reduced glacial melt as the ice mass declines and vanishes.[211] By the end of the century, a future of diminished flows of glacial melt waters will threaten the water supply of Quito, Ecuador; Santiago, Chile; and Lima, Peru; and will reduce hydropower production in Peru.[211] Other regions may see enhanced capacity for hydropower based on increased precipitation and water in-flows to key river systems.[211]

Approximately 29 million people live in the total area of the expanded metropolitan region of Sao Paulo, Brazil.[243] This densely populated megacity and surrounding smaller cities and urban areas is located in a monsoon climate characterized by wet summers with abundant rainfall and drier winters.[244] During the 1950s, extreme precipitation days defined as rainfall greater than 50 mm were rare for Sao Paulo.[211] Since the year 2000, extreme precipitation events have occurred two to five times per year and future climate change is expected to further increase and perhaps double the frequency of these events.[211] Severe climate change is predicted to increase the vulnerability of Sao Paulo to the damaging effects of urban flash floods and landslides.

The drainage basins of the Rio de la Plata and the Amazon are the largest in the world and drain the massive water flows from the Amazon River system and the river systems in the vast areas of south and central Brazil, Paraguay, Uruguay, and northern Argentina. Buenos Aries and other coastal cities within the Rio de la Plata are vulnerable to extreme flooding driven by climate change-dependent sea level rise and storm surges.[211]

The Mesoamerican Barrier Reef System stretching along the coasts of Belize, Guatemala and Honduras may collapse by mid-century if current trends of ocean warming and acid-ification continue.[211] The coral reefs off the coast of Brazil are under similar threats.[211] As is the case for other tropical coastal regions of the world, climate change has the potential to cause catastrophic damage to the coral reefs of Central and South America with severe consequences to marine biodiversity, fish stocks, and coastal livelihoods that are dependent upon the reefs.[211]

Climate variability and extreme weather events are known to increase outbreaks of vector and water-borne disease such as malaria, dengue fever, schistosomiasis, Hantavirus infection, leishmaniosis, and cholera.[211] Under a future scenario of extreme climate change, regional poverty, flooding, and a lack of access to clean water will increase the risk of infectious disease. Low GDP per capita countries of Central and South America have limited capacity for adaptation to extreme surface warming. Populations within these countries, and in particular the elderly or those otherwise predisposed to illness, are vulnerable to heat stress related dehydration, respiratory failure, and cardiovascular disease.

Overall, the vulnerability of Central and South America to a future of extreme climate change varies considerably over the whole of the region. The abundance of rainfall in the highly productive agricultural regions south of the Amazon basin provides capacity for adaptation. Further, this area has considerable potential for further intensification of agriculture and expansion of crop production land. The resilience of this sub-region contrasts the threat of extreme climate change to the food security of poorer nations of Central America. A business as usual future scenario of continued excessive rates of emissions threatens the water supply, hydropower generation capacity, and the agricultural productivity of the drier regions of South America. In addition, as described for other continents, sea level rise is a threat to coastal infrastructure and settlements; ocean warming and acidification threaten coral reefs and marine fish stocks; and extreme surface temperatures are a direct threat to animal and human health.

Excluding emissions arising from forestry and land use, the per capita rate of emissions from the aggregate of countries of Central and South America was 18% below the world average, and lower than any any other region with the exception of Africa.[124] The GDP per capita of South and Central American countries is below that of advanced economies, and this provides a partial explanation for the relatively low rates of emissions. However, the profile of energy consumption differs in comparison to other regions of the world. Almost 70% of the region's electricity is produced by zero-emission hydro power while coal is a minor contributor to energy supply.[211] The difference in emissions per unit of power production between hydro and coal contributes to the relatively low rates of emissions from this region of the world when forestry and land use are excluded from consideration.

Thirty-nine percent of the total GHG emissions from South and Central America come from the forestry and land use sector. When this source of emissions is considered, the per capita rate of emissions exceeds the world average by 26%. As will be discussed, Brazil has made significant inroads to halting deforestation in the Amazon; however, over the entire region, emissions from deforestation and land use practices are substantial and are a significant contributor to the ongoing global accumulation of greenhouse gases in the atmosphere.

3.2.9 Small Islands. The Alliance of Small Island States (AOSIS) was established in 1990 as an intergovernmental organization to provide a unified voice for small, low-lying island nations that share common vulnerabilities to the impacts of climate change.[245] The AOSIS consist of 39 member states, many of which are a part of the group of nations aligned as Small Island Developing States (SIDS).[245] Member states of both the SIDS and the AOSIS come from all regions of the Atlantic, Pacific, and Indian Oceans. The aggregate populations of the AOSIS account for less than 1% of world population, and the diversity of membership is such that there are no defining characteristics of economic and human development.[106] Singapore is a highly advanced nation with a

GDP per capita of $56,000 and an 11th place ranking on the UN Human Development Index.[203,213] In contrast, five of the member states of the AOSIS rank in the Low Human Development category of the index and are among the world's most impoverished nations.[203] A common feature of small island states is a relatively high percentage of low-lying coastal land area and associated populations exposed to the potentially catastrophic effects of sea level rise, storm surges, and ocean warming and acidification.

Climate change risks for small islands mirror those of low-lying coastal settlements of larger nations that are characterized by dependencies on coastal resources and marine ecosystems. However, for small island states, the magnitude of risk is generally greater, and the capacity for adaptation is limited when compared to larger countries.

Small islands are particularily vulnerable to sea level rise, salt water inundation, coastal erosion and flooding.[246] Coastal populations, buildings and other infrastructure are exposed to the potential for direct damage from sea level rise and severe storms surges. Extremes of sea level rise, based on a business as usual scenario of emissions, can lead to flooding of river esturaries, coastal landslides and reshaping of coastlines.[246] The primary effects of sea level rise, and ocean warming and acidification, have the potential to cascade into serious secondary impacts such as damage to coastal and inland ecosystems (including irreversible descimation of coral reefs), salination of fresh water supplies and inland crops, declining production and failures within the capture fish and aquaculture industries, declining tourism, and an overall loss of ocean and coastal dependent livelihoods.[246] Under conditions of extreme surface warming leading to a 1 m rise in sea levels, beach erosion and coastal damage threatens the sustainability of 49–60% of the tourism industry in the Caribbean.[246] Flooding and severe storms can compromise sanitation, fresh water supplies, and health care delivery, leading to disease outbreak.[246] Human health vulnerability to climate change is greatest among low GDP per capita small islands states.

On a per capita basis, when compared to larger nations, small island states have a greater exposure of population and assets to the potentially damaging outcomes of climate change. An analysis of physical exposure to storms in the Asia-Pacific region indicates that 7 of the 9 most at-risk countries were small island states.[246] Construction of coastal protective barriers allows for some degree of adaptation to sea level rise.[246] Capacity for adaptation varies among small island states and is dependent upon available

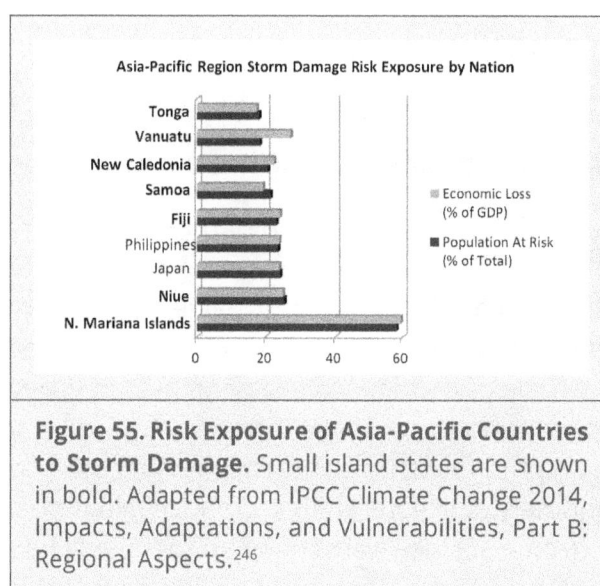

Figure 55. Risk Exposure of Asia-Pacific Countries to Storm Damage. Small island states are shown in bold. Adapted from IPCC Climate Change 2014, Impacts, Adaptations, and Vulnerabilities, Part B: Regional Aspects.[246]

resources and the requirements for the construction of effective sea walls and other structures. In-land migration and abandonment of highly at-risk coastal settlements will reduce mortality and risk exposure.[246] However, for many small islands, entire societies were established based on coastal resources and traditional practices. Willful continuation of excessive GHG emissions by larger advanced economies that result in a future of small island climate change refugees following a forced relocation of coastal populations can only be viewed as a violation of basic human rights.

The Republic of Maldives is a country of 26 coral atolls and 1192 coral islands located in the Indian Ocean southwest of India.[247] Tourism and fishing are the primary industries supporting a population of 400,000.[245] The Maldives are the lowest nation on Earth with over 80% of the land area less than 1 m above sea level.[248] The consequence is that a 1 m sea level rise, by the year 2100, will largely submerge the Maldives. A business as usual future scenario of continued excessive emissions of GHGs, will likely lead to abandonment of the inhabited islands of the Maldives well before the end of the century.

In November 2007, the Maldives hosted a conference of Small Island States to discuss the question of human rights and climate change.[249] The outcome of this meeting was "The Malé Declaration on the Human Dimension of Global Climate Change." The document was the first intergovernmental position statement recognizing the link between anthropogenic climate change and the fundament rights to an environment capable of supporting human society. The declaration noted the particular vulnerability of small island low-lying coastal states to climate change-dependent sea-level rise, ecosystem damage, and increased incidence of natural disasters. Since issuance of the Malé Declaration, the Human Rights Council of the UN has adopted a series of resolutions linking anthropogenic climate change with basic human rights.[250] In December 2015, as part of the Paris Climate Change conference, the United Nations High Commissioner for Human Rights issued a statement calling for urgent and effective action on climate change, based not only on moral obligations but as a necessity to satisfy the duties of states under international human rights law.[248]

3.3 Summary and Conclusions

An indefinite continuation of modern era practices of limitless fueling of the economies of the world by combustion of fossil fuels will reshape the Earth's climate and dramatically alter weather patterns with profound effects on natural ecosystems and human societies. By the year 2100, based on the IPCC RCP8.5 business as usual scenario of continued excessive rates of greenhouse gas emissions, atmospheric CO_2 concentrations will have increased by 3-fold, and methane levels will be 5-fold greater than pre-industrial levels. The net effect will be an average global surface temperature anomaly of greater than 4°C above pre-industrial temperatures. In-land hot spots of extreme surface warming of up to 8°C will have occurred within all continents. Ocean surface water temperatures will have increased by about 3°C, and the oceans will have become increasingly acidic with a drop of greater than 0.4 units of pH. Ocean waters will have risen by 0.75 to 1.2 metres above pre-industrial sea levels. By mid-century, the arctic will have become completely devoid of September sea ice.

By the year 2100, under a business as usual scenario, sea level rise will have increased the potency of storm surges to damage coastal settlements and other infrastructure. Densely populated at-risk coastal cities in Asia and other regions are likely to have been severely damaged by the combination of sea level rise and cyclones. In many coastal regions, physical barriers and other adaptive measures will be in place in an attempt to limit the potential for damage. Vulnerable small island states are likely to have been decimated by the combination of coastal inundation, and severe storm damage. Abandonment of coastal settlements and mass migration of climate change refugees will have occurred in highly vulnerable locations around the globe.

Tropical coral reefs will have undergone irreversible decimation on a worldwide basis with loss of reef dependent ecosystems, coastal protection, fish stocks, fisheries, and human livelihoods.

Pelagic fish will have relocated from tropical to northern waters. Overall, fish stocks will have declined, along with fish body weights. The productivity of tropical capture fisheries will have undergone a dramatic decline. Relocation of commercial fisheries to northern waters will have occurred.

In Africa, a failure to transition to modern intensive agricultural practices in combination with extremes of surface warming, changing precipitation patterns, and a higher frequency of droughts and extreme precipitation events will have had devastating effects on the poorest citizens on Earth. Declining agricultural productivity and a higher frequency of crop failures will have resulted in shortages of food supply, malnutrition, and a dependence on international food aid to avoid famine. Regional flooding, lack of access to clean water, poor sanitation and inadequate medical treatment will have led to increased outbreaks of water and vector borne diseases. Food and water security issues within Sub-Saharan Africa will have impeded economic development, impaired advances in education and fueled regional conflict.

Crop yields will have declined in many parts of Asia including the critical wheat growing regions of India. In arid regions of Asia, current water supply issues will be exacerbated under future conditions of extreme climate change. The frequency of abnormally heavy precipitation events will have increased leading to widespread flood damage and associated disease outbreaks. Severe heat and limited capacity for adaptation among less developed nations in the region will have led to widespread heat-related mortality and morbidity. Conditions of extreme climate change in Asia will have threatened food and water supply, damaged coastal cities and infrastructure, degraded tropical coral reefs, and relocated tropical fish stocks. The damaging effects of extreme climate change will be magnified for less developed Asian nations with limited capacities for adaptation and for populations exposed to severe storm surges and coastal degradation.

Relative to other regions of the planet, Europe, the Americas, and Australia plus New Zealand do not have the same degree of vulnerability to the catastrophic damage projected under a future of extreme climate change. However, extreme surface warming and changing precipitation patterns will have aggravated current water supply issues in densely populated arid regions, and arable lands that are dependent upon irrigation. In arid regions of Australia, Southern Europe, Southwestern US, Northern Mexico and parts of South America west of the Andes, adaptation to diminished water inflow and high evaporation will require extensive, costly, water flow infrastructure and supply management practices. Under extreme conditions of heat waves and droughts, these adaptations will be insufficient, resulting in periodic shortfalls in water supply, reduced agriculture productivity and reduced hydropower production. Property damage from wildfires will be a common occurrence in arid regions. In other regions, a higher frequency of extreme precipitation events will have caused considerable flood damage. Low-lying coastal cities such as Miami will have adapted to the threat of storm surges by erection of costly sea barriers and/or relocating vulnerable populations. Heat stress and water supply issues will reduce crop yields in important agricultural regions such as the US corn belt and areas of Southern Europe. Northern relocation of crop land will have occurred to provide some degree of adaptation to changes in growing conditions. In Mexico and much of Central America, declining crop yields and a higher frequency of crop failure will threaten food supply. Extreme heat will directly affect human health and productivity with the highest incidence of heat stress mortality and morbidity among poorer nations with less capacity for adaptation.

When compared to the business as usual case, an alternate future scenario of aggressive abatements of GHG emissions (RCP2.6) will result in a radically different climate. By the end of the century, atmospheric CO_2 will have stabilized below 450 ppm and methane levels will have declined by 38% due to reduced emissions and chemical conversion in the atmosphere. Global average surfaces temperatures will have likely peaked between 1.5–2.0°C above pre-industrial temperatures. Ocean temperatures and pH will stabilize near current levels. The juggernaut of glacial and sea ice melt that has been set in motion over the past 50 years will continue but at a reduced rate relative to melt flow projections under conditions of extreme climate change. By the end of the century, ocean levels will have risen between 0.50–0.83 metres above pre-industrial sea levels. Extremes of inland surface warming and associated threats to agriculture productivity and water supply are unlikely to occur. The increased frequency and severity of extreme precipitation events and droughts projected under business as usual climate change conditions will not occur, based on a future scenario of aggressive abatement of emissions.

Committed climate change, assuming a future of successful mitigations, will still require adaptations to minimize the potential for damage. Low-lying coastal regions and small islands will remain vulnerable to sea level rise, storm surges, and coral reef degradation. However, the degree of vulnerability and potential for catastrophic damage diminish in correlation with reduced accumulation of atmospheric GHGs. The impacts of climate change under a scenario of aggressive mitigation will be such that cost-effective adaptations can be implemented to minimize actual damage.

Intermediate scenarios of partial mitigation of GHG emissions will result in climate change futures somewhere between the catastrophic projections of extreme climate change and the manageable projections following aggressive abatement of emissions.

Given the quality of the science and predictive capability of climate models, it is difficult to envision the combination of willful ignorance and political incompetence required to carry on a business as usual future of continued high rates of GHG emissions. Climatologists can provide exact numbers for future emissions of GHGs that will limit surface temperatures increases to achievable targets. Much of the replacement technology to transition to a carbon neutral future is understood, and there is considerable capability for further innovation. There is potential to develop science-based regulatory frameworks that would effectively drive the energy supply and other sectors of world economies toward a carbon neutral future.

Implementation of economic mechanisms to transition away from continued accumulation of atmospheric GHGs requires a greater public awareness and understanding of the issues along with effective political leadership with a willingness to act on achieving long-term societal objectives. The outcome of the 2015 Paris Climate Change Conference provides some degree of optimism that the nations of Earth may have mustered the political will to take effective action on climate change.

Chapter 4: PARIS 2015 UNITED NATIONS CLIMATE CHANGE CONFERENCE

At the onset of the 20th century, 51 million people visited the Paris 1900 World Exposition and marvelled at the splendor of the lighting display at the Palace of Electricity.[72] A coal-fired steam engine powered the generator to produce the electricity for the display and over the next 115 years coal use would continually escalate as the world population exploded and societies progressively increased the intensity of energy use. Over time, oil and natural gas provided additional fossil fuel options to power industrialization and economic advancement. Between 1750 and 2011, human activity has been responsible for the release of approximately 2 trillion tonnes of carbon dioxide, of which 40% remains in the atmosphere with the remainder sequestered by ocean and land sinks.[86]

By the year 2015, in contrast to the majority of the metropolises on Earth, the City of Light was no longer illuminated by lamps using electricity generated by the combustion of fossil fuels. When representatives of the 196 nations of the Conference of Parties assembled in Paris in December of 2015 for the 21 annual climate change meeting, the electricity that powered the lighting, communications, audio and visual systems, the elevators and escalators and all manner of appliances and devices required to accommodate the delegates and facilitate discourse was largely produced by 0 emissions sources. By 2012, a full 92% of the electricity consumed in France was generated by a network of nuclear, hydro, wind and solar installations with nuclear power providing 77% of the total supply.[251]

The selection of Paris as the host city of the 2015 UN Climate Conference was fitting in that the meeting took place in the same city that, 115 years earlier, was the sight of a world exposition with a focus on the latest innovations that had evolved out of the Industrial Revolution and on the potential of human inquiry and ingenuity to shape a prosperous future. The meeting took place within the borders of a country that had already begun to decarbonize economic sectors and curtail greenhouse gas emissions while maintaining a high standard of living. As of 2012, the per capita rate of emissions of France was 6.6 tonnes of CO_2 equivalents per person.[124] This figure was 36% of the US per capita rate, 27% of the rate for Canada, and 22% of the per capita GHG emissions rate of Australia.[124]

By 2015, 21 years had passed since universal ratification of the original United Nations Framework Convention on Climate Change (UNFCCC), yet little had been accomplished to effectively address the agreed upon objective of "stabilizing GHG concentrations in the atmosphere at a level that would prevent dangerous anthropogenic interference with the climate system."[252] The Conference of Parties (COP) to the original agreement had met on an annual basis since 1994 to assess progress and to develop international treaties that would set legally binding limits to emissions. However, despite the clarity of the science and the urgency of accumulating atmospheric GHGs, COP was unable to come to a top-down universal agreement that would mandate limits to emissions. China and the United States did not ratify the original Kyoto Protocol and, other than the notable exception of member states of the European Union, the nations of the world had failed to introduce effective policies and regulations to effectively combat climate change. The 2010 Cancun pledges were a step forward toward an inclusive approach among COP nations; however, these pledges did not extend beyond 2020 and thus did not provide a long-term framework to limit surface warming. Since 1990, the annual rate of global GHG emissions had increased by 40%.[124] Looking forward to the year 2030, continued failure to effectively reduce emissions would commit the planet to a future of extreme climate change leading to catastrophic damage to ecosystems and human societies well before the end of the century.

The strength of the Conference of Parties is universal agreement among the nations of Earth as to the reality of current and future climate change driven by anthropogenic greenhouse gas emissions. The weakness of COP is the vast diversity of political and economic national interests and thus an inability to reach common ground on a top-down agreement that would impose specific emissions standards on member states.

4.1 Intended Nationally Determined Contributions (INDCs)

The past failures of the Conference of Parties to establish a means to reduce global emissions called for a new approach to negotiations and an alternate type of agreement. The 2010 COP16 conference in Cancun concluded with an agreement that member states should develop their own targets for the control of future emissions. The concept of nationally determined pledges had a broader acceptance than was the case for the Kyoto protocol and was the first step toward a unified approach to combating climate change.

Overall, the aggregate of projections for year 2020 rates of GHG emissions based on the actual policies and trends for the top 8 emitters matches the sum of the Cancun pledges made by these countries. China and India did not pledge to reduce total emissions and set their Cancun targets on the basis of a formula of reducing the intensity of emissions per unit of GDP.[253] China's Cancun pledge, when considering the projected rate of economic growth, is equivalent to a ceiling on year 2020 emissions that is 36% greater than emissions on record for the year 2012. China is projected to exceed the level of ambition in its Cancun pledge such that emissions should be limited to a 16% increase over 2012 rates.[215] The 2.1 Gt difference in the pledged and projected emissions for China is balanced by the aggregate of the projected shortfalls in achieving Cancun targets for the remainder of the world's top emitters. The United States had pledged to reduce emissions by 17% over the period from 2005 to 2020[122]; however, as of 2015, the US is on a trajectory of increased emissions by the end of the decade.[254]

Overall, the 2010 Cancun pledges of the world's top emitters consist of a set of nationally determined targets that, as an aggregate, pledged to limit the growth in emissions over the coming decade. Achieving Cancun targets will result in a year 2020 emissions rate that is 13% greater than emissions on record for 2012.[122] As a group, the Cancun pledges are lacking in sufficient ambition to adequately control emissions over the short-term period to 2020. Further, these pledges have no provision for progressive longer-term curtailments as required to avoid the catastrophic consequences of severe climate change.

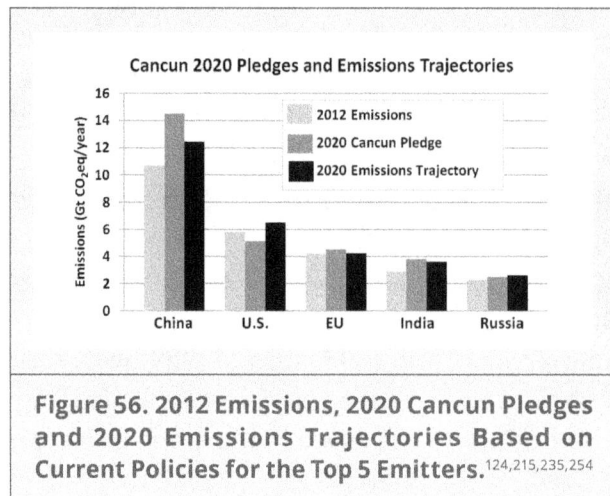

Figure 56. 2012 Emissions, 2020 Cancun Pledges and 2020 Emissions Trajectories Based on Current Policies for the Top 5 Emitters.[124,215,235,254]

The agreements put in place following the COP16 conference in Cancun were clearly inadequate. However, the Cancun pledges did provide the beginnings of a universally acceptable process with the potential to effectively control future global emissions.

Prior to the critical 21st meeting of the Conference of Parties, each member state was asked to develop longer term targets for internally derived commitments to control emissions (Intended Nationally Determined Contributions or INDCs).[254] These commitments were to be compiled and assessed to determine the collective outcome on global emissions and projected surface warming relative to pre-industrial temperatures. INDCs were to form the foundation for the 2015 Paris climate change agreement. The concept was to combine a top-down agreement as to targets to limit surface warming with bottom-up contributions developed by each member state to control their own emissions.

On February 27, 2015, Switzerland became the first member of COP to submit an INDC.[254] Switzerland set a year 2030 target to reduce the rate of emissions of total CO_2 equivalents by 50% relative to baseline emissions from the year 1990.[254] Typically, advanced economies followed the example of the US by targeting 25–30% reduction in emissions for the year 2030 when compared to baseline rates.[254] The European Union extended Kyoto and Cancun pledges and submitted an INDC based on a year 2030 target of a 40% reduction in emissions relative to 1990.[254]

As was the case for 2010 Cancun pledges, China and India deviated from the general approach of the other top emitting nations and used a formula of intensity of GHG emissions per unit of GDP to develop their national contributions. Sustained levels of high economic growth are projected for both countries. The INDCs of China and India were designed to avoid restricting GDP growth, based on the premise that it was not possible to completely decouple emissions from economic development. China has set a year 2030 target of a 60–65% reduction in total greenhouse gases emitted per $ of GDP when compared to the GHG/GDP ratio for the year 2005.[254] India used a similar formula with a 2030 target of a 25–30% reduction in the GHG/GDP ratio.[254] In addition, China set a target of reaching peak emissions on or before 2030, and set a separate ambitious afforestation goal to increase carbon-sequestering forested areas in the country.[254]

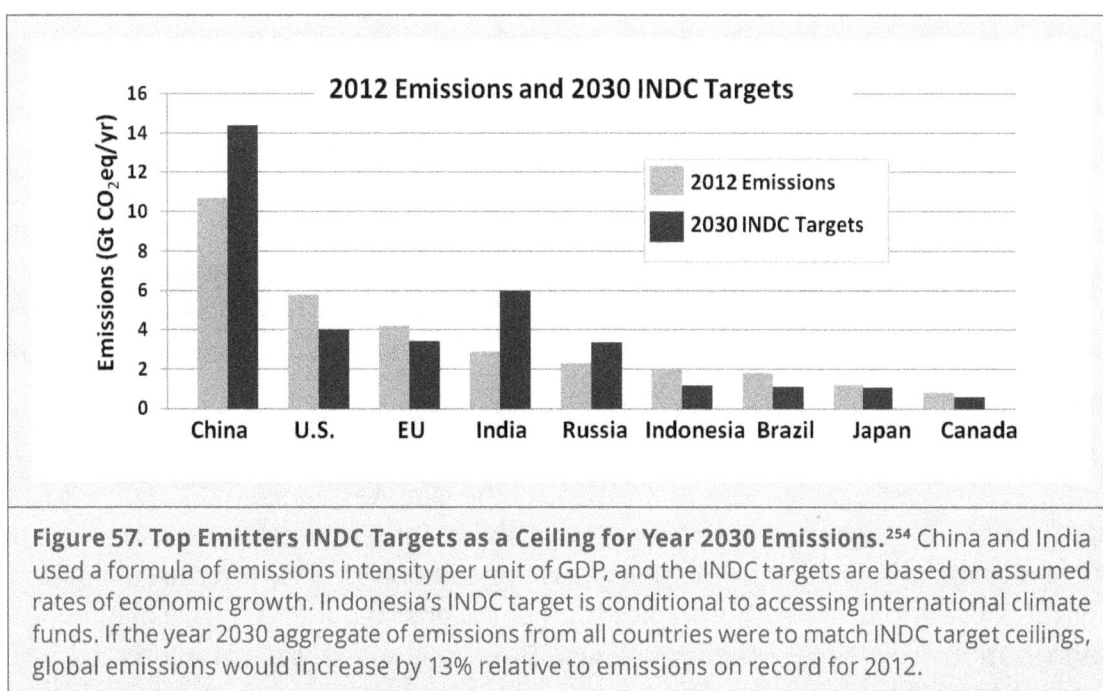

Figure 57. Top Emitters INDC Targets as a Ceiling for Year 2030 Emissions.[254] China and India used a formula of emissions intensity per unit of GDP, and the INDC targets are based on assumed rates of economic growth. Indonesia's INDC target is conditional to accessing international climate funds. If the year 2030 aggregate of emissions from all countries were to match INDC target ceilings, global emissions would increase by 13% relative to emissions on record for 2012.

The formulas used by China and India do not allow for accuracy in predicting the future emissions of these countries. The Centre for Economic and Business Research estimates China's GDP at $34 trillion annually by 2030, which would represent a 15-fold growth relative to 2005.[255] Using the INDC submitted by China and assuming a 62.5% reduction in GHG/GDP ratio, this level of economic growth would result in a catastrophic 39.6 Gt of CO_2 equivalents (84% of the 2012 world total) released by China to the atmosphere in 2030.[254,216] However, a slowdown in the growth of China's economy dramatically reduces emissions predicted by the INDC formula. Further, in recent years, China has made progress in uncoupling emissions from economic growth and is on a trajectory to exceed the 2020 Cancun target of lowering emissions intensity per unit of GDP. In summarizing various predictive models, the United Nations Environment Programme (UNEP) 2015 Emissions Gap Report provides a rough estimate of 14.4 Gt of CO_2 equivalents emitted by China in the year 2030.[122] Based on this estimate, the total emissions of GHGs by China in 2030 will have increased by 35% relative to 2012, and the per capita emissions rate will have approached that of US.

In 2012, China, the US, the EU28, and the other 6 countries that comprised the top emitters, accounted for 67% of worldwide anthropogenic GHG emissions.[124] As an aggregate, the total emissions by the remaining countries of the world is modeled to increase by 19% between 2012 and 2030, which is considerably less than the projected increase in population. Many developing nations submitted INDCs based on a combination of unconditional targets and more aggressive targets conditional to accessing international climate funds to assist in implementing mitigation programs. China, India, Tunisia, Malaysia, and Chile were the only countries that submitted INDCs, using formulas based on a ratio of GHG to GDP.[254]

Many nations included land use and forestry in calculating targets for emissions reductions. China set a target to increase the total forest carbon stock by a staggering 4.5 billion cubic metres through programs of afforestation and reforestation.[254] To achieve this target, forested areas within China would need to be increased by 50–100 million acres (2–4 times the size United Kingdom).[256] This measure alone would reduce world GHG emissions by 2.1% relative to 2012 rates of emissions.[256] The size of the biomass carbon sink proposed by China would be sufficient to sequester nearly all of Japan's projected emissions for the year 2030.[256] China's INDC did not indicate if this massive program of forest management was included in the calculation of overall emissions targets.

Climate Action Tracker (CAT) is the product of a collaboration of 4 leading climate change research organizations.[257] CAT tracks the climate change mitigation efforts of 32 countries responsible for 80% of global emissions. Each country is assessed for current emissions, projected emissions based on national policies prior to submission of INDCs, the impacts of INDCs on projected emissions, and the fair share of intended contributions relative to other countries. Globally, the analysis completed by CAT reached the following conclusions:

1. A business as usual future scenario in the absence of government policy would result in a year 2100 average global surface temperature increase of 4.1–4.8°C relative to pre-industrial temperatures.
2. Successful implementation of the climate change mitigation policies of world governments as they existed prior to submission of INDCs was consistent with a surface warming of 3.3–3.9°C.
3. The pledges contained in the 32 INDCs advanced the level of ambition to reduce emissions; however, in agreement with the United Nations Environment Programme, CAT analysis revealed a substantial gap between the summation of future emissions, based on achieving INDCs pledges and the actual reduction in emissions required to limit surface warming to well below 2°C and ideally close to 1.5°C.

CAT has developed a ratings system of classifying the climate change action plans of countries as role model, sufficient, medium or inadequate. Role model countries have put in place programs that are more than consistent with limiting surface warming to below 2°C. Countries classified as "sufficient" have submitted INDCs and developed national action plans that are fully consistent with limiting surface warming to below 2°C. "Medium" countries do not have INDCs or programs that are consistent with a fair share contribution to limit future surface warming to less than 2°C. The projected rates of emissions from medium countries are such that deep compensatory cuts in emissions will be required from other countries. Countries classified as "inadequate" have submitted INDCs that, along with an assessment of actual policies and practices, would elevate surface temperatures by 3–4°C if all nations were to follow a similar level of ambition to combat climate change. Where appropriate, CAT has provided recommendations to strengthen INDCs and align pledges with the concept of a fair share contribution to adequately control future global emissions.

The Climate Action Tracker assessment of the INDC submitted by the Philippines provides an indication of the depth of analysis employed. In 2012, the Philippines had a population of 95 million people, a GDP per capita that was 25% of world average, and a tiny per capita GHG emissions rate of 1.63 tonnes of CO_2 equivalents per person.[124,213] The INDC submitted by the Philippines states an unconditional target to cut emissions by 70% in the year 2030 relative to current rates.[254] A simplified projection based entirely on this INDC would indicate that the Philippines will release only 47 million tonnes of CO_2 equivalents in 2030 at a microscopic per capita emissions rate of 0.37 tonnes per person.[124] This rate of per capita emissions would be 6% of world average and 3% of the projected per capita emissions rate of the US. Based on these numbers, the Philippines would be a model nation in the global effort to combat climate change. However, a detailed analysis of the energy programs in the Philippines revealed intentions to construct additional coal-fired power plants and CAT concluded that, based on actual government energy policy, it was unlikely that the Philippines could achieve the ambitious target stated in the INDC.[215] CAT classified the INDC submitted by the Philippines as "medium" but indicated that if government policy clearly supported low emissions energy supply and aggressive emissions reduction targets were applied to the industry, energy and agriculture sectors, the Philippines would have been given a "sufficient" ranking.

The combination of national climate policies and INDCs submitted by China, the EU, India, Brazil, Indonesia, the Philippines and 5 other countries were rated as medium and not consistent with fair share contributions to limit global warming. CAT rated the INDCs and climate action plans of the US, Japan, Russia, Canada, Australia, Argentina, and 9 other countries as inadequate and far removed from fair share contributions.[215] The only nations rated as sufficient were Bhutan, Costa Rica, Ethiopia, Morocco and The Gambia.[215] There are no role model nations in the CAT classification system.

Table 5. Climate Action Tracker Ranking of National Climate Action Policies[215]	
CAT Classification	**Countries***
Role Model	None
Sufficient	Bhutan, Costa Rica, Ethiopia, Morocco, The Gambia
Medium	**China**, **EU**, **India**, **Brazil**, **Indonesia**, **Mexico**, Kazakhstan, Norway, Peru, Philippines, Switzerland
Inadequate	**USA**, **Russia**, **Japan**, **Canada**, Argentina, Australia, Chile, New Zealand, Saudi Arabia, Singapore South Africa, South Korea, Turkey, UAE, Ukraine
*Top 10 emitter shown in bold	

Clearly, the INDCs of all major emitting nations must be strengthened and the level of ambition to combat climate change increased if end-of-century average surface warming is to be less than 2°C. However, it should be noted that the global level of ambition contained in the aggregate of INDCs represents an advance relative to projections based on policies in place prior to the submission of intended contributions.

The INDC process provides the first draft of how a universally accepted system of developing bottom-up nationally determined emissions targets would merge with a globally derived, top-down, over-arching objective to limit future surface warming.

4.2 Paris 2015 Conference of the United Nations Framework Convention on Climate Change (COP21)

On Friday, November 13, 2015, two weeks prior to what was scheduled to be history's largest most inclusive gathering of world leaders, a series of coordinated attacks occurred in Paris. At the conclusion of the carnage, 130 people had been killed, 368 were injured, and seven attackers died.[258] The attacks were the deadliest to occur in France since the Second World War, and the government imposed a state of emergency that lasted for 3 months. In light of the terrorist attacks, questions were raised as to the safety and security of the COP21 climate talks and the possibility of cancellation. Laurent Fabius, France's Foreign Minister and President of the Conference of Parties, emphatically declared that the conference would continue as planned but under heightened security.

The timing of the attacks was likely unrelated to the Paris climate change conference, and the underlying motives are beyond the scope of this narrative. However, a future characterized by extreme climate change would exacerbate hunger, poverty, inequality, and lack of education within the poorest nations. A failure to adequately control future emissions would undoubtedly result in a generation of disaffected impoverished climate refugees, and thereby create conditions ripe for civil unrest, terrorism, and international conflict. The repercussions of COP21 extended beyond an attempt to reach an international agreement to limit global warming.

Over 36,000 participants assembled in Paris for the 21st meeting of the Conference of Parties from November 30 to December 11 of 2015.[259] Representatives of the 195 nations on Earth plus the European Union that comprise COP were joined by delegates from intergovernmental and non-governmental organizations. In his opening remarks Laurent Fabius stated that the primary objective of the 11 day conference was to reach a "Universal, ambitious agreement that will be differentiated, fair, sustainable, dynamic, balanced and legally binding, and will need to ensure that the global temperature does not rise by 2°C — or even 1.5°C — compared to the pre-industrial era because of greenhouse gas emissions."[260] The goal of Mr. Laurent and his team was to facilitate an ambitious compromise that did not deviate from the core objective but remained acceptable to the vast diversity of economic, social, and political agendas of COP member states.

Mr. Fabius concluded his opening remarks with broad a reference to poverty and terrorism in the context of climate change:

> *I believe in success. I believe because I hope for it. I believe because all of us know that combating global warming is more than just an environmental matter. It is an essential condition to provide the whole world with food and water, to save biodiversity and protect health, to combat poverty and mass migration, to discourage war and foster peace, and, at the end of the day, to give sustainable development and life a chance.*

> *Here and now, as 2015 comes to an end, it is France's responsibility to help address two of the greatest challenges of the century: combating terrorism and fighting climate change. Today's generations are calling upon us to act, while tomorrow's generations will judge our action. We cannot hear them yet, but, in a way, they are already watching us. The word "historic" is often a hyperbole. Today, it is not. Together, let us make the Paris Climate Conference the historic success the world is waiting for.*

The congress began with the largest gathering of world leaders ever assembled. The 2 day Leaders Event consisted of statements from over 150 heads of states and was designed to empower the subsequent process of negotiations. Individual statements were read by leaders from the most powerful nations to the smallest of developing islands states at greatest risk of damage from climate change. Overall, these statements provided an indication of unanimity among world leaders as to the present reality and future threat of climate change, and that a universal agreement to limit surface warming was an essential outcome from the conference. The Marshall Islands is one of the smallest nations on Earth with a population of 72,000 and a GDP per capita that is 33% of world average. President Loeak of the Marshall Islands gave an emotional statement that was among the highlights of the Leader's Event:

> *I address you today, not only as a president but as a father, as a grandfather, as a custodian of my culture, and as a representative of a nation that lies just two metres above sea level and risks being submerged by the rising waves.*
>
> *Everything I know, and everyone I love, is in the hands of all of us gathered here today.*
>
> *The climate we have known over many centuries has, in a matter of three short decades, changed dramatically before our very eyes.*
>
> *We are already limping from climate disaster to climate disaster, and we know there is worse to come.*
>
> *For us, COP21 must be a turning point in history, and one that gives us hope.*
>
> *Our Paris Agreement must set a path for the safe climate future we all strive for. We all know, and must acknowledge, that the targets on the table now are not enough to limit warming to below 1.5 degrees, although they are a start in the right direction.*
>
> *Therefore, if it is to deliver the end we all seek, the Paris Agreement must be designed for ambition.*
>
> *It must send a message to the world that if we're to win the battle against climate change, the fossil fuel era must draw to a close, to be replaced by a clean, green energy future, free of the carbon pollution that is harming our health, stunting our growth, and suffocating our planet.*
>
> *It must set a rhythm for our action that sees us ratcheting up our national targets every five years.*
>
> *And it must assure countries as vulnerable as mine that the world's helping hand will be there when climate change, unfortunately and unavoidably, unleashes its devastating impacts.*
>
> *This is a time for human solidarity. But this is also a time for action. And this is a time for us to be the leaders that we were elected to be. Let's get it done.*
>
> **H.E. Christopher J. Loeak — President of the Republic of the Marshall Islands, Opening Remarks of the UN Framework Convention on Climate Change , 21st Conference of Parties, Paris France, Nov 30, 2015.**[261]

The greatest danger is not that our target is too difficult and that we miss it. The greatest danger is that it is too easy, and that we hit it.

Francois Hollande — President of France, Opening Remarks of the UN Framework Convention on Climate Change , 21ˢᵗ Conference of Parties, Paris France, Nov 30, 2015.[262]

The Leader's Event was followed by 11 days of marathon negotiations and diplomacy at all levels. Among the key developments over the course of the meetings was the formation of a "coalition of high ambition" that was initially forged by small island states and the EU. The coalition was joined by many of the least developed nations and later by the US, Canada, and Australia and expanded to over 100 countries that could negotiate together on key issues.[263] In the end, no single nation wished to disrupt the process. Difficult and contentious issues such as climate change-dependent loss and damage liability, differentiation of responsibility between developing countries and advanced economies, resource transfer and economic aid, and the impact of a 1.5°C limit to surface warming on the economic growth of China and India were addressed to the satisfaction all parties. Shortly after 7:00 pm on December 12, 2015, COP President Fabius made the announcement of universal acceptance of the entirety of the final version of the Paris Agreement to combat climate change.[264]

4.3 The Paris Agreement

The agreement is a relatively concise 16-page document comprised of 29 articles that is centered on an overarching objective of achieving a specific limit to global surface warming.[265] By design the agreement is short on specifics of how to achieve the objective and avoids any attempt at top-down imposition of universal emissions rates on member states. The core of the agreement is contained in articles 2–15 and can be summarised as follows:

Article 2 is the center piece of the agreement and states that the global average surface temperature increase should be held well below 2°C above pre-industrial temperatures, and that all efforts should be made to limit the temperature increase to 1.5°C. The article goes on to state that adaptation to climate change and transitioning to low GHG emissions should not threaten food production, and the flow of finance should be consistent with reducing emissions and building climate change resilience.

Article 3 directs all parties to develop and communicate nationally determined contributions with the view of achieving the global objective of limiting surface warming as stated in article 2. The contributions are to be progressive over time and recognize the need to support developing countries.

Article 4 states that parties should aim to reach peak GHG emissions as soon as possible and transition to a rapid reduction in emissions, so as to achieve a balance between anthropogenic emissions by source and removals by GHG sinks in the second half of the century. Transition to carbon neutral economies should be in accordance with best available science, on the basis of equity and in the context of sustainable development and eradication of poverty. The article acknowledges that developing countries may take longer to reach peak emissions. A list of the 19 points under article 4 and can be summarized as follows:

- Nationally determined contributions are to be basis for how each nation is to achieve internally derived targets for reduced emissions. These contributions are to be ambitious and successive. Each successive contribution will be a progression beyond the previous contribution and reflect the highest possible ambition. Contributions are to reflect common but differentiated responsibilities and capabilities, in light of different national circumstance. The special circumstance of developing and small island nations may be reflected in the contributions developed by these countries.

- Contributions are to be submitted to the UNFCCC at 5-year intervals with enough detail to provide clarity, transparency and understanding of emissions targets. National contributions can be modified at any time with the intent of increasing the level of ambition to reduce emissions.

- Developed countries shall take the lead by undertaking economy-wide absolute emissions reduction targets. Developing countries are encouraged to move toward economy-wide reduction, in light of different national circumstances.

- Support shall be provided to developing countries in recognition that this will allow these countries to achieve more ambitious emissions reduction targets.

- Parties forming joint integrated economic organizations shall be responsible for their own emissions levels as defined by their individual nationally derived contribution and for emissions levels defined under the contribution of the joint organization.

Article 5 directs parties to conserve and enhance carbon sinks and specifically develop regulations, policies and incentives to reduce deforestation and forest degradation and to promote sustainable forest management practices with the goal of enhancing forest carbon stocks.

Article 6 addresses voluntary international cooperative approaches that would allow for an overall higher ambition in reducing emissions. The Article establishes the authority of COP to supervise internationally transferred mitigations such that an overall increase in reductions is achieved. Specifics of governance are to be worked out for each agreement. Parties cannot use emissions reduction based on the transfer of mitigation to a host party if the host party is claiming these emissions reductions.

Article 7 references climate adaptations and the need to implement adaptation to protect livelihoods, ecosystems and human health dependent on the specific degree of vulnerability. Each party is asked to submit and periodically update adaptation communications. The need for international cooperation and sharing of information on adaptive measures is addressed, along with the specific adaptation requirements of developing and high at-risk countries.

Article 8 addresses loss and damage associated with climate change and references the Warsaw International Mechanism for Loss and Damage functioning under the authority of COP.

Article 9 tackles the contentious issue of financial resource transfer from advanced economies to assist developing countries in climate change mitigation and adaptation efforts. Every 2 years, high GDP per capita countries are asked to communicate quantitative and qualitative information as to projected levels of financial support provided to developing countries.

Articles 10 and 11 reference the pressing need to develop and enhance technologies to reduce GHG emissions and to achieve efficient transfer mechanisms for broad scale implementation of technology to achieve worldwide objectives.

Article 12 outlines the need for building capacity within developing nations to implement adaptation and mitigation plans and the need for advanced economies to assist in this process.

Article 13 outlines a transparency framework to facilitate effective and accurate communication of information by each party on GHG emissions by source, removal by sinks, and information necessary to implement and achieve nationally determined contributions. The transparency framework includes the requirement for advanced economies and developing nations to provide information on international transfer of funds, technology and capacity building.

Article 14 implements a global stockade tracking procedure and thus is a critical component of assessing the progress of the aggregate of member states in achieving the overarching targets of limited global warming as outlined in article 2. On 5-year intervals beginning in 2023, COP will assess the collective progress of the parties toward achieving the purpose of the agreement. The outcome of the global stockade will contribute to the ongoing process of successive enhancement of national contributions by member states and other efforts to enhance international cooperation to combat climate change.

Article 15 establishes a mechanism to facilitate implementation and promote compliance with the overall provisions of the agreement. An expert committee operating under the guidance of COP will be formed to continually monitor progress and will report on an annual basis to COP. The committee is to be facilitative in nature and function in a transparent, non-adversarial and non-punitive manner.

The remaining articles in the agreement are largely governance provisions. The agreement was open for signature and ratification over a 1-year period beginning April 22,

2016 and would come into force on the 30th day after 55% of the parties accounting for at least 55% of total global greenhouse gas emissions have formally signed on.

Shortly after the announcement of the Paris Agreement, the archway of the Eiffel Tower was illuminated with a simple message that read "1.5 degrees." The tower was constructed 126 years earlier using structural grade iron produced by the skill and brute strength of industrial era iron puddlers. At the time of construction, GHG levels were, at most, slightly above pre-industrial levels and had not affected surface

Figure 58.

temperatures. The tower stood during the whole of the 20th century as the world underwent a population explosion and advances in industrialization and development that elevated atmospheric CO_2 concentration from 280 to 400 ppm and increased global average surface temperatures by 1°C. Barring unforeseen events, the tower will continue to stand for the duration of the 21st century as surface temperatures continue to climb. What remains uncertain is the actuality of a 1.5°C limit to global surface warming from the time of construction of the tower to the year 2100.

The elation and tears shed among the delegates at the conclusion of the Paris conference were a rightful acknowledgment of a tremendous achievement in diplomacy and negotiations. The success of the Paris Agreement was based on a realization among the leadership of COP as to what could and could not be accomplished, coupled with considerable skills in compromise and negotiation. China, the United States, the EU, India, Russia, and other major emitters around the globe were placed in a position whereby scuttling the deal would have been seen as an act of denial based on short-term self-interest with dire consequences to future generations and to the most vulnerable, poorest citizens on Earth.

The Paris Agreement states that future surface warming should be held to well below 2°C and ideally 1.5°C above pre-industrial temperatures with nations devising their own targets and programs to reduce emissions. Under the agreement, peak emissions are to occur as soon as possible with transition to a carbon neutral world by the second half of the current century. Nationally derived emissions targets are to be renewed in 5-year intervals in a progressive manner to ratchet up the overall process of completing the transition. Advanced economies are to take the lead in emissions reduction and provide aid and technology transfer to assist developing nations in building their capacity for climate change adaptation and mitigation. The entire process is to be monitored by COP to allow for continued optimization and adjustment of national contributions and international efforts to combat climate change.

The Paris Agreement has no mention of punitive trade embargoes or any other action to be taken against laggard nations that fail to develop meaningful INDCs or fail to achieve targets within their INDCs. The agreement has come under criticism based on the absence of specific mandated top-down emissions standards, and the lack of punitive consequences for countries that fail to adequately curb emissions. However, the agreement is the best possible outcome of climate change negotiations given the diversity of nations involved in the process. Realistically, the Paris Agreement is the first tentative step forward in what will be a demanding journey undertaken by the unified family of nations to take ownership and responsibility for anthropogenic accumulation of greenhouse gases. Given the time squandered over the past few decades, the long journey to a carbon neutral future must now be completed over a short time span of 50–60 years. There is concern among economists and the scientific community as to the practically of achieving a 1.5°C limit to future surface warming in this reduced period.

On November 4, 2016, thirty days after conditions of national ratification had been met, the Paris Agreement came into force and Intended Nationally Determined Contributions became Nationally Determined Contributions.[266]

4.4 The Paris Agreement and the World Carbon Budget

The IPCC has calculated that the total quantity of anthropogenic carbon dioxide (excluding non-CO_2 GHGs) emitted since the beginning of industrialization should not exceed 3 trillion tonnes if there is to be a 66% chance that future surface temperatures are to be less than 2°C above pre-industrial temperatures.[*,86] As of 2011, approximately 2 trillion tonnes of CO_2 had been emitted, leaving 1 trillion tonnes remaining in the less than 2°C global warming carbon budget.[86] Emissions beyond 1 trillion tonnes between 2011 and 2100 are likely to drive surface temperatures above 2°C.[86]

At the current rate of emissions, the remaining CO_2 budget will be exhausted over the next 25 years. Cost-effective models with the highest probability of success in limiting future emissions within the constraints of the less than 2°C budget all require that global emissions peak prior to 2020 and then decline at a steady rate that arrives at total GHG carbon neutral status sometime between 2060 and 2080.[267]

The Paris Agreement states that surface warming should be held well below 2°C and closer to 1.5°C above pre-industrial temperatures. A 1.5°C limit to global warming has not been extensively modeled; however, there is some suggestion that to achieve this

---

* Long-term carbon budgets are based on carbon dioxide emissions and do not include methane, nitrous oxide other greenhouse gases. Methane and nitrous oxide have defined chemical half-lives in the atmosphere and thus the global warming potential for a unit of these gases is dependent upon the time frame from release. CO_2 does not undergo chemical transformation in the atmosphere. Non-CO_2 GHGs are important contributors to global warming and must be included in emissions calculation at any given time. However, the budgeting of emissions since the onset of Anthropocene and going forward to the end of the current century is generally restricted to the subset of CO_2 emissions.

level of ambition, world economies must complete the transition to carbon neutral status 10–15 years ahead of the schedule outlined for a threshold of 2°C in surface warming.[122]

Assuming that peak emissions stabilize close to 2014 rates and begin to decline after 2020, emissions rates should approximate 42 Gt CO_2eq/year by the year 2030 to limit global warming to less than 2°C and ideally 39 Gt/year to limit surface temperatures increases closer to a 1.5°C maximum.[122]

Based on the current NDCs and other policy indicators, the UNEP 2015 Emissions Gap Report predicts that 54–56 Gt of CO_2 equivalents will be released to the atmosphere in 2030. This rate of emissions equates to a 14% increase over 2012 rates.[122] A substantial 12–14 Gt/year gap exists, between projected year 2030 emissions and the emissions rates aligned with the highest probability to limit future surface warming to less than 2°C. The actual size of this emissions gap is dependent on the use of conditional or unconditional NDCs in the calculations. If emissions were to track the ceilings contained in unconditional NDCs, the emissions gap would be 14 Gt. A 12 Gt emissions gap exists for the more ambitious set of NDCs that are conditional to lower income countries accessing international climate funds and technical cooperation. The emissions gap

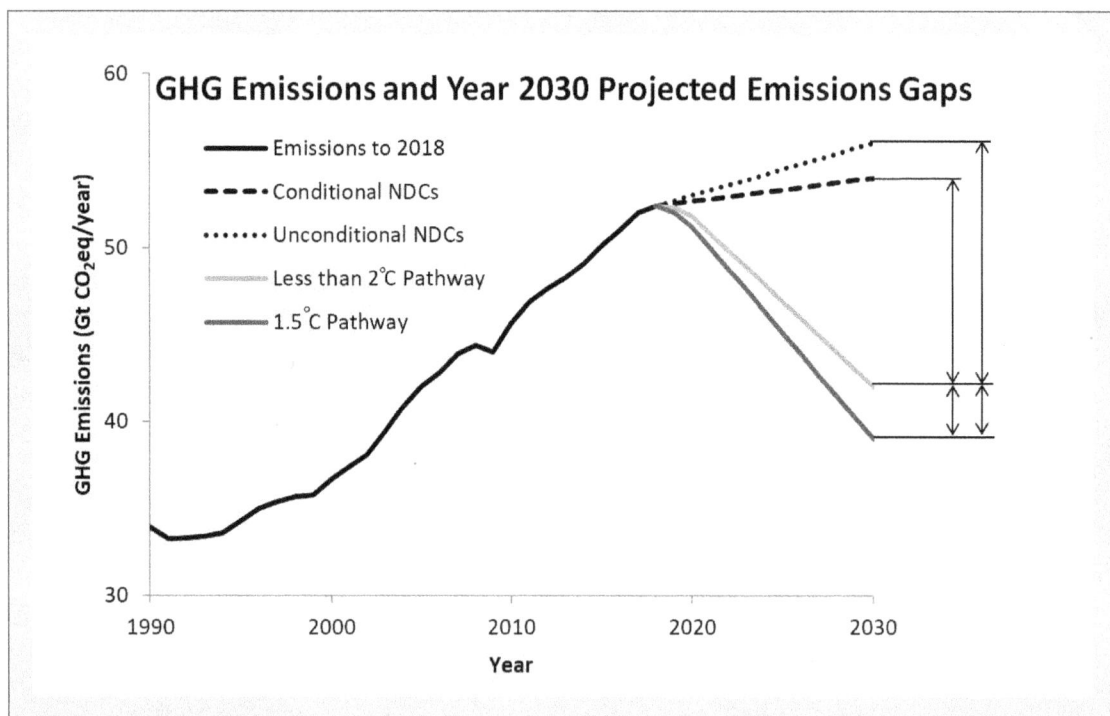

Figure 59. Global Emissions since 1990 and Projected Emissions to 2030. Less than 2°C and 1.5°C pathways are the emissions trajectories aligned with the most cost efficient and highest probability pathway to limit end-of-century surface warming. Emissions gaps are indicated as the difference between year 2030 less than 2°C and 1.5°C pathway emissions and the aggregate target ceiling of all unconditional NDCs or conditional NDCs.

extends to 14–16 Gt/year for pathways that would limit surface warming to 1.5°C above pre-industrial temperatures.

If emissions were to follow current NDCs, another 600–700 billion tonnes of CO_2 will have been added to the atmosphere between 2015 and 2030. By 2030, the economies of the world would have begun to deviate from the RCP8.5 business as usual scenario but would remain on a path toward serious global warming. Based on full implementation of NDCs, the UNEP 2015 Emissions Gap Report predicts that, by the end of the century, the Earth's surface will have warmed by an average of 3–4°C above pre-industrial temperatures.[122] Future generations will have incurred the extreme costs and damages of severe climate change resulting from an excessive accumulation of atmospheric greenhouse gases that were willfully emitted by their parents and grandparents.

The current set of NDCs is grossly inadequate and not aligned with the objectives specified in the Paris Agreement. The 2010 Cancun pledge of the United States stated an ambition to pass legislation that would place the US on a trajectory toward a 30% emissions reduction by the year 2025 advancing to a 42% cut in emissions by 2030 and an 83% curtailment by mid-century relative to a 2005 baseline.[268] The current NDC of the US has pulled back from these Cancun targets and is based on a 27–28% emissions reduction by the year 2025.[254] An ambitious year 2030 target of a 42% reduction in emissions by the world's largest economy would have provided the political incentive for other nations to increase the level of ambition of their own NDCs. Around the globe, with a few exceptions, the intended contributions of advanced economies and developing countries lack the ambition to adequately control future emissions. Member states must revise their NDCs such that the aggregate of contributions begins to align with the agreed upon target of limiting surface warming to well below 2°C above pre-industrial temperatures.

Table 6. Future Emissions Projections*			
Scenario	2030 Emissions Gt CO₂eq	Warming (°C)	Outcome
Business As Usual Emissions increase with population and GDP growth	65	4-6	•Catastrophic climate change damage •Food and water supply failures •Little capacity for adaptation
Current low ambition NDCs Emissions are not aligned with with Paris Agreement objectives	54	3-4	•Severe climate change damage •Threats to food and water supply and security •High adaptation cost and limits to capabilities
Revised high ambition NDCs Emissions aligned with Paris Agreement objectives	39-42	1.5-2	•Moderate climate change damage •Cost-effective adaptations to limit damage
*Projections for emission rates and surface temperature warming from the UNEP 2015 Emissions Gap Report.[122]			

Chapter 5: HOW WE GET THERE – TECHNICAL PATHWAYS TO A SUSTAINABLE FUTURE

The Stone Age did not end for lack of stone, and the Oil Age will end long before the world runs out of oil.

Sheikh Zaki Yamani — Former Oil Minister of Saudi Arabia

The International Energy Agency (IEA) produces an annual in-depth analysis and perspective on current trends and future projections for the production and use of primary and secondary energy sources. This information is used by stakeholders for long-term planning purposes. In recent years, the IEA reports have included a "450 Scenario," whereby an end-of-century atmospheric CO_2 concentration of 450 ppm becomes a determining factor for short and long-term energy projections. The climate model used by the IEA for the 450 Scenario is based on a remaining global emissions budget of 1,200 Gt of CO_2.[172] Complete release of this budget would lead to an end-of-century accumulation of about 450 ppm of CO_2 in the atmosphere with a 50% chance of limiting surface warming to 2°C above pre-industrial temperatures.

The core objective of the Paris Agreement is to limit surface warming to well below 2°C and, ideally, 1.5°C relative to pre-industrial temperatures. The level of ambition contained in the Paris Agreement exceeds the assumption of the IEA 450 Scenario. The precise meaning of "well below 2°C" is not provided; however, the IEA assumes that a 50% chance of limiting future surface warming to less than 1.84°C would qualify.[172] The IEA concludes that, as of 2014, about 950 Gt of CO_2 remained in an emissions budget that is consistent with limiting future surface warming to well below 2°C.[172] This conclusion aligns with the IPCC carbon budget with a 66% chance that surface warming will be held to less than 2°C.

IEA World Energy Outlook 2016 estimates future use of fossil fuels, nuclear, hydro, bioenergy and renewables to the year 2040.[172] The report is heavily influenced by the Paris Agreement and is focused on a New Policies Scenario that incorporates both existing policies in place as of 2016, and climate pledges contained in NDCs submitted prior to the 2015 Paris climate change conference. Comparisons are made between the main New Policies Scenario and the 450 Scenario. In addition, some discussions of the more ambitious Well Below 2°C and 1.5°C Scenarios are included in the report.

In the 2014 Fifth Assessment Report of the IPCC, Working Group III assembled a vast data base of over 900 climate change mitigation models and then compiled this information to generate a set of future scenarios that outline implementation requirements, costs, constraints and limitations in reducing emissions and minimizing surface warming over the course of this century.[269] The focus of this narrative will be on IPCC and IEA mitigation pathways to limit surface warming to less than 2°C.

Cost-effective pathways to safely dispense with the remaining emissions budget can be divided into a short-term immediate action phase that extends to 2030, a mid-term mitigation period between 2030 and 2050, and a final long-term phase over the second half of the century. The time frame of the short-term mitigation period corresponds to the 2030 target dates found in most NDCs. Short-term climate change actions are based on implementation of available known cost-effective technologies. Economies will continue to transition to progressively lower emissions practices over the mid-term mitigation period as higher cost options come into play. By 2050, deep decarbonization of world economies must be in place, and the extent of emissions reduction achieved will largely define end-of-century surface temperatures. Over the long-term, global emissions must eventually arrive at a balanced equilibrium that avoids any further accumulation of atmospheric GHGs. Emissions will continue from the agriculture sector and perhaps to a limited extent from other sectors extending beyond the end of the century. These residual emissions must be balanced by a combination of natural and anthropogenic carbon sinks.

Many climate change mitigation models incorporate a slight overshoot in atmospheric CO_2 concentration above final end-of-century targets.[270] In these models, the balance between release of GHGs and sequestration of atmospheric carbon dioxide reaches net zero-emissions and then transitions to negative emissions status during the second half of the century. After this transition, atmospheric CO_2 levels slowly decline over decades to about 450 ppm.

Cost-efficient, high probability, pathways to limit end-of-century surface warming to less than 2°C require that global rates of total GHG emissions peak prior to 2020 and then decline by about 1 Gt of CO_2 equivalents per year.[122] This rate of curtailment will lead to a 20% cut in emissions by 2030 and a 59% reduction by mid-century relative to peak emissions. Over the long-term mitigation period, emissions would continue to decline to net zero status sometime around the year 2070 followed by a gradual transition to a state of net negative emissions of about 3 Gt/year by the end of the century. In this scenario, the less than 2°C carbon budget would have been effectively utilized to facilitate an orderly decarbonization of world economies. An over expenditure of the carbon budget would have been avoided and surface warming would have been held to less than 2°C.

If global warming is to be limited to 1.5°C, ideally GHG emissions should be reduced from peak emissions by about 26% in 2030 and 85% by mid-century.[122] The transition to zero-emissions should be completed about 15 years ahead of the less than 2°C timeline and net withdrawal of about 5 Gt/year of atmospheric CO_2 equivalents should be in place by 2070.[122]

All climate change models designed to limit end-of-century surface warming to less than 2°C incorporate net atmospheric carbon dioxide reduction (CDR) during the second half of the century. In these models, the rate of CO_2 withdrawal from the atmosphere exceeds residual CO_2 emissions and balances continued emission of non-CO_2 greenhouse gases primarily from the agriculture sector.

By mid-century, in most climate change mitigation models, programs of afforestation and reforestation have established a world forest biomass with a carbon sequestration capacity well in excess of CO_2

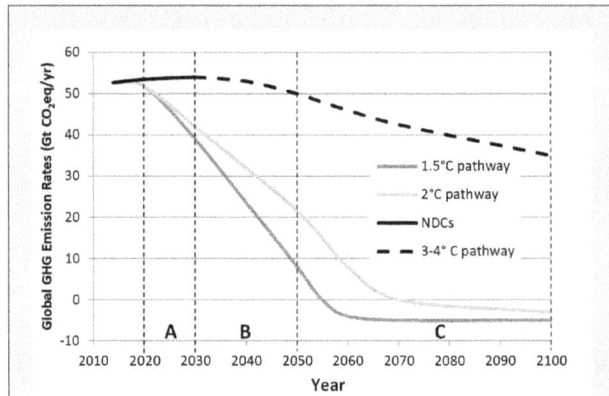

Figure 60. Emissions Reduction Pathways over the Current Century. A — Short-term mitigation phase to 2030. **B** — Mid-term pathways to 2050. **C** — Long-term pathways to final equilibration of emissions. Dashed line indicates the likely long-term outcome based on the level of ambition to curtail emissions contained in the aggregate of NDCs submitted under the Paris Agreement.

emissions from agriculture forestry and other land use (AFOLU).[270] The AFOLU sector begins to offset GHG emissions from other economic sectors. The electricity supply sector rapidly decarbonizes and reaches net zero-emissions between 2050 and 2060. Emissions from the transportation, buildings and industry sectors continually decline from the year 2020 peak; however, these sectors do not arrive at a final equilibrated state of zero or very low emissions until the second half of the century.

Most long-term climate change mitigation models use a combination of forest biomass and industrial bioenergy production with carbon capture and storage (BECCS) to achieve the level of carbon dioxide removal required to fully offset continuing GHG emissions in the latter part of the century.[270] Future BECCS power plants would produce electricity from the combustion of biomass. During the growth stage, forest biomass or energy crops grown on plantations, would sequester and thus withdraw atmospheric CO_2. The biomass would then be harvested and combusted to generate electricity at facilities with carbon capture technology. The captured carbon would be transported to sites for permanent isolation in geological formations. With the implementation of BECCS, the production of electricity transitions to a net negative carbon balance during the second half of the century.

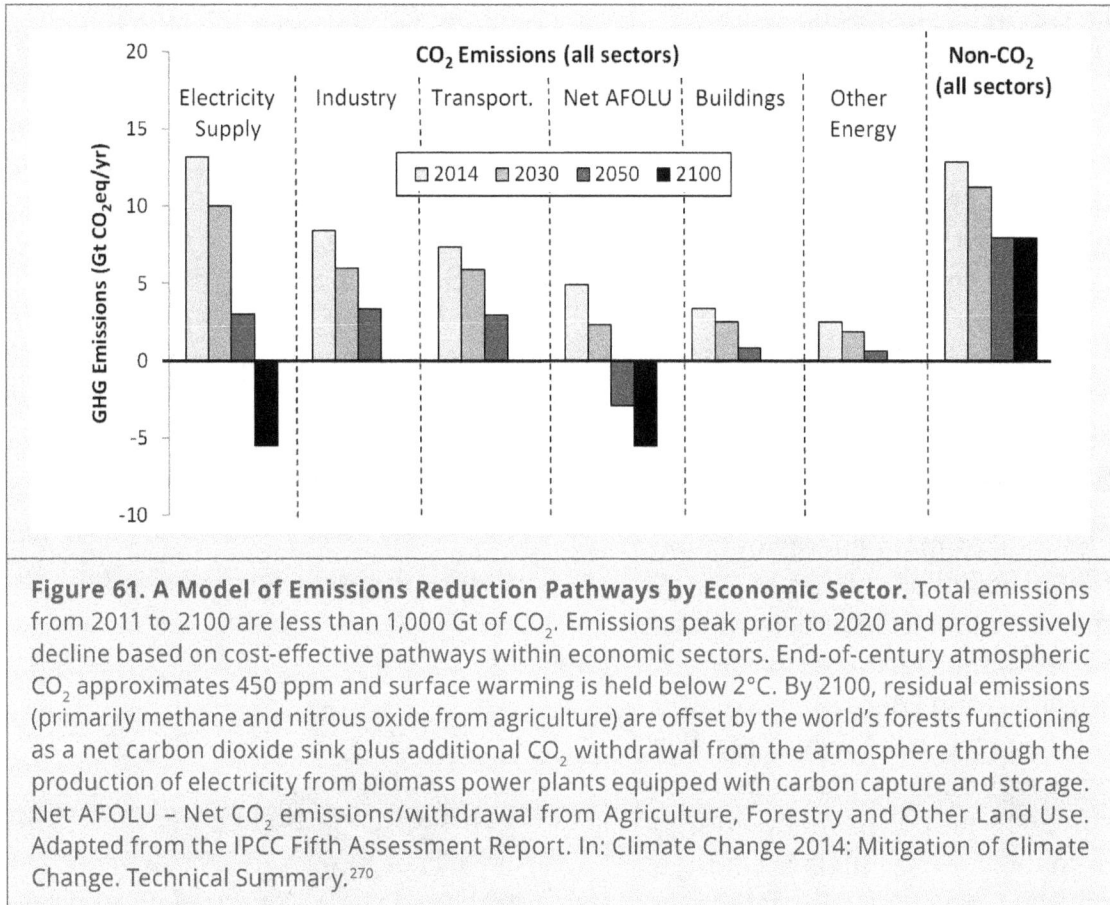

Figure 61. A Model of Emissions Reduction Pathways by Economic Sector. Total emissions from 2011 to 2100 are less than 1,000 Gt of CO_2. Emissions peak prior to 2020 and progressively decline based on cost-effective pathways within economic sectors. End-of-century atmospheric CO_2 approximates 450 ppm and surface warming is held below 2°C. By 2100, residual emissions (primarily methane and nitrous oxide from agriculture) are offset by the world's forests functioning as a net carbon dioxide sink plus additional CO_2 withdrawal from the atmosphere through the production of electricity from biomass power plants equipped with carbon capture and storage. Net AFOLU – Net CO_2 emissions/withdrawal from Agriculture, Forestry and Other Land Use. Adapted from the IPCC Fifth Assessment Report. In: Climate Change 2014: Mitigation of Climate Change. Technical Summary.[270]

The IPCC rates BECCS as a technology under development. There are uncertainties as to the costs, complexities, and land use practicalities of the sequence of technologies required to implement the process. In the absence of BECCS, forested areas of the planet must provide sufficient carbon sequestration capacity to balance residual GHG emissions.

The particular route to achieve the core objectives of the Paris Agreement is dependent on the time frame to peak emissions, the rate of emissions reduction from peak to carbon neutral status, the magnitude of any overshoot in atmospheric CO_2 concentrations, and the capacity of carbon dioxide reduction to drawdown residual emissions over the long-term. Delays in the implementation of effective climate change action plans will push back the date of peak emissions. If peak emissions are extended to 2030, a substantial portion of the total carbon budget will be spent prior to meaningful cuts in emissions. To compensate, an abrupt and costly decarbonization of world economies would be required, along with long-term reliance on vast programs of carbon dioxide removal.

With an expanding world population, the demands of agriculture for land use will limit areas available for forest expansion and thus limit the sequestration capacity of forest biomass, as such, a heavy reliance on carbon dioxide removal would require a vast fleet of BECCS installations. Given the uncertainties of broad scale implementation of BECCS, an over-reliance on carbon dioxide removal could easily become a pathway to a future of severe climate change damage.

The global route with the lowest cost and highest probability of success to achieve the objectives of the Paris Agreement is to complete an orderly process of decarbonization of global economies over the next 50 years without over-expending the remaining budget of emissions. Carbon dioxide removal will be an important component of the overall global program but must be developed alongside of aggressive measures to improve efficiencies, and transition to zero and very low emissions practices.

5.1 Emissions Reduction Pathways in the Energy Supply Sector

The total primary energy demand (TPED) equals all sources of energy that are either used directly or converted to secondary energy carriers. The IEA World Energy Outlook 2016 provides a breakdown of the mix of energy sources that comprised the 2014 global primary energy demand.[172] Fossil fuels accounted for 81% of the total. Bioenergy met about 10% of demand with most of this energy used for cooking and heating purposes in developing countries. Nuclear and hydro

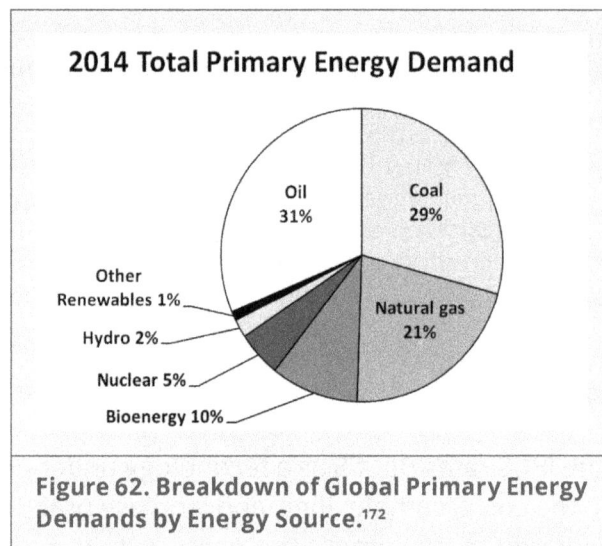

2014 Total Primary Energy Demand

Oil 31% · Coal 29% · Natural gas 21% · Bioenergy 10% · Nuclear 5% · Hydro 2% · Other Renewables 1%

Figure 62. Breakdown of Global Primary Energy Demands by Energy Source.[172]

accounted for 7% of the TPED with other renewables providing the remaining 1%.

The IPCC estimates that about 27% of primary energy is used directly to fuel thermal energy processes within the buildings and industry sectors. Direct energy combustion is divided between natural gas (40%), coal (27%) and biomass (33%).[271] The remaining 73% of the total primary energy supply is converted to electricity and other secondary energy carriers prior to end use.

The energy supply sector is defined as the extraction, conversion, storage, transportation, and distribution of primary energy. The sector includes conversion of primary energy to electricity and all aspects of supply of electricity and other primary and secondary energy carries used by the industry, buildings and transportation sectors. Crude oil is classified as an input to the energy supply sector with refined petroleum products as outputs. The sector includes petroleum products that are used for non-energy purposes. In 2010, the energy supply sector accounted for 35% of total emissions and was by far the most significant contributor to anthropogenic GHG accumulation.[271] The sector operates at about a 56% efficiency with 29% of the TPED consumed in the production and supply of electricity and distributed heat, and in the extraction, transport and refining of primary fuels to other secondary energy products.[271]

5.1.1 Emissions Reductions within the Fossil Fuel Production Sub-Sector. In 2010, an estimated 6% of global GHG emissions were attributed to the extraction and processing of oil and gas prior to end use of the products.[272] In the US, methane release through venting and fugitive gas losses from the oil and gas sector accounted for 2.7% of total GHG emissions in 2015.[273] Gas flaring at oil production sites alone contributes an estimated 1.2% of

Figure 63. Summary of 2010 Energy Flows from Total Primary Energy Demand.[271]

total world CO_2 emissions.[274] These high rates of emissions are exacerbated by recent advances in hydraulic fracking and other unconventional processes to access oil and gas reserves. These processes are more energy intensive and come with higher levels of release of fugitive gas, such that, relative to conventional deposits, fossil fuel products from unconventional sources come with an elevated complete life cycle of GHG emissions.

Countries and sub-national regions with considerable oil and gas activities have some of the highest per capita rates of GHG emissions on Earth. Initiatives such as the recent US-Canada agreement to target a 40–45% cut in methane emissions from the oil and gas sector by the year 2025, and the World Bank initiative to eliminate routine flaring by 2030 are important short-term climate change mitigation actions.[275,276] Global implementation of measures to reduce fugitive gas release and to limit venting and flaring during extraction, processing, distribution and supply of fossil fuels can significantly reduce emissions from the energy supply sector.

5.1.2 Emissions Reduction through Improved Efficiencies in the Supply and Use of Electricity. The production of electricity is a straightforward process whereby the energy from a primary source is used to spin the turbine of a generator or, in the specific case of photovoltaic cells, directly converted to electricity. Provision of electricity to end users is a complicated balancing act whereby power is produced in the absence of storage capability and then distribution across a network or grid in response to variable demands. Adequate generation capacity must be in place to meet peak demands, and the system must have the inherent flexibility to balance supply and demand and redirect power flows to meet localized, often unplanned, demand

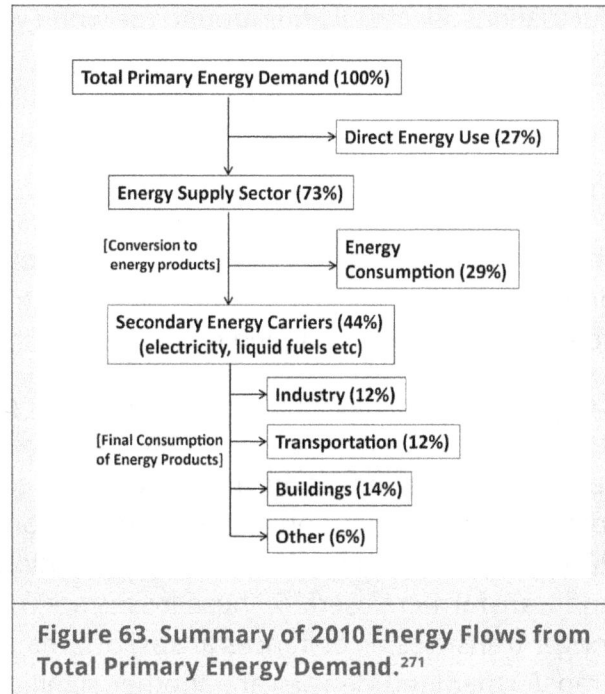

fluctuations. Electrical grids around the world were built around high capacity, centrally located power stations using primary energy sources and processes that allowed for modulation of supply to meet demand requirements. Hydropower is an exception in that the production site is dependent upon the location of the energy source and often involves long distance transmission from a remote hydro dam to more heavily populated regions. The size of electrical grids, the degree of interconnections within these grids, the technical sophistication of power flow management, and the policies and pricing mechanisms for end use of electricity are based on the historical use of coal, natural gas, nuclear and hydro as fuels in large power plants.

Conversion of primary energy to electricity, distribution of power supply, and final consumption by end users are steps in an inefficient series of processes. For steam turbine power production, about 60% of the energy is lost as waste heat.[271] Power losses during distribution of electricity are considerable and can vary between 6.5 to 20%, and are dependent upon the geographical layout of electrical grids and the lines and transformers used.[271] These losses are mainly due to cable resistance such that longer transmission distances are associated with lower efficiencies in power supply. Transformer inefficiencies are another significant source of power loss during distribution. Finally, end user inefficiencies are substantial, and vary with the particulars of industrial motors, appliances, electronics, refrigeration, lighting, heating, and cooling.

Improved transformer design, use of high voltage direct current transmission, a greater degree of interconnection and an overall modernization of the electrical distribution system will reduce distribution losses and increase the efficiency of power supply. These upgrades will reduce emissions per kilowatt hour of electricity delivered to the end user.

Combined heat and power (CHP) or co-generation plants retain the waste heat energy output from steam turbines. This energy is generally used to heat water that is then distributed to buildings on a district heating grid. Tri-generation or building cooling, heating and power (BCHP) plants incorporate a refrigeration unit and have the flexibility to provide district heating and cooling outputs. CHP plants generally operate at 58% net efficiency of conversion of a primary energy source to the combination of electricity and useable waste heat.[271] Promotion of well-designed decentralized BCHP systems has the potential to efficiently heat and cool residential areas and industrial complexes and thus reduce or eliminate regional energy demands for these purposes.

Modern natural gas combined cycle (NGCC) power plants make use of steam and gas turbines in the same facility. Waste heat from the gas turbine is directed to power generation from steam, and these plants operate at 50–60% efficiency in energy conversion to electricity.[271] When complete life cycle accounting of GHG emissions is used, emissions from an NGCC power plant are about half of the emissions per unit of power generated from a coal-fired power plant.[271] Power output from gas turbines can be readily modulated to control supply in response to fluctuations in demand. Further, smaller NGCC plants can be used to provide distributed energy within electrical grids.

These small plants can provide power to residential and industrial areas located near the plant and thus avoid the inefficiencies associated with long distance distribution of electricity from large centralize plants. Overall, switching from coal to natural gas power plants will achieve substantial emissions reductions; however, natural gas power production can only be considered as a bridge technology to an eventual complete decarbonization of power production. The plant life-span economics and GHG emissions for the installation of future natural gas power plants must be fully evaluated in comparison to zero-emissions options for power production.

Improved efficiencies in the demand side use of electricity can reduce per capita power production and can help to ameliorate supply side balancing and flexibility requirements. Smart metres, smart grids, and pricing mechanisms that favour power consumption during non-peak hours are effective tools in smoothing the volatility of demand and thus reducing the costs associated with providing capacity for peak demand and balancing demand volatility.

Improved efficiencies of demand side power consumption, in combination with grid updates, fuel switching from coal to natural gas and, where applicable, installation of heat recovery systems to power plants, can enhance the overall performance of the power supply chain and achieve a significant reduction in GHG emissions. Near-term optimization of the energy supply sector can be a significant contributor toward achieving a year 2020 goal for peak emissions and will be an ongoing contributor to declining emissions during the critical period of infrastructure transition as world economies progressively decarbonize.

5.1.3 Decarbonization of the Electricity Supply Sector. Improved efficiencies in the energy supply sector will not adequately reduce total GHG emissions as required to meet the future surface temperature limitations specified in the Paris Agreement. Effective climate change mitigation requires a fundamental restructuring of the energy supply sector to achieve net 0 or very low emissions status by 2050–2060.[269] Generation of electricity accounts for 70% of total emissions from the overall sector and thus deep decarbonization of power production by mid-century is an essential component of all effective climate change action plans.[269] To achieve this goal, power must be produced by zero or very low emissions energy sources or incorporate carbon capture and storage (CCS) technology to maintain low emissions standards.

Achieving such an ambitious reconfiguration of power production is technically feasible, and, along with improved efficiencies in energy use across all economic sectors, this reconfiguration will have the greatest impact among the full basket of measures required to decarbonize world economies.

Hydro, nuclear, solar, wind and geothermal are classified as commercially viable zero-emissions energy sources for power production. The intensity of emissions from these sources is less than 40 and, for nuclear and wind, down to 4 g of CO_2eq/kWh of electricity produced.[271] Emissions assessments are based on a full life cycle accounting of atmospheric withdrawal and emission of GHGs associated with all aspects of equipment manufacture and plant construction and operation. When these standards are applied, the production of electricity from bioenergy combustion (without CCS), and from fossil fuel plants equipped with CCS are in the range of 150–250 g of CO_2eq/kWh.[271] These technologies, while emitting considerably fewer GHGs than traditional fossil fuel-based power production, are classified as low emissions processes. Fossil fuel power plants equipped with CCS without enhanced oil recovery have yet to reach commercial implementation.[271]

The capacity factor of a given energy source is a measure of the average power output over time relative to designed power production under ideal conditions. Some renewable energy sources such as solar and wind have low capacity factors with variable power production and limited flexibility to modulate output. In contrast, bioenergy, nuclear and fossil fuels are flexible, dispatchable power sources with high capacity factors. The capacity factor of hydro power varies with the specifics of water flow and the design of the reservoir but is generally between that of dispatchable and variable energy sources. Hydro power output can be subject to seasonal variations and changing precipitation patterns.

In general, existing grids can accommodate up to 20% of total power input from wind or solar without issue.[277] Higher levels of grid penetration by variable power sources are possible. On average, wind turbines generate 42% of Denmark's total power production.[278] Distributed power from solar installations and wind turbines offers the advantage of local use and decreased transmission losses; however, over and under production requires further management of flows from distributed sources to the larger grid. High levels of penetration by variable renewables require an installed capacity well in excess of peak demands and extensive grid interconnections over larger distances. As an example, by linking wind parks in separate regions, it becomes possible to smooth variabilities in wind power production and maintain the balance of supply and demand at considerably higher rates of wind energy penetration than could be accomplished using the same total installed power in a single region.[172] Investment in transmission capacity can allow high demand, densely populated areas to access wind, solar and hydro rich regions that may otherwise may be lacking in market opportunities. Ultimately, a combination of grid infrastructure and sophistication in both supply and demand management will be required for optimum utilization of zero-emission, low capacity factor renewables.

A robust commercially viable technology for the storage of electricity produced by wind and solar installations would be a game changer. Potentially, electricity could be stored directly in banks of advanced batteries and then dispatched to the grid on a need basis. Alternatively, excess power produced by variable renewables could be converted to an intermediate energy storage product. As an example, hydrogen could be produced from electricity through hydrolysis of water. Hydrogen can be readily stored and used as a high capacity factor fuel to produce power as needed with zero-emissions.[172] As of this writing, hydrogen and other robust electricity storage technologies with broad applications are far from commercial readiness.

As described in chapter 3, foregoing the replacement of fossil fuel-based power plants with nuclear or hydro options based on environmental, safety and human health concerns is without foundation. In contrast, a switch away from fossil fuels to the production of electricity using these well-established, dispatchable, zero-emission technologies will markedly improve air and water quality, reduce human illness and mortality associated with non-greenhouse gas emissions, and improve the overall safety of the energy sector. Electrical grids can readily accommodate a high penetration from nuclear and hydro plants without incurring the costs and complexity associated with extensive power supplies coming from variable energy sources.

The US Energy Information Administration has projected the Levelized Cost of Electricity (LCOE) for various power plant options assuming a year 2020 and year 2040 installation.[279] LCOE accounts for the total lifespan costs of building and operating a power plant and then calculates the cost per unit of electricity produced. Geothermal is by far the most economical source of power and has the advantages of a high capacity factor and zero GHG emissions. However, the accessibility of geothermal hotspots limits the potential contribution of this energy source to about 1% of the world's total primary energy supply.[280] The remaining zero-emissions options for power production do not have primary energy supply constraints. Nuclear power is the only zero-emissions source for the production of dispatchable electricity with a high capacity factor.

Table 7. Comparison of Options for Power Production						
	Capacity Factor (%)	LCOE ($/MWh)*		Emissions Intensity **		
		2020	2040	High	Low	Zero
Dispatchable Technologies						
Advanced Coal	85%	$116	$106	812		
Advanced Coal with CCS	85%	$144	$128		228	
Advanced Natural Gas	87%	$73	$79	488		
Advanced Natural Gas with CCS	87%	$100	$106		164	
Nuclear	90%	$95	$89			4
Geothermal	92%	$48	$57			40
Biomass	83%	$101	$94		232	
Non-Dispatchable Technologies						
Hydroelectric***	54%	$84	$90			20
Wind (Onshore)	36%	$74	$75			4
Wind (Offshore)	38%	$197	$176			4
Solar (concentrated thermal)	20%	$240	$197			20
Solar (photovoltaic)	25%	$125	$107			40
*Levelized Cost of Electricity. Accounts for all aspects of capital costs, plus variable and fixed operating costs to produce electricity over the lifespan of the plant.[279]						
**Emission Intensity as g CO_2eq/kWh of electricity produced.[271]						
***Capacity factor of hydro assumes seasonal variations in water flow.[279]						

In recent years, the LCOE from wind, solar and biomass has dropped to levels that are competitive with well-established primary energy sources including coal and natural gas. In contrast to wind and solar, power production from a biomass power plant is readily controlled. However, in the absence of CCS, the production of electricity from biomass is encumbered by net emissions of GHGs.

Carbon capture and storage can potentially extend the utility of coal and gas-fired power plants by reducing the intensity of emissions; as such, CCS can provide some flexibility in the energy mix to produce electricity. However, as will be discussed, there are considerable economic and logistic barriers to the application of CCS to coal and gas power plants. In the longer term extending to 2050–2060, an orderly and complete worldwide shut-down of fossil fuel power plants with replacement by zero-emissions nuclear, hydro, wind, solar and geothermal energy sources will likely be necessary to limit end-of-century surface temperature elevation to less than 2°C.

The configuration of future electricity supply chains around the world will be dependent upon local options for primary energy inputs in combination with public perceptions and political will to implement change. In France, a heavy reliance on nuclear power production has resulted in near complete transition to a zero-emissions supply of electricity. In 2015, the consumer cost of electricity in France was among the lowest in Europe at $0.09 (US dollar cents)/kWh.[281] The experience in France provides a blueprint for cost-effective transition to zero-emissions power production for nations that will accept an expansion of the nuclear energy industry. Countries with access to geothermal hotspots or substantial water flow resources have the opportunity to build zero-emissions geothermal or hydro power plants. Regions that are rich in solar and wind energy are likely to rely more upon these low capacity factor renewables in their electricity supply network. To control costs, optimally designed networks of power supply from zero-emissions sources will be based on a suite of complementary technologies whereby dispatchable energy is used to compensate for the variable power output from wind and solar.

In Germany, there is little public support for nuclear power, and the complete phase out of the nuclear industry is scheduled for 2022.[282] Germany has established an "Energiewendeenergy" or energy transformation policy with the goal of achieving an 80–95% reduction in GHG emissions by 2050 relative to the nation's 1990 Kyoto baseline emissions.[282] The core components of the policy are to double the efficiency of energy utilization, and to produce 80% of the nation's electricity from renewables while phasing out fossil fuel and nuclear power plants.[282] Achieving such a high level of penetration by wind and solar, requires a complex, highly interconnected electricity supply network, along with back-up systems of dispatchable power production. These restrictions have limited Germany's success in achieving cost-effective emissions reductions. In 2012, per capita GHG emissions rates in Germany were 60% greater than the per capita emissions rates of France. While in 2015, the consumer cost of electricity in Germany was 70% above the utility rate in France.[124,281] Germany has one of the highest costs for electricity among major economies in the world. There are questions as to the practicality and cost-effectiveness of achieving the mid-century objectives of Germany's energy transformation policy. However, the policy is a driving force for transition and innovation within the world's fourth largest economy. Considerable technical expertise and resources within Germany will be applied to challenges such as the development of energy storage systems and emissions-free dispatchable power supplies that can be integrated with variable renewables.

Over the long-term, a transition of world economies to carbon neutral status will require deep cuts in fossil fuel use within the transportation, buildings, and industry sectors. Electricity produced from zero or negative emissions sources can replace liquid and gas fuels in many applications. New demand will be partially offset by improved efficiencies; however, the future design of electricity supply networks must consider increased electrification of industry and transportation, along with significant grid penetration by variable renewables and sources of small modular distributed power production.

5.1.4 Carbon Capture and Storage. The oil and gas sector operates several large scale commercially viable natural gas processing facilities that incorporate CO_2 capture and storage. The world's second largest carbon capture facility is located in LaBarge, Wyoming and operated by ExxonMobil.[283] Natural gas containing 65% CO_2 is extracted and piped to the facility for processing to capture 7 million tonnes of CO_2 annually.[283] The captured CO_2 is concentrated, piped to nearby oilfield injection sites, and used for Enhanced Oil Recovery (EOR). Prior to 2008, the treatment facility sold 55% of the captured CO_2 for EOR and vented the remainder to the atmosphere.[283] In June 2008, the Oil and Gas Conservation Commission ordered ExxonMobil to curb emissions at the LaBarge facility.[283] In December of 2010, the complete carbon capture facility began operations, resulting in a 50% reduction in CO_2 emissions with the additional captured CO_2 applied to EOR.[283] The viability of the ExxonMobil's LaBarge CO_2 capture facility is based on a combination of low capture costs due to the high concentration of CO_2 in the gas stream, revenue from CO_2 sold for EOR, and environmental legislation mandating reduced emissions from the facility.

Enhanced oil recovery linked to CCS will not contribute to the long-term decarbonization of world economies. EOR is based on injection of captured CO_2 into spent geological formations to displace oil that otherwise would have remained underground. However, CCS with EOR does advance the technology, and the resultant commercial scale experience and cost reductions will apply to future CCS installations. At present, EOR is the only economic driver for commercial scale deployment of carbon capture and storage.

There are no remaining serious technical barriers to implementation of large scale CCS. The process begins with on-site capture of the CO_2 at a power station or industrial site of production. Options for CO_2 capture consist of post-combustion, pre-combustion or oxygen combustion technologies, and each option comes with differences in technical maturity, capital and operating costs, and carbon capture efficiency.[284] Carbon capture is an energy intensive process regardless of the technology employed. Internal energy use by power plants equipped with CCS will reduce the efficiency of power supply such that the cost of electricity is estimated to increase by 28–44%.[284] CO_2 is then separated from impurities using liquid amines (absorption) or adsorption to solids such as activated carbon and converted to a fluid state for transport by pipeline, truck, rail or ocean tanker.[284] Assuming a future of large scale adoption of CCS, pipelines between sites of CO_2 capture and final geological storage are by far the safest and most cost-effective mode of transport. Effective and acceptable use of CCS to mitigate climate change requires final permanent isolation in deep saline aquifers. These aquifers are located beneath layers of impermeable rock at depths of over 800 metres and are isolated from ground and sea water. Total world storage capacity within saline aquifers is uncertain but can be considered as enormous with estimates ranging from 7,000 to 23,000 Gt of CO_2.[285] Deep saline aquifers are relatively well distributed around the world such that storage capacity and location is unlikely to be an impediment to worldwide adoption of CCS.[285]

Effective use of CCS will require a network of pipelines for transport of CO_2 to injection sites over selected deep saline aquifers. A suite of measuring, monitoring and verification (MMV) technologies will be applied to the entire CCS system to ensure and maintain safe, permanent isolation of CO_2 from the atmosphere.[271]

The cost of CCS varies with the purity of carbon dioxide in the source. Excluding storage and transportation costs, the cost of CO_2 capture and separation for coal-fired power plants is estimated at $23–36 per tonne.[284] Costs for separation from the more dilute streams found in natural gas-fired power plants are considerably higher and estimated at $58–112 per tonne dependent on the technology employed.[284] In comparison, the costs of carbon dioxide capture from natural gas can be as low as $5–15 per tonne.[286] Overall, current estimates for the total costs of capture, separation, concentration, pipeline transportation, geological storage, and measuring, monitoring and verifying the end-to-end process of CCS from fossil fuel power plants are in the range of $60–100 tonne of CO_2, with the costs of capture accounting for 70–80% of the total.[284] The US Energy Information Administration estimates the capital costs for building an advanced combined cycle natural gas power plant with CCS will be twice the cost of the same plant without CCS.[287] Installation of CCS to a coal fueled power plant is projected to increase capital costs by 50%.[287]

Greenhouse gas life-cycle assessment studies indicate that application of CCS will reduce emissions by 75–84% from fossil fuel-based power plants.[284] Residual emissions will be higher than competing, zero-emissions technologies. A complete assessment of the costs and efficacy of emissions reduction must be applied to CCS equipped fossil fuel power plants in comparison to other zero on very low emissions energy supply options.

In the absence of EOR, significant financial barriers prevent commercial adoption of CCS technology to the production of electricity from fossil fuels. A recent study concluded that an emissions tax of at least $125/tonne of CO_2 would be required for a CCS equipped natural gas combined cycle power plant to be competitive with a conventional NGCC plant.[288] Given the high capital and operating costs, the reduced efficiency of energy conversion, and the residual GHG emissions, fossil fuel power plants equipped with CCS are unlikely to be competitive with nuclear, hydro, wind, solar, and bioenergy options for the production of electricity.

The end game for CCS extends beyond power production from fossil fuels. As will be discussed, CCS has the potential to reduce emissions from cement manufacture and other emissions-intensive industrial practices. However, the most significant opportunity for the future use of CCS in combating climate change is to achieve atmospheric withdrawal of CO_2 by applying the technology to the generation of electricity from biomass energy.

Bioenergy with carbon capture and storage is the only technology at the developmental stage that can achieve net negative emissions and thus drawdown atmospheric CO_2. Successful development and adoption of BECSS is a game-changer in that a controlled removal of atmospheric CO_2 facilitates the timing and management of the transition of world economies to total GHG carbon neutral status during the second half of the century.

The technical and cost considerations for adoption of BECCS are similar to those of CCS applied to fossil fuel power plants, with the important exception that combustion of biomass is used to produce electricity. As will be discussed under mitigation pathways within the agriculture and forestry sector, upstream management of land use for intensive growth of biomass fuel is a complex issue that must be integrated with food production and other land use applications. Development of BECCS will require a carbon pricing mechanism; however, unlike other technologies, carbon pricing will provide a revenue stream to BECCS operations. With BECCS, the price applied to carbon emissions would be equally applied as payment for the net atmospheric withdrawal and permanent isolation of CO_2 from the environment. High emitting processes and energy sources are less competitive in markets that include a significant carbon pricing mechanism. Economic drivers designed to curb emissions will also function to enhance the profitability and commercial adoption of negative emissions technologies such as BECCS.

Critics of BECCS describe the overall chain of technologies as unwieldly and encumbered by considerable technical, land use, environmental, and cost challenges. As of this writing, BECCS is a pre-commercial concept that requires extensive study and development. Biomass power plants equipped with CCS will be costly, and given the costs of planting, harvesting and transporting biomass, the levelized cost of electricity will be well in excess of the costs to produce power from zero-emissions options.

There are no aggressive models of climate change mitigation designed to limit surface warming to well below 2°C that do not involve atmospheric carbon dioxide reduction. Afforestation, while simpler than BECCS, will be far less efficient and will require considerably more land mass to achieve continued and effective withdrawal of atmospheric

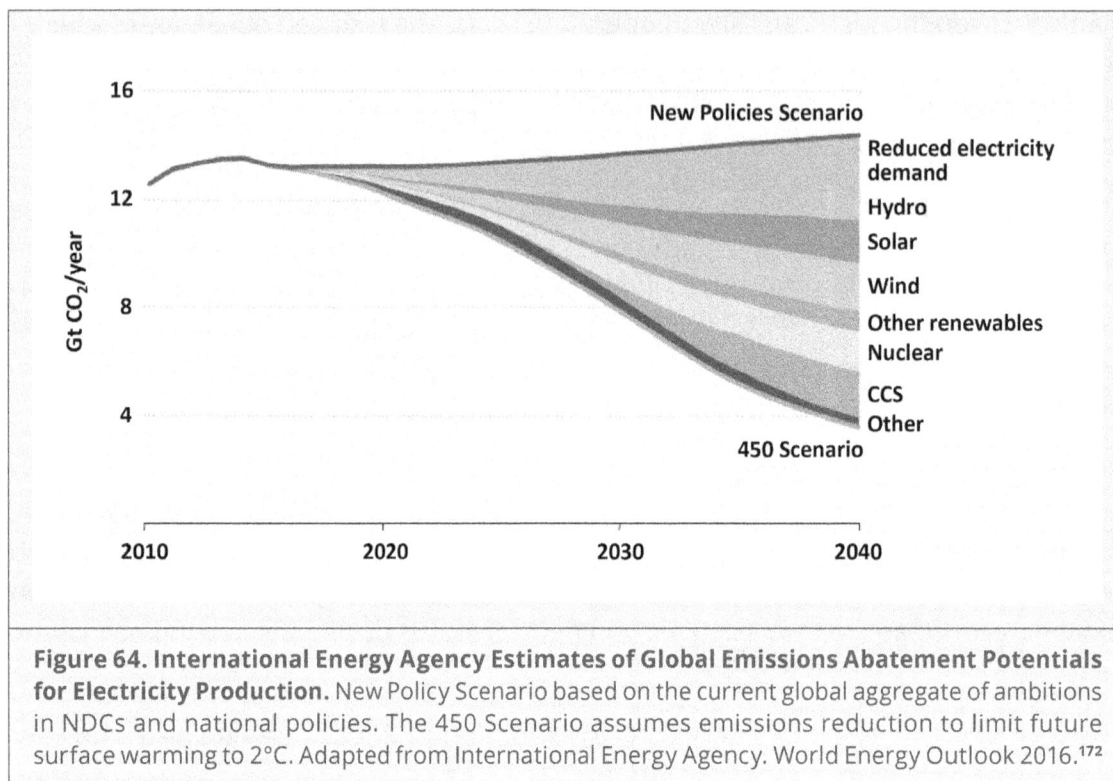

Figure 64. International Energy Agency Estimates of Global Emissions Abatement Potentials for Electricity Production. New Policy Scenario based on the current global aggregate of ambitions in NDCs and national policies. The 450 Scenario assumes emissions reduction to limit future surface warming to 2°C. Adapted from International Energy Agency. World Energy Outlook 2016.[172]

carbon. The practical limits of CDR, based purely on afforestation, are such that the majority of models that achieve end-of-century atmospheric GHG levels of 450 ppm CO_2 assume successful development and deployment of BECCS. Delays in peak emissions beyond 2020 and inadequate emissions reduction over the next 30 years will progressively increase the demands on BECCS or other largely theoretical CDR technologies, to compensate for excess emissions. Given the early stage of development of BECCS and the potential limitation of biomass supply, an over-reliance on industrial processes to remove and sequester atmospheric CO_2 increases the risk of failure to adequately control GHG levels and thus prevent excessive surface warming during the current century.

5.1.5 Summary of Potential Abatements in the Electricity Production Sub-Sector. In the World Energy Outlook 2016, the IEA compared emissions projections for power production under a New Policies Scenario that roughly approximates the level of ambition contained in the global aggregate of NDCs with the 450 Scenario that aligns with a 50% chance of limiting future surface warming to less than 2°C.[172] By 2040, improved efficiencies in the use of electricity across economic sectors will be essential to control demand. The IEA indicates that the potential for savings through efficiency measures more than compensates for progressive electrification of industry and transportation, and can become a significant contributor to emissions reduction from power production. Under the 450 Scenario, increased grid penetration by zero and very low emissions hydro, solar, wind, other renewables and nuclear power will be the major contributors to achieving emissions reduction from power production. Going forward from 2030, the IEA anticipates that CCS will become an increasingly significant contributor to reduced emissions. By the year 2040, the fully realized combination of all abatement potentials under the 450 Scenario could drive down global emissions from the production of electricity by 72%. Given inevitable inconsistencies of implementation, when all abatements are considered under the 450 Scenario, the IEA projects a 55% reduction in emissions by the year 2040. Under the more aggressive Well Below 2°C and 1.5°C Scenarios, emissions cuts could be further advanced. If one of these scenarios were to unfold, the sector would be well positioned for deep decarbonization by mid-century.

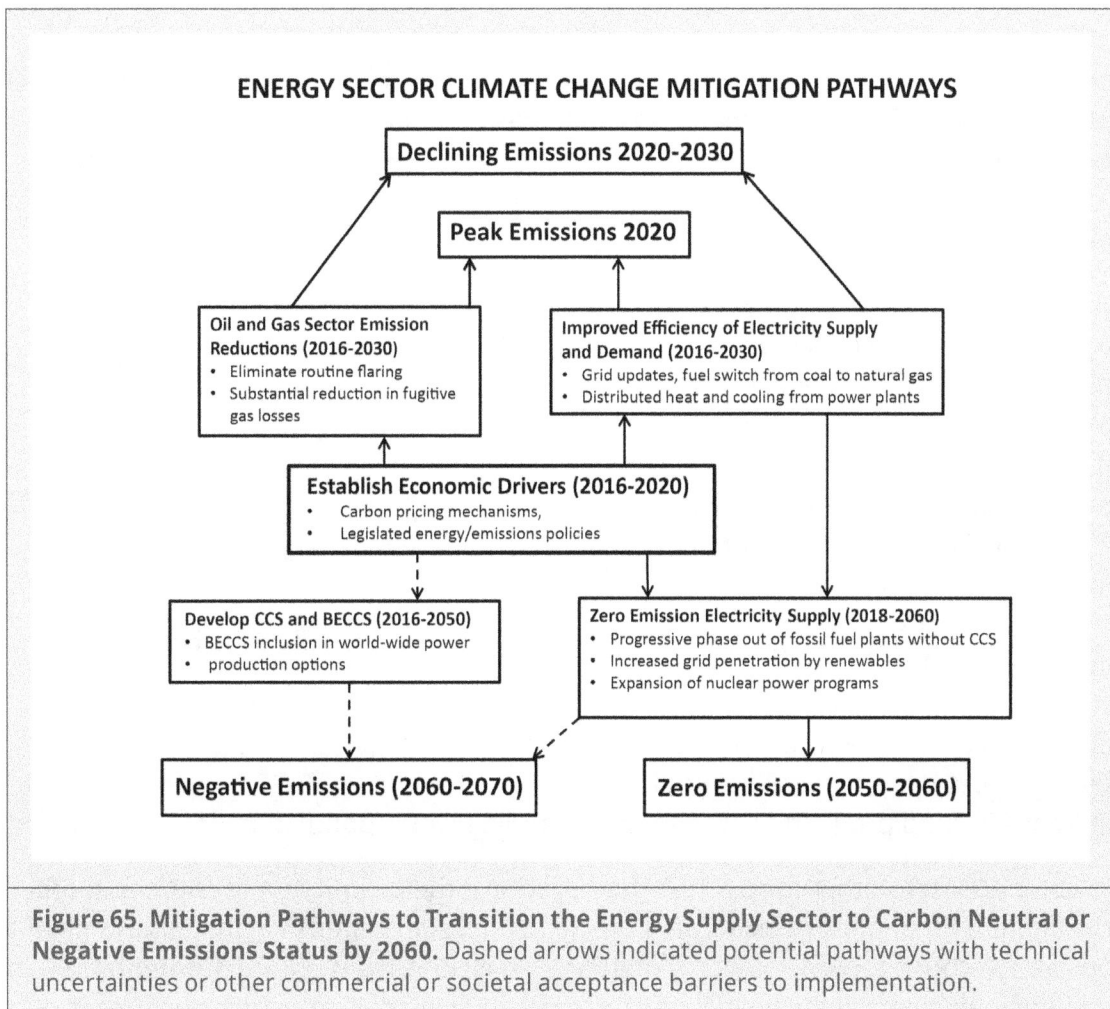

Figure 65. Mitigation Pathways to Transition the Energy Supply Sector to Carbon Neutral or Negative Emissions Status by 2060. Dashed arrows indicated potential pathways with technical uncertainties or other commercial or societal acceptance barriers to implementation.

5.2 Emissions Reduction Pathways in the Agriculture, Forestry and Other Land Use (AFOLU) Sector

The AFOLU sector accounted for 24% of world GHG emissions in 2010.[267] The sector can be divided into the sub-sectors of agriculture, where emissions of methane and nitrous oxide predominate, and forestry and other land use (FOLU). Net emissions from forestry and other land use are calculated as the difference between emissions from deforestation, forest degradation and land use, and CO_2 that is sequestered by plants for photosynthesis and thus withdrawn from the atmosphere.

5.2.1 Emissions from Agriculture. In 2010, methane output from enteric fermentation by farm animals accounted for 4% of global GHG emissions and 36% of the total emissions from the agriculture sub-sector.[107] Release of nitrous oxide from the application of synthetic fertilizers is the second major source of emissions from agricultural practices. In 2010, the cultivation of rice paddies released a quantity of methane equal to 1.2% of world GHG emissions.[107] Burning of biomass as part of agricultural practices, manure management and other sources account for the remaining 2.1% of emissions originating from the agriculture sub-sector.[107]

With improvements in crop genetics and yields, enhanced feed efficiency of farm animals, and other advances in agriculture, the intensity of GHG emissions per unit of product has declined over the decades. Between 1990 and 2012, world population increased by 33%,[289] and the rate of annual total GHG emissions increased by 40%.[124] However, over the same period, emissions from the agriculture sub-sector were up by only 20% despite a substantial increase in food production to feed an expanding world population.[124]

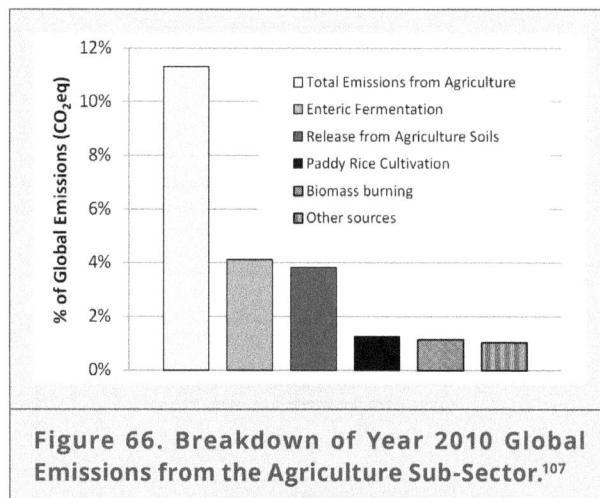

Figure 66. Breakdown of Year 2010 Global Emissions from the Agriculture Sub-Sector.[107]

In comparison to other food production systems, the production of cattle and other ruminant animal meat remains an inefficient, emissions intensive process. Ruminant animals account for 75% of world methane emissions from enteric fermentation. In addition, two thirds of the emissions from manure deposited on grazing pastures comes from ruminant animals.[107] For each metric tonne of cattle meat produced, 5.5 tonnes of CO_2 equivalents are emitted to the atmosphere.[107] The intensity of emissions per unit of cattle meat dwarfs that of all other food products. The UN Food and Agriculture Organization estimates that 26% of the world's ice-free land is used for livestock grazing and 33% of the total cropland on Earth is devoted to animal feed production.[290] Maintaining the current dedication of land resources and world GHG emissions budget to the production of animal protein, and specifically cattle meat, will be a challenge in transitioning world economies to full carbon neutral status over the next 60 years.

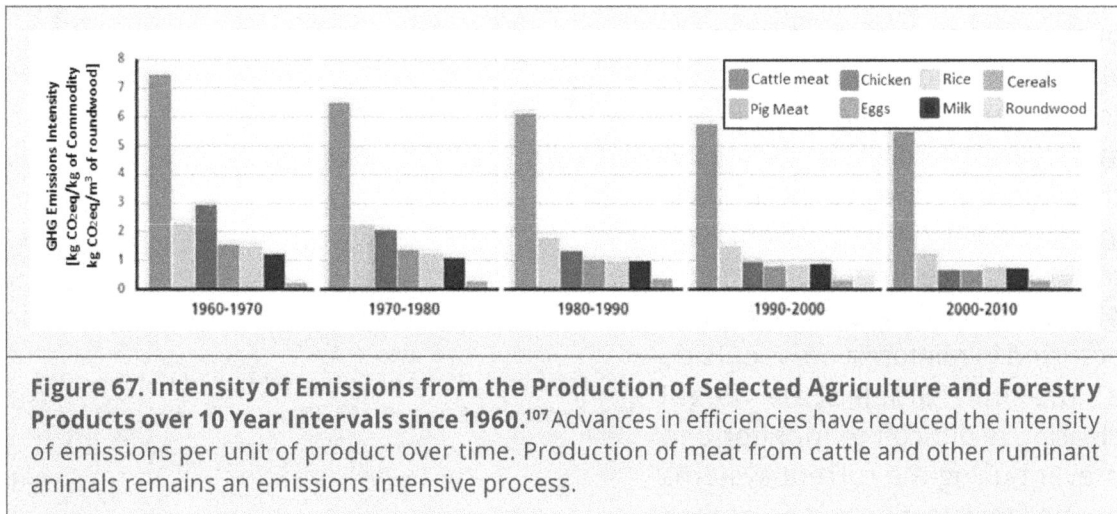

Figure 67. Intensity of Emissions from the Production of Selected Agriculture and Forestry Products over 10 Year Intervals since 1960.[107] Advances in efficiencies have reduced the intensity of emissions per unit of product over time. Production of meat from cattle and other ruminant animals remains an emissions intensive process.

5.2.2 Emissions from Forestry and Other Land Use (FOLU). Between 1990 and 2010, deforestation in South America, led by deforestation of the Amazon, accounted for 70% of the total loss of world forest area[208] and 78% of the worldwide decline in forest-sequestered carbon.[209] During the peak years of deforestation, the narrow sector of forestry and other land use in Brazil accounted for 45% of the nation's GHG emissions and 2.5% of world emissions.[124] The loss of vegetative biomass from deforestation of the Amazon was, and continues to be, both a major source of emissions, and an ongoing loss in capacity for sequestration and draw-down of atmospheric carbon. Historically, deforestation in South America was driven by a complex array of unsustainable logging practices, mining, illegal forest exploitation, and most significantly, conversion of forest area to expansive low productivity pastures for cattle grazing.[240] Specifically, an expansion of the soybean crop area in Brazil displaced cattle ranchers leading to a northern extension of grazing pasture lands into, what was, the edges of the Amazon forest.[240]

In response to the devastating rate of rainforest depletion, Brazil implemented the "Action Plan for Prevention and Control of Deforestation in the Amazon."[291] Over 200 measures have been introduced since the launch of the plan including the creation of protected areas, implementation of satellite monitoring systems and improvements in policing practices. Under Brazil's forest management law, all public forests are covered and have either been transformed to protected lands, allocated to indigenous populations or given over to land use concessions for monitored sustainable economic use.[240] Approximately 48% of the Amazon is under protection, and systems designed to detect and prevent illegal forest exploitation are in place.[240] The actions of the Brazilian government have effectively eliminated large scale descimation and reduced the overall rates of deforestation of the Amazon by 82% relative to the 2004 peak of deforestation.[292]

Forest management and conservation of the Amazon can be described as a work in progress. Forty percent of the total area of the Amazon is within the boundaries of rainforest countries other than Brazil, and overall deforestation rates in these neighbouring countries have increased over the period from 2004 to 2012.[291] Since 2012, approximately 40% of annual deforestation has occurred in rainforest areas outside of Brazil.[291] Small scale illegal clearing is difficult to monitor and prevent using the current systems of satellite detection and on-ground law enforcement. Further advances in policy development, monitoring, and policing over the entirety of the Amazon are required to eliminate deforestation. Potentially, forest protection could extend to stimulation of recovery and reforestation.

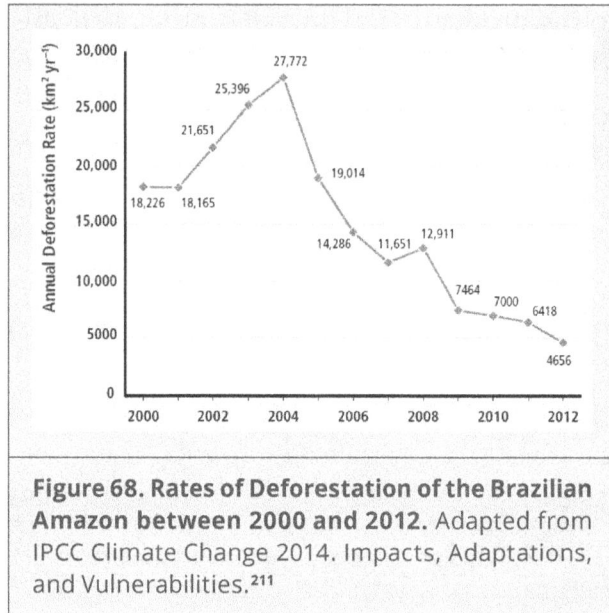

Figure 68. Rates of Deforestation of the Brazilian Amazon between 2000 and 2012. Adapted from IPCC Climate Change 2014. Impacts, Adaptations, and Vulnerabilities.[211]

Conservation of the Amazon rainforest and future growth in agricultural productivity in the expansive savannah region south of the Amazon are not mutually exclusive. Arable land use in the sub-region consists of 25% crop production and 75% expansive low productivity pasture land, much of which is defined as degraded pasture land.[293] In the future, a well managed and regulated conversion of low productivity pasture land and potentially arable non-forested land to intensive agriculture can facilitate an increase in overall agricultural productivity independent of deforestation of the Amazon.[293]

Globally, during the decade from 2001–2010, GHG emissions from deforestation plus biomass fires, and the draining of peatlands accounted for 10.5% of world emissions.[107] These emissions were partially offset by forest biomass added to the planet through forest management programs. Atmospheric withdrawal of CO_2 following reforestation and afforestation was equivalent to 3.8% of world emissions such that the net effect of the forestry and

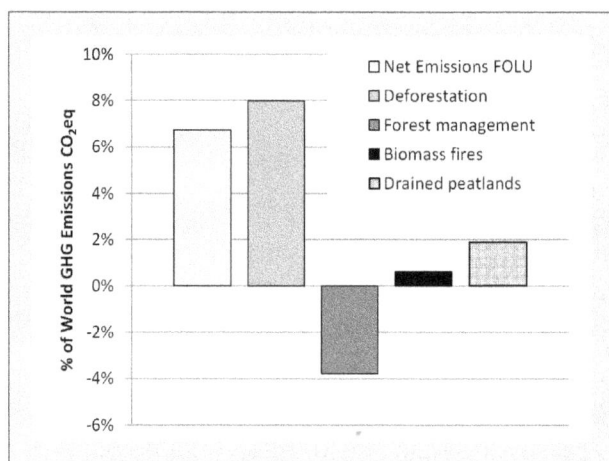

Figure 69. Breakdown of Emissions from the Global Subsector of Forestry and Other Land Use (FOLU) from 2001 to 2010. Forest management consists of programs of forest preservation, reforestation and afforestation to increase forest carbon stocks and atmospheric withdrawal of carbon dioxide.[107]

other land use sub-sector was a 6.7% contribution to global GHG emissions. Since 2003, net emissions from FOLU have been cut by 28% primarily due to the declining rates of deforestation in the Amazon.[294]

All aggressive climate change mitigation models require that the FOLU sub-sector transition to net withdrawal of atmospheric CO_2 by mid-century, with further increases in forest biomass and carbon dioxide reduction capacity going forward over the long-term mitigation period. In the absence of development and deployment of BECCS, substantial forested areas will be required to provide the carbon dioxide removal capacity that is required to prevent atmospheric GHG accumulation and limit surface warming within the specifications of the Paris Agreement.

5.2.3 Reducing Emissions from Deforestation and Degradation (REDD) Plus Enhancement of Forest Carbon Stocks (REDD+). The concept of incentivized programs to reduce emissions from deforestation and degradation of forests and peatlands was first discussed at the 11th session of the UNFCCC Conference of Parties.[295] These discussions eventually lead to the 2007, "Bali Action Plan."[296] This initiative identifies reducing emissions from deforestation and forest degradation, conservation and sustainable management of forests, and enhancement of forest carbon stocks as eligible activities under REDD and REDD+ programs.

As part of the Bali Action Plan, the United Nations REDD Programme was established with the goal of providing assistance to developing nations to achieve a state of readiness to effectively participate in REDD and REDD+ mechanisms. By 2014, membership in the UN-REDD Programme had grown to 49 participants.[297] Thus far, UN-REDD programs are at the development stage, and the efforts of participant nations have focused on establishing technical capabilities including national forest monitoring systems for measurement, reporting and verification (MRV).[107] Ultimately, results based payments are to be made to developing nations that have successfully implemented REDD eligible activities.

In addition to the UN-REDD Programme, the Forest Carbon Partnership Facility (FCPF) was formed as part of the Bali Action Plan. The FCPF consists of 36 developing nations as REDD country participants in partnership with a group of donor nations consisting of Canada, the US, Japan, Australia and most of the high GDP per capita nations of Europe, along with the European Commission.[298] Many of the participant countries in the FCPF also take part in the UN-REDD Programme. The FCPF has established a Readiness Fund and a Carbon Fund. Participant nations can access the Readiness Fund to aid in developing capabilities and then apply for access to the Carbon Fund. In 2013, Costa Rica became the first FCPF participant nation to successfully apply to the Carbon Fund.[299] The FCPF signed a letter of intent to negotiate emissions reduction payments to Costa Rica upon achieving REDD milestones to conserve 340,000 hectares of forests, regenerate degraded lands, and scale up agro-forestry systems. The forest management opportunity, along with an abundance of zero-emissions hydro power resources*, is such that Costa Rica has established a national objective of attaining carbon neutral status by 2021.[298,300]

In addition to early stage, large, multi-party REDD programs, several nation-to-nation REDD partnerships have formed. In 2008, a presidential decree created the Amazon Fund in Brazil with the express purpose of funding REDD activities in the Brazilian portion of the Amazon rainforest.[107] The Brazilian, Norwegian and German governments, along with Brazil's largest oil company, have pledged over 1 billion USD to the fund.[107] The Norway-Indonesia REDD Partnership is based on an agreement between the two countries whereby Indonesia would receive 1 billion USD from Norway upon achieving verified emissions reduction targets following implementation of programs to reduce deforestation, peatland draining and peat fires.[107]

The responsibility to curtail emissions is global, but some of the most cost-effective and efficient abatement opportunities are found in tropical regions in South America, Sub-Saharan Africa and Southeast Asia. Internationally funded REDD and REDD+ programs have considerable potential to transition forestry and land use in developing countries from a significant source of emissions to net carbon dioxide reduction. Potentially, advanced economies could make good use of these international abatement opportunities by contributing either directly to REDD programs or to climate funds that can be accessed by REDD programs within developing countries. Some funding arrangements could see a transfer of international carbon credits to donor countries. Performance-based economic incentives will be required for developing nations to effectively implement REDD programs.

5.2.4 Changing Food Consumption Patterns. By mid-century, world population will exceed 9 billion, and current value global average GDP per capita is projected to increase. Under a business as usual scenario, this combination of population growth and increased affluence will increase the demand for meat products.

In the near-term, crop yields and animal feeding efficiencies are expecting to increase with adoption of Green Revolution agriculture practices in Sub-Saharan Africa and other regions of the developing world. By mid-century, intensification of agriculture in the developing world, along with continued advances in seed genetics and crop production practices, are projected to increase world crop yields by 54%.[107] These advances will largely offset the impact of an increase in the demand for food on the land area required to produce this food.[301] A business as usual scenario that incorporates projected improvements in efficiencies anticipates that a 10% expansion of agricultural lands would be sufficient to meet mid-century food production requirements.[301] However, by 2050, the demands to feed a more affluent global population diets that are rich in animal protein has the potential to increase cattle numbers from 1.5 to 2.6

---

* Costa Rica uses renewable energy to generate 98% of the country's electricity. Hydro power provides 74% of the total power production capacity. For 250 days over the year 2016, Costa Rica did not burn fossil fuels to produce electricity.[300]

billion, and increase the number of sheep and goats from 1.7 to 2.7 billion.[302] Methane and nitrous oxide emissions from enteric fermentation and manure deposits from this massive population of ruminant animals would double emissions from the agriculture sub-sector up to 11.9 Gt per year of CO_2 equivalents. If this were to transpire, future emissions from ruminant animals would approximate the year 2010 rate of global GHG emissions from the production of electricity.[301]

A recent study compared land use, greenhouse gas emissions, and climate change mitigation costs based on year 2050 models of a world average business as usual diet; a diet without ruminant meat but with pork, poultry and fish; a diet without meat; a diet devoid of animal products; and the Harvard Medical School "Healthy Diet" that is low in meat but does allow limited quantities of beef protein consumption.[301] When compared to the business as usual diet, removal of ruminant meat resulted in an 80% decrease in agricultural grasslands and an overall 56% reduction in total land use by agriculture. Production of fish, poultry and pork are landless, intensive, highly efficient processes. Complete removal of meat or animal products had a marginal effect on land requirements when compared to diets without ruminant meat. The study concluded that 29% less land would be required for agriculture if the consumption of beef followed the limits outlined in the Harvard Medical School Healthy Diet. Elimination of ruminant meat resulted in a 50% cut in GHG emissions from the agriculture sub-sector, relative to the business as usual diet scen-

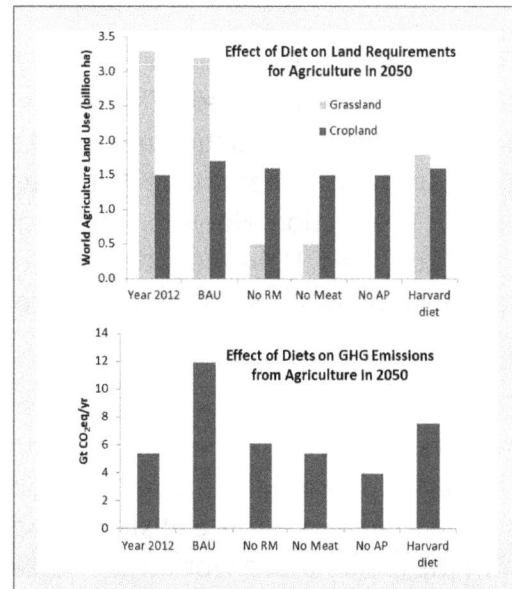

Figure 70. Effect of Model Global Average Diets on Year 2050 Land Use and GHG Emissions from the World Agriculture Sector.[301] Business as usual average global diets (**BAU**). Global consumption of ruminant meat increases with future population and affluence. Average global diet with no ruminant meat (**No RM**). Diets without meat (**No Meat**) or animal products (**No AP**). Average global diet based on the Harvard Medical School "healthy diet" with a lower content of ruminant meat (**Harvard diet**).

ario, while restricting ruminant meat consumption as recommended in the Harvard Medical School Healthy Diet reduced emissions by 37%.

Mid-century GHG emissions following a business as usual scenario of increased per capita ruminant meat consumption will require an equivalent carbon dioxide removal capacity. The land area for the biomass required to sequester the emissions from vastly expanded global herds of cattle, sheep and goats will conflict with requirements to raise these animals and maintain world food production. Emissions from agriculture will grossly exceed the global carbon dioxide removal potential of forest biomass such that an extensive fleet of costly BECCS power plants and associated biomass plantations would be required. Going forward to mid-century, the sustainability of a business as usual population and affluence driven increase in the global consumption of ruminant meat is highly questionable. Overall, abatement costs to limit future surface warming to less than 2°C while maintaining vast herds of ruminant farm animals are estimated to be double the costs that would be incurred under an alternate scenario whereby consumption of ruminant meat was limited to the healthy diet recommendations of the Harvard Medical School.[301]

There may be additional opportunities to increase the density of cattle per unit of land and to increase feed efficiency through breeding and diet beyond what is anticipated in animal production models. Additives such as hormones and antibiotics can increase growth rates and feed efficiency, and thus will reduce emissions per unit of protein production. However, consumer preferences may exclude the future use of these additives. Ultimately, the physiology of ruminant farm animals will limit what can be accomplished to further reduce emissions and improve the efficiency of meat production.

Consumption of ruminant products is deeply entrenched in the dietary practices of many cultures around the world. It is difficult to envision the degree of societal acceptance and political will required to abolish the farming of cattle, sheep and goats and transition to a world free of steaks and burgers. This statement is seemingly at odds with the reality that a worldwide increase in the per capita consumption of ruminant meat is unsustainable.

Carbon pricing mechanisms applied without exception across all sectors including agriculture will drive household economics away from the purchase of ruminant products toward poultry, pork, fish and other less carbon intensive foods. This transition will not compromise human health and will increase world food production capacities. A realistic outcome of successful climate change mitigation pathways within the agriculture sector would be a reduction in per capita consumption of ruminant products on par with the healthy diet advocated by the Harvard Medical School. Under this scenario, the size and frequency of steaks on the barbeque will decrease and some households will choose not to purchase ruminant animal products in favour of lower cost food options.

A steady state of anthropogenic release of methane and nitrous oxide from the agriculture sub-sector will continue after all other emissions abatements are in place. These residual emissions must be balanced by carbon dioxide reduction. The costs and ultimate practicality of CDR will be largely dependent on the intensity of continued GHG emissions from agriculture and other sectors. Assuming successful decarbonization of the electricity supply, industry, transportation, forestry and buildings sectors, by the second half of the century, agriculture will become the dominant source of future emissions. Successful application of CDR to manage continued global emissions of greenhouse gases may well be dependent upon the size of ruminant animal herds on the world's farmlands.

5.2.5 Food Waste Reduction. An estimated 30–40% of food produced in the world is lost as waste in the supply chain from harvest to consumption.[107] The FAO estimates that, in 2011, food wastage accounted for approximately 8% of the global GHG emissions.[303] Waste at the consumption stage accounts for 35% of world food waste emissions with the remainder divided among the production, storage, processing, and distribution stages of the food supply chain.[303] Asia accounts for half of world food waste emissions; however, consumer and food service industry wastes are significant sources of emissions in Europe and the Americas.

The United Nations has set a year 2030 sustainability goal of reducing world food waste by 50%.[303] Achieving this target would bring down global GHG emissions by 3% and, by improving the efficiency of the entire food chain, would decrease land requirements for agriculture.

5.2.6 Bioenergy Production. In 2008, biomass provided 10% of world's total primary energy supply.[304] Traditional use of wood, straws, charcoal and dung for cooking, heating and lighting in developing countries accounts for 60% of the total biomass energy consumption.[304] The World Health Organization estimates that approximately 3 billion people are dependent upon traditional use of biomass for domestic cooking and heating.[305] Modern bioenergy processes consume 22% of the global biomass energy. These processes consist of biomass combustion to generate electricity and heat, plus conversion of biomass to liquid transportation fuels.[304] The remaining biomass energy supply is allocated to unaccounted informal sectors.[304]

Biomass harvested from the world's forests for traditional uses approximates the industrial harvest of forests for roundwood products. Gathering of firewood and other biomass for domestic cooking and heating is generally an informal, unregulated process that contributes to unsustainable deforestation in many parts of the world. Traditional domestic cook stoves or open fires are highly inefficient with energy losses of 80–90%.[304] Incomplete combustion of biomass in combination with poor ventilation is the major source of indoor air pollution among the poorest nations on Earth, leading to an estimated 3.3 million early deaths per year.[305] Over the short-term mitigation period, conversion to low-cost, higher efficiency, improved cook stoves, in combination with sustainable forest management, has considerable potential to reduce deforestation and GHG emissions in developing nations. Co-benefits of upgrades to traditional biomass cooking and heating are improved indoor air quality and less dependence on gathering large quantities of firewood and other solid fuels.

The International Energy Agency estimates that just 10 countries in Sub-Saharan Africa and developing Asia account for 75% of the global population that relies on traditional solid biomass for fuel.[306] Rural electrification and poverty alleviation, along with modernization of agricultural practices, in these countries will be required over the next 30 years to transition away from traditional biomass to more efficient energy carriers for domestic heating and cooking. This transition will be key to successful implementation of REDD programs in least developed countries.

As discussed in chapter 3, bioethanol and biodiesel produced from conventional crops such as oilseeds, corn and wheat are transitory liquid fuels that accomplish little to reduce GHG emissions and cannot sustainably replace conventional gasoline and diesel fuels at high levels of inclusion without impacting upon land use for food production. Ethanol production from sugarcane has advantages over bioethanol from corn and wheat or biodiesel from oilseeds, based on the efficiency of production, land requirements and life cycle assessment of GHG emissions. Ultimately, cost-effective liquid biofuels must be produced from second generation feedstocks consisting of low valued agriculture and forestry by-products (stems, hulls, woodchips, skins, bark and pulp and leaves) or sustainably grown fuel crops such as switchgrass, poplar, and willow. Bioenergy plantations dedicated to the efficient production of rapidly growing fibrous biomass could effectively minimize land use for liquid fuel production. However, the chemistry and physical processes for conversion of fibrous plants to ethanol is complex and must overcome considerable technical and cost barriers. The future of liquid biofuels is uncertain given the potential of electric or hydrogen fueled zero-emissions vehicles. However, biofuels may have niche applications in aviation and other transportation sub-sectors that require portable, high energy density liquid fuels.

Biomass fueled power plants, combined heat and power plants, and heat plants are classified as low intensity emitters of GHGs. Carbon dioxide emitted during combustion of biomass to produce electricity and heat had been sequestered from the atmosphere during plant growth. However, based on the GHG life cycle assessment of the complete process of biomass harvesting, transport and combustion, industrial bioenergy plants are a source of low intensity emissions.[271] In addition, biomass heat and power plants emit significant levels of non-GHG pollutants that must be factored into decisions of deployment.[304] Coal power plants can be converted to biomass combustion often at considerable cost-saving compared to green field construction of new power plants.[307] Where the logistics of biomass fuel supply and the economics for power generation are favourable, there is potential to replace coal combustion with biomass and achieve a 70% reduction in GHG emissions while retaining the flexibility of dispatchable power production.[271] Decisions to build new biomass power or heat plants or to convert coal plants to biomass must be based on verified sustainable forest management or biomass plantation practices with complete life cycle assessments of GHG emissions. The net benefits of land use for carbon dioxide removal following afforestation and reforestation must be compared to the benefits of land use for biomass energy production on a case by case basis.

Equipping biomass power plants with carbon capture and storage technology converts the production of electricity from a source of emissions to an active process of carbon dioxide withdrawal from the atmosphere. Potentially, trees or energy grasses selected specifically for rapid growth could be produced on bioenergy plantations. These plantations would function as highly efficient atmospheric CO_2 sequestration sites. Harvested biomass would be replaced with new growth such that the plantation would operate at a sustainable steady state of continuous carbon sequestration. Plant sequestered CO_2 would then be captured at the power plant and piped to injection sites for permanent isolation from the environment in deep saline aquifers. In theory, well managed programs of BECCS will be a far more effective use of land area to achieve carbon dioxide removal than programs based entirely on afforestation, reforestation and maintenance of forest carbon stocks. However, BECCS is far from commercial reality and will require carbon pricing mechanisms as an economic driver for development and eventual broad scale deployment.

5.2.7 The Impact of Land Use on AFOLU Climate Change Mitigation Pathways. A business as usual scenario for the AFOLU sector would see a continuation of emissions from deforestation and land degradation. Emissions abatements in the agriculture sub-sector would be restricted to current projections for improved efficiencies and yields in food production. This reference baseline excludes the potential for damage to food production systems and food security from unmitigated climate change. Under a business as usual scenario, mid-century emissions from agriculture, forestry and other land use would total about 15 Gt/year, and the planet would be committed to future surface warming well in excess of 2°C, regardless of abatements in other economic sectors.[107]

Without diet change or reductions in food waste, the land area required to produce food and maintain the massive future global herd of ruminant animals would limit the carbon dioxide withdrawal capacity of forest biomass to about 6.1 Gt/year.[107] This scenario assumes that the complete theoretical potential for afforestation and reforestation is in place as of mid-century. Under these conditions, total GHG emissions as CO_2 equivalents from the global AFOLU sector would be cut by 62% relative to the year 2012.[107] However; excessive emissions from the agriculture sub-sector would be well in excess of the carbon dioxide reduction capacity of forest biomass.

The business as usual scenario of future diets assumes an increase in the per capita consumption of ruminant meat. If, however, consumption patterns followed the

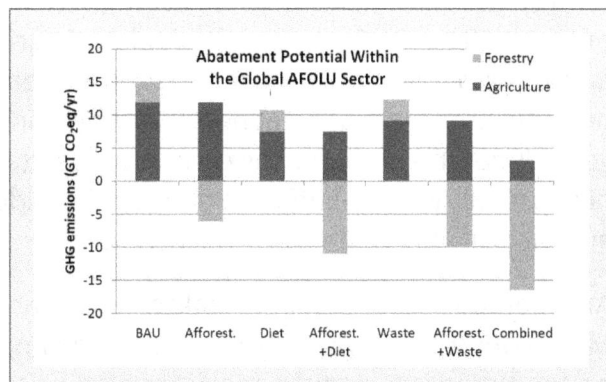

Figure 71. Emissions Abatement Opportunities within the Global Agriculture, Forestry and Other Land Use Sector.[107] Business as Usual (**BAU**) case for future emissions assumes a 54% increase in global agriculture productivity without further abatements. Afforestation (**Afforest.**) based on an optimal expansion of global forested areas. Abatement potential of global diet change (**Diet**) assumes that the consumption of ruminant meat follows the Harvard Medical School "healthy diet" recommendations. Food waste abatement potential (**Waste**) based on achieving UN Food and Agriculture Organization targets to limit future food waste.

recommended limits of the Harvard Medical School Healthy Diet, a substantial cut in direct GHG emissions would be realized and additional land area would become available for expanded programs of afforestation and reforestation. The combined fully realized potential of diet change plus increased forest biomass equates to a net atmospheric withdrawal capacity of 3.5 Gt of carbon dioxide equivalents per year from the global agriculture, forestry and other land use sector.[107,301]

Achieving UN targets for reducing food waste by 50% would curtail food waste GHG emissions and further improve efficiencies in agriculture such that more land area would be available for forest growth. Without diet change, a fully realized combination of food waste reduction and increased forest biomass would approximate net zero-emissions.[107]

Potentially, advances in crop yields and animal feeding could exceed predictions under the baseline scenario. These advances would further increase food production efficiencies and thus open up additional lands for forest expansion.

Implementation of multiple abatements within the AFOLU sector will have overlapping effects on reducing agriculture land use and does not result in additive gains in land availability for forest growth. If the full potential of forest management, diet change, food waste reduction, and further improvements in the efficiencies in agricultural productivity were to be realized, global agriculture would meet food production quotas with minimal land use. Under this idealized scenario, the AFOLU sector could potentially operate at a net negative emissions rate of 13.4 Gt of CO_2 equivalents per year. Under this scenario, the carbon sequestration capacity of the future global forest biomass will likely be sufficient to balance residual emissions of total global GHGs after all practical abatements are in place.[107]

An idealized scenario of fully realized abatements within the AFOLU sector is unlikely. Many countries simply may not accept the imposition of carbon pricing mechanisms on food products and the resultant high prices that would be assigned to ruminant meats. The efficacy of afforestation and reforestation programs will vary between countries and will not approach the full theoretical potential to expand global forest biomass. Significant further advances in the efficiencies of food production beyond current expectations may not be realized. Based on practical considerations, most aggressive climate change mitigation models designed to limit future surface warming to less than 2°C use a combination of forest biomass and BECCS to fully offset residual emissions from agriculture and other economic sectors over the long-term mitigation period extending through the second half of the century.

5.2.8 AFOLU Sector Mitigation Pathways Summary and Conclusions. There are few technical barriers in transitioning the world AFOLU sector to carbon neutral or negative emissions status by mid-century. In recent years, the rate of deforestation in the Amazon basin has dropped dramatically in response to initiatives from the Brazilian government. Incentivized, results-based REDD and REDD+ programs may well be the most cost-effective means for partnerships between developed and developing nations to achieve near-term reductions in GHG emission. Adoption of Green Revolution agriculture practices in Sub-Saharan Africa and other regions of the developing world will increase the efficiency of agriculture to feed an expanding population without impacting land use. Carbon pricing mechanisms can function to limit ruminant animal meat inclusion levels in world average diets without impacting human health or food supply. Programs to limit food waste in the supply chain and at the consumption level can further increase the efficiency of agriculture. These advances in food provision will optimize land use for agriculture. Programs of afforestation, reforestation and forest carbon stock maintenance can be applied to available land such that carbon dioxide removal by forest biomass can offset GHG emissions from agriculture. Continued progress in furthering the efficiency of agriculture and reducing the size of the global herd of ruminant animals combined with continued expansion and maintenance of forested areas can transition the overall integrated sectors of agriculture, forestry and other land use to negative emissions status during the second half of the century.

Globally, a business as usual increase in per capita consumption of ruminant meat is unsustainable. However, in many countries, societal barriers to diet change may be such that carbon pricing mechanisms and other measures to limit ruminant meat consumption are not implemented. By mid-century, an intermediate outcome may result in an average global consumption of ruminant meat somewhere between the business as usual case and the recommendations of the Harvard Medical School Healthy Diet. Under this scenario, the land use requirements and GHG emissions from the global herd of ruminant animals will likely exceed the practical limits for carbon dioxide reduction by forest biomass.

As needed, land use management can accommodate the production of feedstocks for conversion to second generation liquid biofuels and biomass for power and heat production. Successful development and broad scale deployment of bioenergy with carbon capture and storage may well be an important component of global programs to achieve the surface warming limitations specified in the Paris Agreement.

AFOLU SECTOR CLIMATE CHANGE MITIGATION PATHWAYS

Declining Emissions 2020-2030

Improved Agricultural Efficiency in Developing Regions (2016-2030)
- Adoption of intensive agricultural practices
- Enhanced food production with minimal land use

Reduce Deforestation and Land Degradation (2016-2030)

Land Use for Bioenergy Crops (2016-2060)
- Solid biomass fuels
- Second generation liquid biofuels

Establish Economic Drivers (2016-2020)
- Carbon pricing mechanisms, REDD+ programs
- Legislated land use emissions policies
- Agriculture technology transfer to developing nations

Optimization of Food Supply (2016-2050)
- Limit ruminant meat consumption
- Limit food waste
- Optimize land use for agriculture

Afforestation, reforestation and maintenance of Enhanced forest carbon stocks (2016-2050)
- Net atmospheric CO_2 withdrawal by forest biomass
- Carbon neutral status for AFLOU sector

Electricity Production Sector (2016-2060)
- Biomass replaces fossil fuel power stations
- BECCS (negative emissions from power supply)

End of Century Low Or Zero Emissions Status

Transportation Sector (2016-2060)
- Liquid biofuels replace petroleum products

Figure 72. Mitigation Pathways to Transition the Integrated Sectors of Agriculture, Forestry and Other Land Use to Carbon Neutral or Negative Emission Status. Dashed arrows indicated potential pathways with technical uncertainties or other commercial or societal acceptance barriers to implementation.

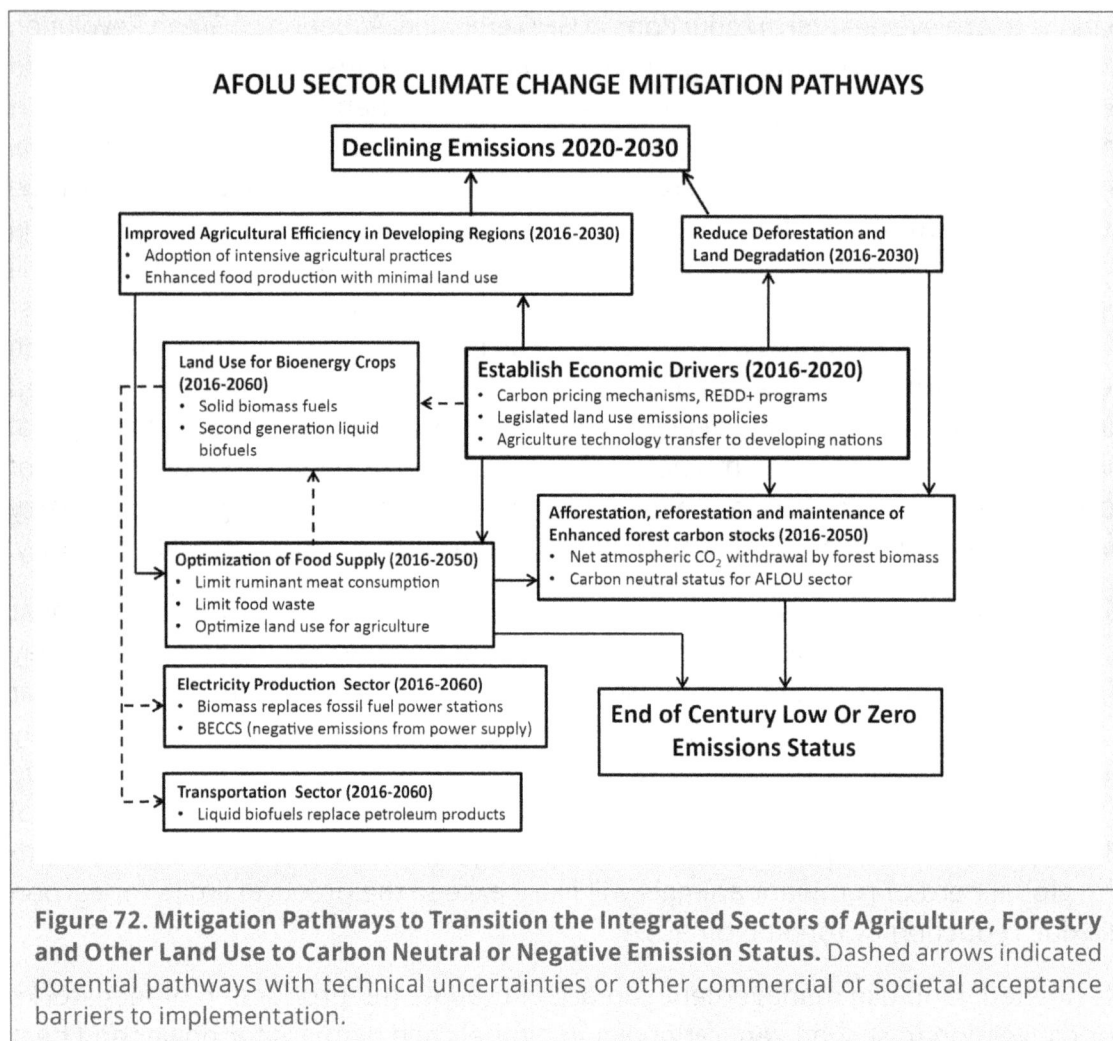

5.3 Emissions Reduction Pathways in the Industry Sector

The IPCC defines the industry sector as the complete supply chain for the manufacture of physical products for human use.[308] The overall sector is subdivided into extraction, materials, manufacturing and construction, and waste industries. Extraction industries such as mineral mining, function to supply raw inputs to the energy intense materials sub-sector. Iron and steel and cement production are examples of industries that refine ores and other raw extracts to intermediate materials that are used by the manufacturing and construction industries to produce final products, buildings and infrastructure. The processing and conversion of primary agriculture outputs to food and animal feed products is included in the sub-sector of manufacturing and construction industries. Waste industries encompass all handling, processing, recycling and disposal of waste from extraction, materials, and manufacturing industries and end-of-life products retired from use. In 2010, the industry sector accounted for 21% of the world total of direct GHG emissions and consumed 45% of the global supply of electricity.[269] When indirect emissions from the generation of electricity used by industry are included, the industry sector is responsible for over 30% of world emissions.[269]

In 2010, direct emissions of carbon dioxide from the combustion of fossil fuels to provide thermal energy for industrial processes accounted for 11% of world GHG emissions.[308] Steel manufacture is a thermally intense process, and this single industry produced 4.5% of global emissions.[308] The manufacture of cement is another energy and emissions intensive process accounting for 4.1% of total emissions.[308] Production of ammonia fertilizer via the Haber-Bosch process consumes 1.2% of the world primary energy supply and is a major component of direct emissions from thermal energy use in the manufacture of chemicals.[308] The remaining direct emissions from fossil fuel combustion within the industry sector come from the mining of raw materials, food and feed processing, pulp and paper industries, textiles manufacturing, construction, plus a diverse array of other manufacturing industries.

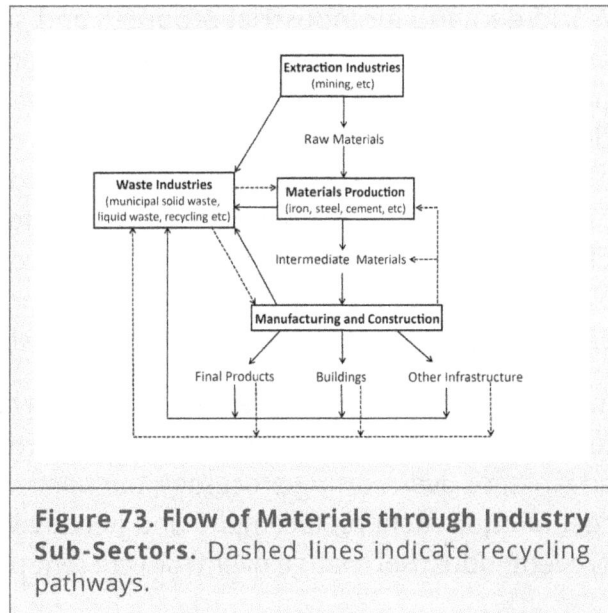

Figure 73. Flow of Materials through Industry Sub-Sectors. Dashed lines indicate recycling pathways.

In 2010, 5.3% of world GHG emissions originated from CO_2 released from chemical reactions during industrial processes.[308] Cement production accounted for half of this total. The remaining in-process CO_2 release came from the manufacture of chemicals (1.0%), refining and processing of steel (0.5%), and from the production of lime, non-ferrous metals and other processes (1.2%).[308]

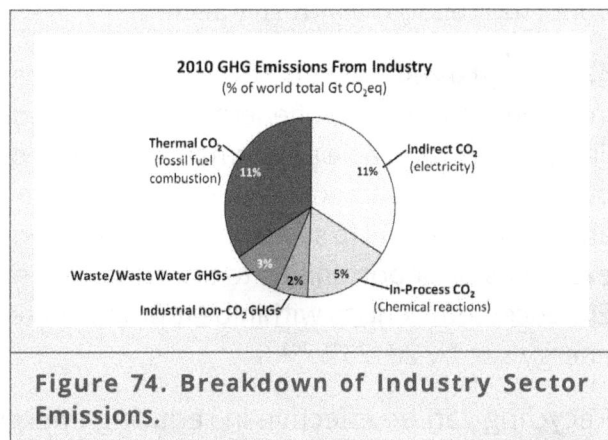

Figure 74. Breakdown of Industry Sector Emissions.

Non-CO_2 GHG emissions from industrial processes other than waste processing and wastewater treatments accounted for 1.8% of world emissions.[308] In 2010, industrial release of hydrofluorocarbons made up 1.3% of total GHG emissions when expressed as CO_2 equivalents.[308]

Emissions from untreated wastewater, landfills and dumps accounted for 3% of global emissions with methane release comprising 90% of this total.[308]

World demand for industrial products and services is expected to double by 2050 based on projections for increased population and GDP per capita. A continuation of current industrial practices and a linear increase in emissions is unsustainable from a climate change perspective. Effective models of climate change mitigation incorporate a curtailment in industrial sector emissions of 15–20% by year 2030 and 50–70% by mid-century.[122] These models arrive at close to zero-emissions during the second half of the century.[122] Based on a future scenario of increased output while reducing total emissions, abatement measures within the industrial sector must achieve substantial and progressive cuts in the intensity of emissions.

Industrial technologies are classified as "best available technologies" and "best practiced technologies". Best available technologies often offer savings in efficiency and other advantages over existing processes, but barriers to implementation such as the capital cost of equipment replacement, and perceived or actual adoption risks, can delay or prevent industrial change over from current practices.

The efficiencies of industrial processes can range between poor, average and best available performance. Investments and installation of new equipment are generally focused on improving the efficiencies of processes that are currently in practice with the goal of achieving best available performance.

The IPCC has identified improved efficiency in energy use, reduced emissions intensity, reduced material use in product design and production, reduced demand through lower consumption, and longer more intensive use of products as emissions abatement opportunities in the industry sector.

5.3.1 Improved Energy Efficiencies. Overall, considerable opportunity exists to improve the efficiency of energy use within industrial processes toward the theoretical limits of best available performance. The use of exhaust heat to generate electricity or offset process heat requirements, installation of high efficiency electrical equipment, and improvements to systems of heat retention, heat recovery and heat exchange are examples of opportunities to improve the energy efficiencies of existing processes. Enhanced efficiencies within best practiced technologies has the potential to reduce energy use by 20–25%.[308]

Recycling can be effective in reducing energy demand to produce new products. Bulk metals are efficiently recycled with considerable saving in energy inputs for the production of steel and non-iron metals. However, recycling programs must be fully evaluated on a case-by-case basis for life cycle assessment of GHG emissions. In many cases recycling programs accomplish little and can sometimes increase emissions relative to disposal and waste treatment.

Improved efficiencies within the industry sector will contribute to reducing industrial GHG emissions over the short-term mitigation period. However, these optimizations will reach a technical limit based on application to best practiced technologies. Successful mitigation of climate change requires progressive decarbonization of the global industry sector over the next 60 years. Additional, more impactful, and ambitious transitions to zero or very low emissions from the best available technologies will be needed to limit future surface warming as specified in the Paris Agreement.

5.3.2 Reduced Intensity of Emissions. Complete decarbonization of the electricity supply sector will eliminate indirect emissions coming from the generation of electricity used by industry. For many industrial processes, electricity or district heat produced from zero or low emissions energy sources can displace on-site use of fossil fuels.

In 2008, coal and oil provided 42% of the energy used by global industries.[308] By 2030, the International Energy Agency predicts that fuel switching to natural gas and electricity plus direct use of renewables will reduce the fraction of total energy supply coming from coal and oil down to 30%.[308]

Currently, 70% of the world's steel is manufactured from pig iron that is produced using coke or coal treatment of iron oxide in large blast furnaces followed by reduction in oxygen blown converters. This is a thermal energy and emissions intensive process that has changed little since the Industrial Revolution. Assuming a future supply of emissions-free electricity, electric arc furnaces can cut emissions per unit of steel produced by over 80%.[308] A progressive transition to best available steel manufacturing technologies can cut emissions by 30 to 60% in 2030, and by 60 to 80% in 2050, compared to emissions from global steel production in the year 2010.[308]

Thermal energy intensive processes in the cement, chemicals, and pulp and paper industries require either on-site combustion of fuels or a local supply of intense heat, and as such, emission-free electricity or renewable energy sources are of limited to no value in displacing fossil fuel use in these applications. Biofuels are an exception and potentially can be combusted to produce heat at lower rates of emissions. As discussed, extensive reliance on biofuels has land use implications and requires sustainable biomass production practices with complete life cycle assessment of GHG emissions. Combustion of municipal and other wastes can provide thermal energy

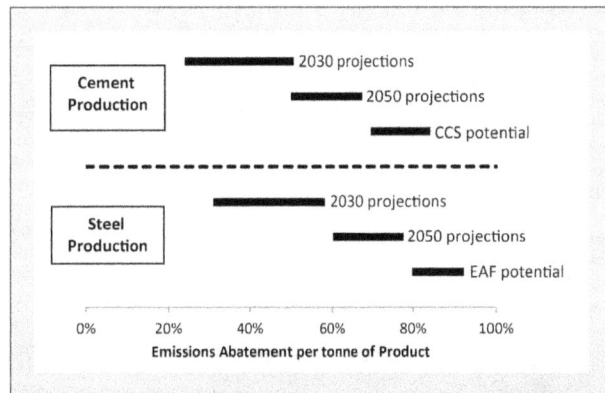

Figure 75. Year 2030 and 2050 Emissions Abatement Potentials for Cement and Steel Production. Electric Arc Furnace (**EAF**) with scrap metal technology applied to steel production. Carbon capture and storage (**CCS**) incorporated into cement production processes. Both technologies assume complete decarbonization of electricity supply. Adapted from IPCC Climate Change 2014. Mitigation of Climate Change.[308]

to industry. In the Netherlands, combustion of waste provides 83% of the energy used to manufacture cement.[308] Emissions of non-GHG pollutants and life cycle assessment of GHGs emissions must be fully evaluated on a case by case basis to determine the net benefits of waste combustion by heavy industry. Combustion of hydrogen is a zero-emissions process that could potentially meet the future thermal energy needs of industry. However, fuel switching to hydrogen requires the development of a commercial scale, cost-effective, zero or very low emissions process to produce hydrogen.

Capture and storage of carbon dioxide formed by chemical reactions in the manufacture of cement, chemicals and steel has the potential to abate 80–90% of in-process CO_2 emissions.[308] CCS will be required to achieve deep decarbonization of the overall industry sector. As is the case for the production of electricity, the capital and ongoing operating costs of carbon capture will be high and commercial implementation will be dependent on effective carbon pricing mechanisms. The chemicals industry may be positioned for early adoption of CCS given the high CO_2 concentration in vented gas streams and thus lower capture costs.

5.3.3 Materials Efficiency in Production and Product Design. Losses in the complete chain of extraction, raw material processing, and final manufacturing are such that 25% of steel and 50% of aluminum are scrapped and internally recycled.[308] Process innovation can reduce this internal inefficiency and thus reduce the intensity of energy use and emissions per unit of product output. Where applicable, replacing emissions-intensive materials such as steel with plastics and other materials can reduce emissions from construction and product manufacture. Improved efficiency in materials use in production and in product design can provide modest emissions reduction; however, these savings are additive to improved efficiency of energy use, fuel switching, and other changes in practice to reduce the intensity of emissions from the industry sector.

5.3.4 Industrial Products and Services. Extension of product lifespan use and overall reduced consumption of emission-intensive industrial products and services will reduce GHG emissions from the industry sector. Consumer behaviour is dependent upon a complex array of cultural and wealth driven motivations. Voluntary change in consumption patterns in the absence of financial drivers is unlikely to affect industrial production. Carbon pricing mechanisms will be effective in driving consumption and leisure activities toward lower emissions options. Unlike production of ruminant meat within the agriculture sector, optimization of efficiencies, adoption of best available technologies, decarbonization of electricity supply, fuel switching, and application of carbon capture and storage can transition the production of industrial goods and services to very low or zero-emissions status. In a future scenario of a decarbonized industry sector, affluence, travel, and enjoyment of products and leisure activities can be sustained without compromising the environment or imposing the risk of climate change catastrophe on future generations.

5.3.5 Solid and Liquid Wastes. In 2010, methane release from wastes accounted for 2.7% of the world emissions of carbon dioxide equivalents.[308] Emissions from municipal solid wastes (MSW) comprised half of this total with the remainder coming from untreated wastewater. Nitrous oxide release from wastewater contributed another 0.27% of global emissions.[308]

Global production of MSW is estimated at 1.5 billion tonnes annually with over 53% discarded in non-sanitary landfills or dumps.[308] The remainder is recycled (20%), treated with energy recovery (13%) or discarded to sanitary landfills (13%).[308] Methane is produced from MSW in unsanitary landfills and dumps by microbial growth under anaerobic conditions. Since 1990, per capita global emissions from MSW have decreased by 5% following the implementation of landfills directives and programs of energy production from waste primarily in the EU and the US.[308]

In developed countries, most wastewater is processed through centralized facilities with secondary aerobic treatment to convert methane to CO_2. The difference in GHG potency between the two gases is such that highly efficient complete conversion to CO_2 in centralized facilities can reduce the immediate global warming potential of methane from wastewater by 80–90%. Alternatively, under controlled conditions of anaerobic digestion, wastewater treatment facilities can produce methane that is then combusted to produce electricity or heat. In developing countries there is often no collection and treatment of wastewaters and thus methane is released to the atmosphere following natural processes of anaerobic fermentation of organic compounds in the waste.

Aerated lagoons and wetlands are effective means of treating wastewater to reduce or eliminate methane release. When available, wastewater application to wetlands can often provide additional ecological benefits, along with a simple, low-cost, method of reducing GHG emissions.

Multiple existing technologies and processes can be applied to the treatment of MSW and wastewater to dramatically reduce methane emissions. Composting and other aerobic processes can breakdown solid wastes without methane production. Anaerobic

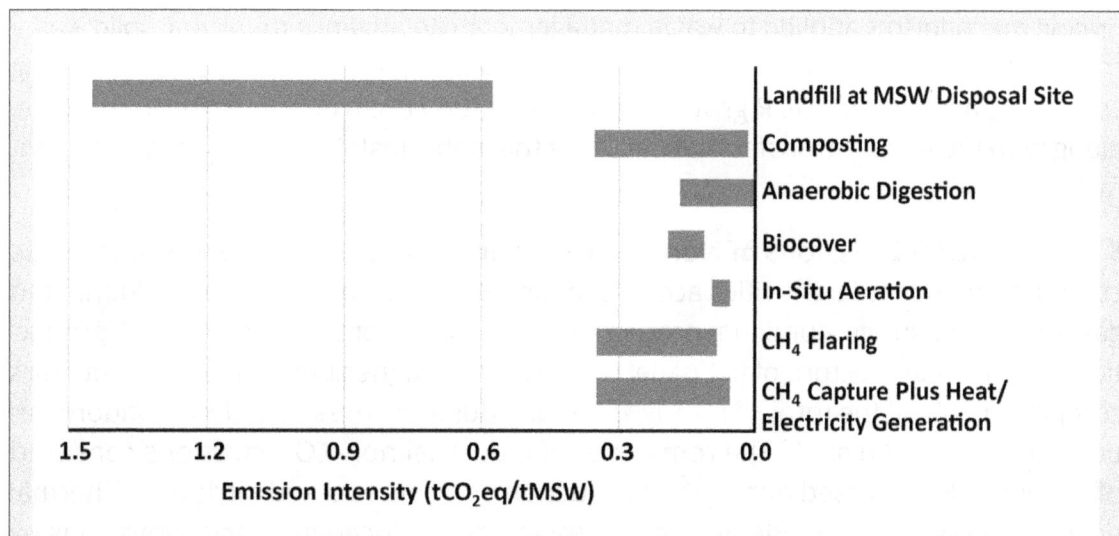

Figure 76. Emissions Abatements from Municipal Solid Waste Treatment Options. Adapted from IPCC Climate Change 2014. Mitigation of Climate Change.[308]

digesters can be used in solid waste to energy facilities. Methane can be collected from landfills and, either flared off to release CO_2 as a far less potent GHG gas or combusted to produce electricity or heat.

Installation of centralized collection and treatment facilities on a global basis can eliminate methane release from wastewater. Where the economics are favourable, waste to energy technology, based on anaerobic digesters, can be implemented.

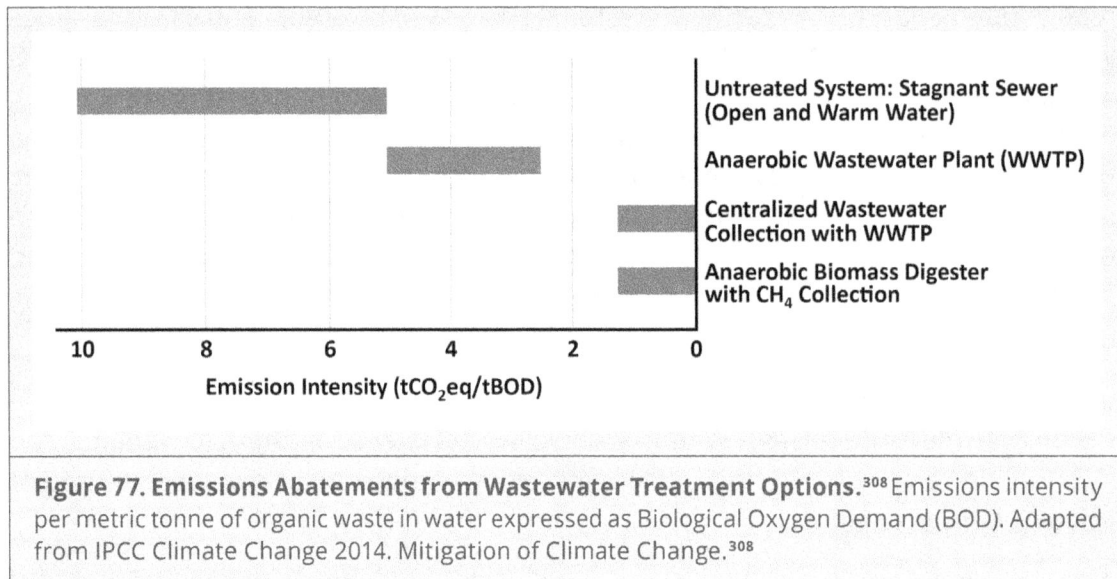

Figure 77. Emissions Abatements from Wastewater Treatment Options.[308] Emissions intensity per metric tonne of organic waste in water expressed as Biological Oxygen Demand (BOD). Adapted from IPCC Climate Change 2014. Mitigation of Climate Change.[308]

There are no technical barriers to achieving near complete elimination of GHG emissions from wastes. However, financial structures are required to accelerate implementation in developed and developing nations. In the developing world, infrastructure to allow waste transport to centralized facilities is often lacking. Emissions legislation and carbon pricing mechanisms applied to waste management can advance municipal solid waste treatment and waste to energy programs in advanced economies. Among developing countries, accelerated GDP growth and associated construction of infrastructure, along with technology transfers can facilitate the global installation of waste treatment facilities.

5.3.6 Industrial Emissions of Non-CO$_2$ Greenhouse Gases. In 2010, non-CO_2 GHGs coming from industrial activities accounted for 1.8% of world emissions.[308] Fluorinated gases released during aluminum processing, chlorodifluoromethane (HFCF-22) production, and the manufacture of flat panel displays, and magnesium and semi-conductors comprised 78% of the total.[308] Nearly 100% of industrial emissions of hydrofluorocarbons come from China.[308] The remainder of industrial non-CO_2 emissions consisted of nitrous oxide released during the production of adipic acid and nitric acid. Thermal destruction of nitrous oxide during processing has reduced industrial emissions by 50% since 1990.[308] Implementation of thermal and other technologies designed to destroy specific pollutants can be applied to reduce or eliminate industrial release of fluorinated gases.

5.3.7 Summary and Overview of Mitigation Pathways to Reduce GHG Emissions from Industry.

Complete transition of the industry sector to zero-emissions status is complicated by in-process release of carbon dioxide and on-site thermal energy demands coupled with limited options for fuel switching. The International Energy Agency projects a 50–70% reduction in emissions relative to a year 2010

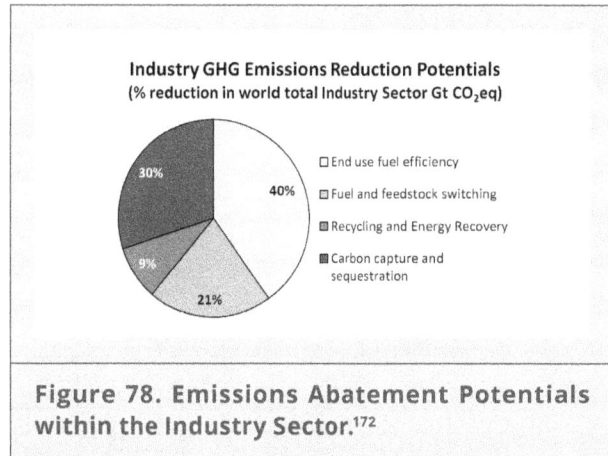

Industry GHG Emissions Reduction Potentials
(% reduction in world total Industry Sector Gt CO_2eq)

30% | 40% | 9% | 21%

☐ End use fuel efficiency
☐ Fuel and feedstock switching
▣ Recycling and Energy Recovery
▪ Carbon capture and sequestration

Figure 78. Emissions Abatement Potentials within the Industry Sector.[172]

baseline as the mid-century climate change mitigation potential of the industry sector.[172] Implementation of best available technologies within the world steel manufacturing industries has the greatest potential to reduce emissions. Well-designed clusters of industry, waste treatment, and energy supply can provide distributed heat and power, along with shared piping and infrastructure to improve the efficiencies of energy intensive industries.

Eventual complete transition of the industry sector to zero-emissions is dependent upon achieving the full potential of capture and storage of in-process CO_2 emissions. CCS applied to industry is in the early stages of development. Cost-effective, models of climate change mitigation do not forecast a transition of industry to carbon neutral status until sometime in the second half of the century. An analysis of global, sector-wide mitigation pathways, concludes that, with implementation of best available technologies, overall industrial emissions can be reduced by 40% from end use fuel efficiency, 21% from fuel and feedstock switching, 9% from recycling and energy recovery and 30% through implementation of carbon capture and storage.[172]

Emissions reduction mitigation pathways within the industry sector face considerable monetary barriers to implementation. Financial analysts are unlikely to recommend costly replacement of equipment if these investments are not required by law or do not meet targets for return on investment. Carbon pricing mechanisms or legislated emissions policies must be applied to industry to induce the sweeping changes in practice required to reduce industrial GHG emissions and limit surface warming to well below 2°C above pre-industrial temperatures.

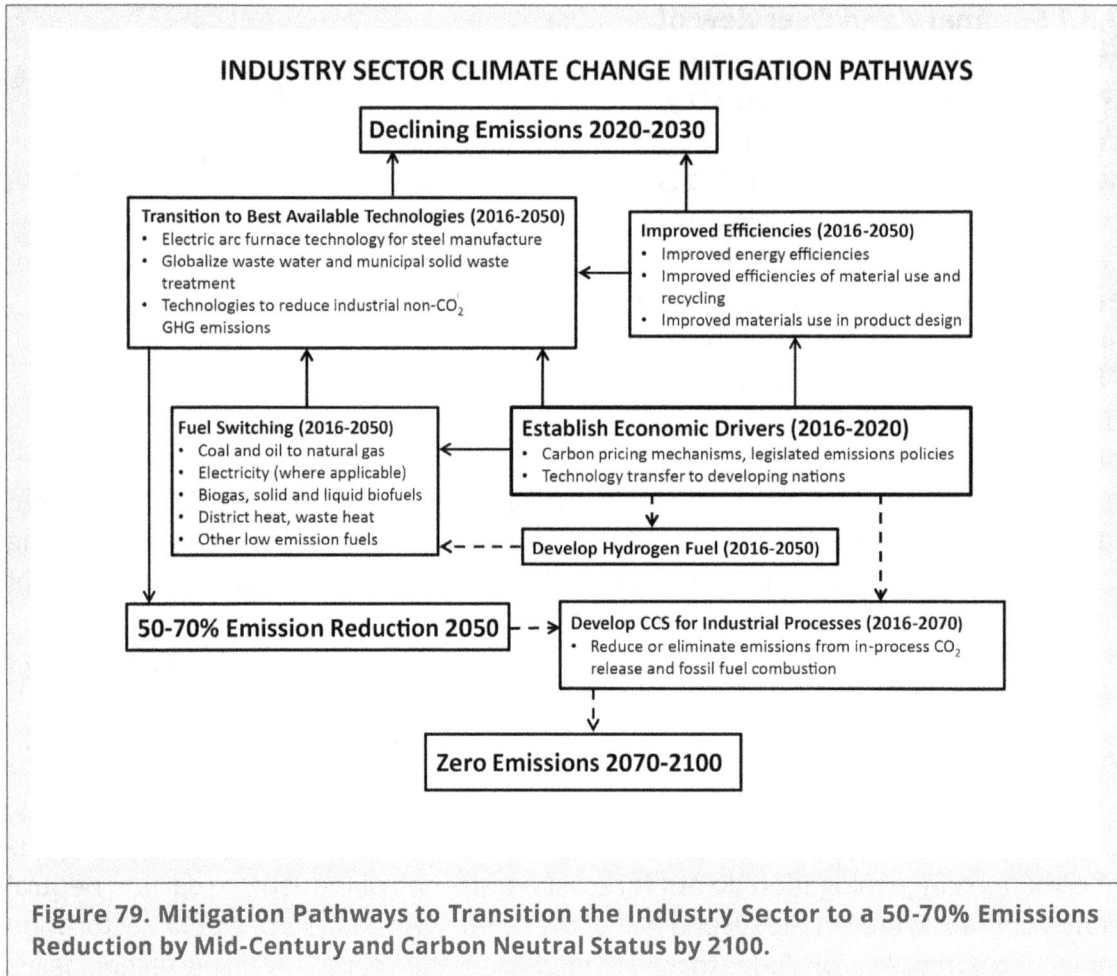

INDUSTRY SECTOR CLIMATE CHANGE MITIGATION PATHWAYS

Declining Emissions 2020-2030

Transition to Best Available Technologies (2016-2050)
- Electric arc furnace technology for steel manufacture
- Globalize waste water and municipal solid waste treatment
- Technologies to reduce industrial non-CO$_2$ GHG emissions

Improved Efficiencies (2016-2050)
- Improved energy efficiencies
- Improved efficiencies of material use and recycling
- Improved materials use in product design

Fuel Switching (2016-2050)
- Coal and oil to natural gas
- Electricity (where applicable)
- Biogas, solid and liquid biofuels
- District heat, waste heat
- Other low emission fuels

Establish Economic Drivers (2016-2020)
- Carbon pricing mechanisms, legislated emissions policies
- Technology transfer to developing nations

Develop Hydrogen Fuel (2016-2050)

50-70% Emission Reduction 2050

Develop CCS for Industrial Processes (2016-2070)
- Reduce or eliminate emissions from in-process CO$_2$ release and fossil fuel combustion

Zero Emissions 2070-2100

Figure 79. Mitigation Pathways to Transition the Industry Sector to a 50-70% Emissions Reduction by Mid-Century and Carbon Neutral Status by 2100.

5.4 Emissions Reduction Pathways in the Transportation Sector

In 2009, light-duty road vehicles (LDV) consumed 52% of the total energy supply to the global transportation sector. The remaining energy use was divided among heavy-duty road vehicles (25%), aircrafts (11%), waterborne vessels (10%), and rail transport (2%).[309] After consolidating all modes of transport, transportation of passengers accounted for 57%, and freight transportation 43% of the total energy use within the sector.[309]

Petroleum products in the form of liquid fuels for internal combustion

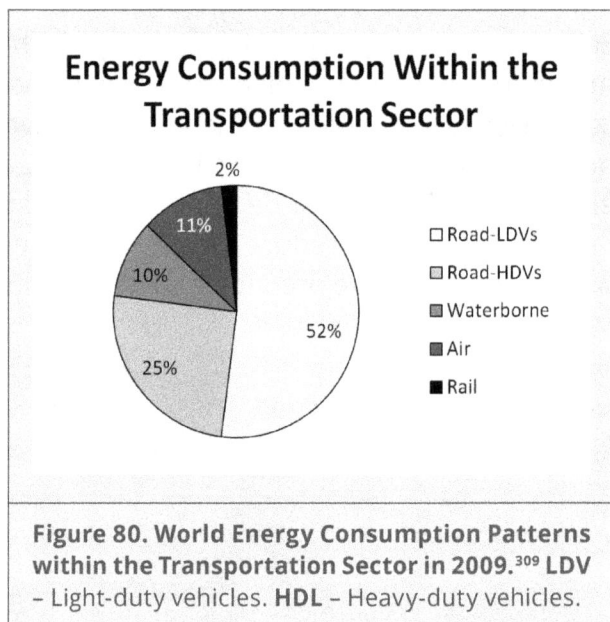

Energy Consumption Within the Transportation Sector

- Road-LDVs
- Road-HDVs
- Waterborne
- Air
- Rail

2%
11%
10%
52%
25%

Figure 80. World Energy Consumption Patterns within the Transportation Sector in 2009.[309] LDV – Light-duty vehicles. HDL – Heavy-duty vehicles.

engines (ICEs) provided 94% of total energy supply to the transportation sector.[309] ICEs are relatively inefficient in that only 36% of fuel energy is converted to mechanical energy that propels the vehicle.[309] The remaining 64% of the energy of combustion is dissipated as heat losses.[309]

In 2010, the global transportation sector accounted for 14% of world GHG emissions.[267] On a national basis, the intensity of transportation and emissions from the sector are strongly correlated with affluence and GDP per capita. In advanced economies, transportation can account for up to 30% of total emissions, while in low income countries, transportation is a minor contributor often amounting to less than 3% of national emissions.[309] Historically, this relationship has been a major contributor to the coupling of economic development and GHG emissions. In 2014, only 17.5% of the world's population resided within the 34 member states of the Organization for Economic Co-operation and Development. However, 64% of world GDP and 46% of GHG emissions from the global transportation sector came from the advanced economies that comprised OECD countries.[106,216] Globally, 10% of the world's population accounted for an estimated 80% of motorized kilometres travelled by passengers in 2010.[309]

Over the next few decades, the projected growth in population and GDP per capita in the developing world will increase the demand for passenger mobility and transportation of goods. Economic development will lead to an expansion in public transportation networks, increased freight transport, and an increase in private ownership of LDVs. The number of LDVs on the planet is projected to triple between 2010 and 2050, based on an expanding more affluent world population.[310] Most of these additional vehicles will operate on new roads in what are currently classified as developing countries. If the transportation infrastructure, urban design, and modes of transport that typify advanced economies were to be replicated in non-OECD countries, the end result would be a catastrophic 75% increase in global transportation sector GHG emissions by 2050.[309]

The challenge to world economies will be to achieve deep cuts in emissions from the global transportation sector without impeding economic growth in developing nations. As will be discussed, recent advances in zero and low emissions road vehicles, along with opportunities to build low emissions infrastructure, provides a significant opportunity to progressively decouple the transportation of passengers and freight from GHG emissions.

5.4.1 Improved Fuel Efficiencies. The Global Fuel Economy Initiative (GFEI) has established a year 2030 goal of doubling the average fuel economy of new LDVs powered by internal combustion engines relative to a 2005 baseline.[310] Technically, there is considerable opportunity to improve the fuel efficiency of light-duty road vehicles through reduced weight and power in combination with improvements in drivetrain design and vehicle aerodynamics, reduced rolling resistance, and improved efficiency of energy use by non-drivetrain vehicle components.[309] However, substantial improvements in the fuel efficiencies of internal combustion engines may well be limited by public preferences for larger more powerful passenger vehicles.

The 1973 Arab oil embargo triggered a shift toward compact, lower powered, more fuel efficient light-duty vehicles. These changes were reflected by a 70% improvement in LDV fuel efficiency between 1975 and 1987.[311] However, since 1985, consumer demand has led to a steady increase in the horsepower and weight of the average vehicle introduced to the market. These changes have offset continued optimization of engine and vehicle design.[311] Meaningful reductions in emissions from gasoline and diesel fueled LDVs will require carbon pricing mechanisms or other legislative means to fundamentally shift consumer preferences in vehicle ownership and operation. In the absence of alternatives to fossil fuel powered internal combustion engines, public acceptance of government policy that would substantially reduce the size and power of privately owned vehicles is questionable.

Fuel efficiency and emissions per tonne of freight hauled by heavy-duty road vehicles can be improved using heavier capacity and multiple trailers vehicles. Further improvements in engine design, truck aerodynamics and vehicle maintenance have the potential to improve fuel efficiency in HDVs by 30–50%.[309]

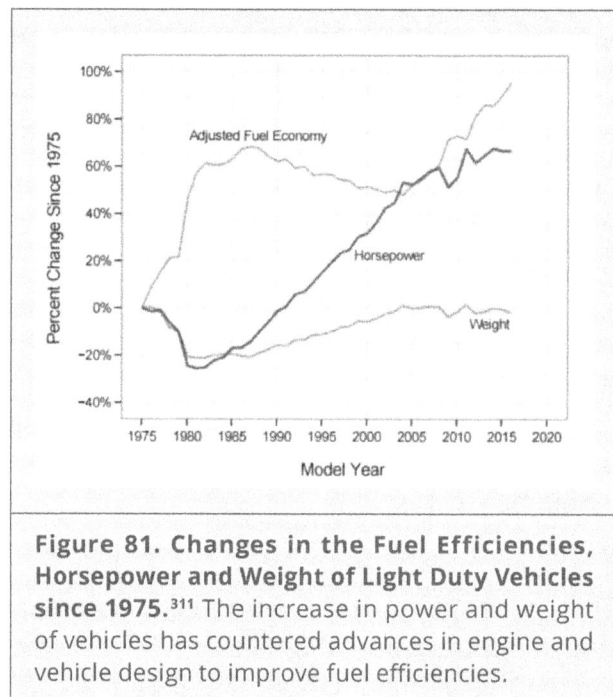

Figure 81. Changes in the Fuel Efficiencies, Horsepower and Weight of Light Duty Vehicles since 1975.[311] The increase in power and weight of vehicles has countered advances in engine and vehicle design to improve fuel efficiencies.

Rail and waterborne crafts are relatively efficient modes of transporting goods and passengers. Incorporation of advances in drivetrain design, weight reductions, waste heat recovery systems, along with improved hydrodynamics, can potentially improve fuel efficiencies in new build ships by 5–30%.[309]

Aircraft fuel efficiency has continually improved with development and application of ongoing engine technologies and design concepts. Potentially, aircraft fuel efficiency could be increased by another 40–50% through weight reduction, improved aerodynamics and further advances in engine performance and system designs.[309] However, the sunk cost into existing aircraft is a significant barrier to a rapid replacement of equipment with more fuel-efficient options.

Overall, significant opportunities exist to improve fuel efficiency and thus reduce the intensity of emissions for all modes of transport. However, implementation of an effective worldwide program to cut emissions, based purely on improved efficiencies in the use of petroleum products, can only serve to partially offset emissions from the massive expansion in passenger and freight transportation projected for the developing world over the next 40 years.

5.4.2 Liquid Fuel Switching. Sugarcane derived bioethanol, along with second generation biofuels produced from plants oils, lignocellulosic energy crops, and waste materials has potential for use as drop-in fuels to reduce GHG emissions from the road transportation sub-sector. However, the plant feedstock must be sustainably grown within overall land use practices that balance food production requirements and the need to preserve and enhance forest carbon stocks. There are serious questions as to the practicality of producing the massive quantities of biofuels that would be required to achieve meaningful replacement of gasoline and diesel fuels. Further, the opportunity for propulsion system changeover may bypass the need for liquid carbon fuel replacement for most road vehicles. Second generation liquid biofuels are in the development stage, and the extent of future use will be defined by demand, production costs, sustainability issues and actual effectiveness in reducing life cycle GHG emissions. These fuels may have niche applications in long haul heavy-duty trucking where internal combustion engines using onboard liquid fuels could have practical advantages over electric or fuel cell propulsion systems.

Aviation will continue to use high energy density liquid fuels for the foreseeable future. Potentially, second generation biofuels with properties similar to those of conventional kerosene-type jet fuels could be used in the aviation sub-sector. The chemistry for conversion of plant oils and other biomass to jet fuel is complex and will likely delay significant fuel switching. The European Commission has targeted a 40% inclusion rate for sustainably produced low carbon aviation fuels by the year 2050.[312]

Most ocean-going vessels are propelled by diesel engines fueled by low-cost marine oil. Second generation biodiesel fuels will likely be produced at a considerable cost disadvantage and thus fuel switching to reduce emissions from marine shipping will likely take place over an extended period. Fuel switching may be surpassed by vessel change over to alternate low emissions propulsion systems. The European Commission has set a year 2050 goal of achieving a minimum 40% reduction in CO_2 emissions from combustion of EU maritime bunker fuels.[312]

5.4.3 Propulsion System Changeover. Battery electric vehicles (BEVs) do not produce tailpipe emissions and, when charged by electricity from zero-emissions sources, these vehicles will operate with near zero life cycle GHG emissions. BEVs are highly efficient in that 80% of the electrical energy is converted to mechanical energy to propel the vehicle.[309] Historically, consumer acceptance of BEVs has been limited due to high costs, limited range, and limited access to recharging stations. In the 2014 IPCC fifth assessment report, commercially available BEVs were stated to have a range 100–160 km, to require 4 hours or more of recharging time and to cost comparatively more than internal combustion engine cars due to the high cost of batteries.[309] The report was limited by the technical status of BEVs at the time of writing. BEV technology has advanced rapidly in the last few years and is beginning to overcome barriers to widespread acceptance. As of this writing, the Tesla Motor Corporation has received nearly 400,000 pre-orders for the Model 3 electric vehicle scheduled for release in 2017.[313] Model 3 specifications, including battery pack details, have not been released; however, the price is projected to be $35,000 USD before incentives, and the vehicle will operate with a minimal range of 350 kilometres.[313] Currently, a total of 632 Tesla supercharger stations are located in North America, Europe and parts of Asia, and vehicles can be recharged to 80% capacity in 30 minutes using a Tesla supercharger.[314] Charging stations in the US are situated to facilitate long distance cross-country travel along interstate highways and other major roadways.

Fuel cell vehicles (FCVs) use compressed hydrogen to provide continuous on-board generation of electricity that propels the vehicle. Approximately 60% of the hydrogen energy is converted to mechanical propulsion energy in an FCV.[309] The efficiency of energy use is superior to that of an internal combustion engine, but FCVs are considerably less efficient than battery electric vehicles. Within the fuel cell, hydrogen and oxygen react to form water with zero tail-pipe emissions and FCVs are often erroneously classified as zero-emissions vehicles. FCVs can be quickly refueled at hydrogen filling stations and thus offer both range and refueling advantages relative to a BEV. Currently, 95% of industrial hydrogen is produced from fossil fuels with in-process release of CO_2 during the reaction.[315] Based on complete life cycle assessment of GHG emissions, operating a hydrogen fuel cell vehicle will result in a 45% reduction in emissions when compared to a similar-sized gasoline fueled ICE vehicle.[309] With current hydrogen production practices, FCVs run at reduced emissions but are not zero-emissions vehicles. Potentially, carbon capture and storage could be applied to hydrogen production to markedly reduce life cycle GHG emissions from FCVs. Alternatively, hydrogen could be produced by electrolysis of water with zero-emissions. Zero or very low emissions commercially viable processes for the production of hydrogen are in the early stages of development with considerable uncertainty as to practicality and costs. Compelling arguments can be made as to the efficiencies and advantages of simply recharging a BEV when compared to using on-board hydrogen to produce electricity to propel the vehicle.

Plug-in-electric hybrids function as zero-emissions vehicles when operating in battery mode. Battery range is much shorter than a BEV and plug-in hybrids release emissions when the internal combustion engine is engaged. Operation of a plug-in hybrid will reduce emissions somewhere within the broad range of 20–60% with actual emissions dependent on the percentage of travel time running in battery mode.[309]

Considerable opportunity exits to use battery power propulsion to reduce emissions from heavy-duty vehicles running in urban settings or used for short and medium distance freight hauling. The operation of transit buses, garbage collection trucks, and delivery vehicles is characterized by frequent stops within a limited range. Conversion of urban service trucks and buses to battery electric vehicles offers the co-benefit of improved air quality within cities by reducing emissions of non-GHGs pollutants.

Emissions from water-borne crafts are dominated by combustion of marine oil in the large diesel engines that propel ocean-going vessels. In 2010, marine shipping accounted for approximately 1.4% of global GHG emissions.[309] Potentially, emissions from the marine subsector could be reduced through a changeover to alternate propulsion systems based on cleaner burning natural gas or more advanced systems such as hydrogen fuel cell driven electric engines. The costs for installation of low emissions propulsion systems in new build ships will be considerably higher than installing conventional diesel engines and will require economic drivers such as effective carbon pricing mechanisms or legislated limits to vessel emissions.

5.4.4 The Transition of the Auto Industry to the Production of Zero and Very Low Emissions Vehicles. On December 3, 2015, during the Paris COP21 climate change meeting, the International Zero-Emissions Vehicle Alliance (ZEV Alliance) issued an initial announcement of rational, vision and actions.[316] The alliance consists of Germany, the UK, Norway and The Netherlands plus 8 US states including California, New York, and Massachusetts, along with the Canadian provinces of Quebec and British Columbia.[316] The stated objective of the ZEV Alliance is that all light-duty vehicle sales in member jurisdictions are to transition to zero-emissions vehicles as quickly as possible and with complete transition no later than 2050. The alliance specifically referred to battery-electric, plug-in-hybrid and fuel cell vehicles with either zero or near zero tailpipe emissions to replace sales of conventional ICE vehicles.

The world's largest automotive company, Toyota, has established a year 2050 goal of achieving a 90% reduction in new vehicle GHG emissions relative to vehicles produced in 2010.[317] Over the next 34 years, Toyota plans to completely phase out production of internal combustion engine vehicles and transition to the production of hybrid, plug-in-electric hybrid, battery electric and fuel cell vehicles. Toyota has stated a willingness to cooperate with stakeholders in the widespread provision of infrastructure that will be required for adoption of electric and fuel cell vehicles.

The Nissan Leaf was released in 2010 as the first mass production highway-capable all electric vehicle. By January of 2017, world sales of these automobiles had surpassed 250,000.[318] Nissan has stated a year 2050 corporate objective of achieving a 90% emissions reduction in new vehicles when compared to vehicles produced in the year 2000.[319]

The corporate websites of Toyota and Nissan make direct reference to IPCC reports and science-based correlations between GHG emissions and global warming. Publicly, these large corporations have aligned their long-term strategies with progressive decarbonization of world economies as required to meet the stated objectives of the Paris Agreement.

Almost all of the major automakers and new players such as Tesla have either released or will be releasing competitively priced BEVs and FCVs in the next few years. Unlike vehicles propelled by internal combustion engines, the size, power and performance of true zero-emissions BEVs and FCVs are unrelated to GHG emissions. Market penetration by BEVs will lead to an expanded infrastructure of public charging stations and continued improvements in battery and vehicle technology. A transition of the auto industry to the production of zero-emissions vehicles that coincides with decarbonization of the electricity supply sector will dramatically reduce road vehicle emissions by mid-century.

5.4.5 Urban Design and Transportation Infrastructure. Population density, transportation infrastructure and urban design features such as the distances between dwellings, workplaces and amenities are significant contributing factors to the overall intensity of emissions from the transportation sector. Established cities in high income nations were often constructed to facilitate passenger transport by personal light-duty vehicles with less emphasis on public transit. The lifespan of these infrastructures ranges from 50 to 100 years.[309] In the US, the successful development of the interstate highway network locked in road vehicle passenger transport while restricting railway options. Barriers to change are substantial and include the sunk cost of low density suburban housing and road transportation infrastructure, along with societal preference for personal ownership and use of LDVs.

Within the core areas of major cities, investments in public transport, higher density housing, and cycling and walking infrastructure, along with constraints on the use of LDVs, can shift modes of urban transportation to lower GHG emissions options. The European Commission has established goals of achieving a 50% reduction in the use of conventionally-fueled LDVs within cities by the year 2030 with complete phase out by 2050.[312]

Construction of new urban infrastructure provides considerable opportunity to coordinate population density with efficient and accessible public transit and to promote zero to low emissions modes of personal transportation such as bicycles, electric cycles and scooters and walking. Future population growth in developing nations will occur in cities that are generally at the early stages of urban design and infrastructure placement. The opportunity exists to build these cities based on low emissions transportation infrastructure and urban design features.

Ultimately, with increased affluence in the developing world, the flexibility and freedoms associated with private ownership of LDVs will lead to construction of roadways and an increase in the number of passenger vehicles on these roads. However, this trend can be balanced and partially offset with upfront investment in passenger rail service between urban centers.

The European Commission has set a year 2030 target to triple the length of existing high-speed rail networks.[309] By mid-century, passenger railways should become the preferred option for the majority of medium distance travel within Europe. In addition, the EC has targeted a 30% shift in long haul transportation of freight from road to rail by the year 2030.[309] By mid-century, "green corridors" in Europe should account for over 50% of long haul freight transportation.[309]

5.4.6 Summary and Overview of Emissions Abatement Opportunities in the Transportation Sector. Direct release of carbon dioxide from the combustion of petroleum products accounts for almost 100% of GHG emissions from the transportation sector. Potentially, hydrogen or low life cycle GHG second generation liquid biofuels could replace gasoline and diesel products. However, these fuels are in development and, in the case of biofuels, the land requirements to produce feedstocks will limit the potential for wholesale fuel switching across the transportation sector. Second generation liquid biofuels may have niche applications in long haul trucking and in the aviation sub-sector. Achieving zero or very low emissions status for hydrogen fuel will require the development of a practical, low emissions method of production.

Efficient transportation of passengers and freight is integral to the economic growth of developing countries. Growth in population and GDP per capita will increase the demand for private ownership of light-duty vehicles. From a climate change perspective, a future scenario whereby transportation sector GHG emissions are coupled to economic growth in developing countries is unsustainable.

In the developing world, new infrastructure builds can include upfront investment in rail and other public transit options, along with high density urban planning to favour low emissions transportation options. Established cities in high income nations can reduce and eventually eliminate internal combustion engine vehicles within core areas while facilitating efficient use of public transport and zero to low emissions modes of personal mobility. Road vehicle use for medium and long-distance travel can be reduced by expanding networks of freight and high speed passenger rail service.

Road vehicles account for 72% of the energy consumption within the world transportation sector, and there is considerable opportunity to achieve deep cuts in emissions from the road transportation sub-sector by mid-century. There are no serious technical or cost barriers to replacing the majority of internal combustion engine road vehicles with zero or low emitting battery electric, plug-in hybrid or hydrogen fuel cell options over the next 35 years. Electric vehicles will be cheaper to operate based on the efficiency of energy use, the lower cost of electricity, and lower maintenance costs. Further, there is no correlation between GHG emissions and the size and power of the vehicle. The transition of the electricity supply sector to carbon neutral status could follow similar timelines to road vehicle fleet change-over to battery electric propulsion systems. Reduced emissions from long distance inland freight transport may require increased rail use or fuel switching for heavy-duty long-haul trucks.

Achieving significant emissions reductions from the aviation subsector is dependent upon further improvements in aircraft and engine design and fuel switching. The sunk costs into existing fleets will prevent rapid adoption of advances in aircraft design. Biofuels suitable for jet aircraft are in the early stages of development. These factors will delay substantial emissions reductions from the aviation subsector.

Ocean-going cargo ships have a life-span of 20-30 years and adoption of design improvements and new propulsion systems will be delayed by the slow turnover of existing fleets.[304] The low-cost of conventional marine diesel fuel presents a considerable barrier to the use of second generation biofuels and possibly hydrogen. A curtailment in GHG emissions from the marine transportation subsector is likely to be a prolonged process that will extend well into the second half of the century.

Given the lack of significant barriers to an efficient changeover to zero and very low emissions propulsions systems for the majority of road vehicles, and other opportunities to improve energy efficiencies, deep cuts of 70-80% in world transportation sector emissions can be achieved by mid-century. Decoupling GHG emissions from the development of efficient passenger and freight transportation systems will facilitate sustainable economic development in non-OECD nations without compromising the core objective of the Paris Agreement to limit future surface warming.

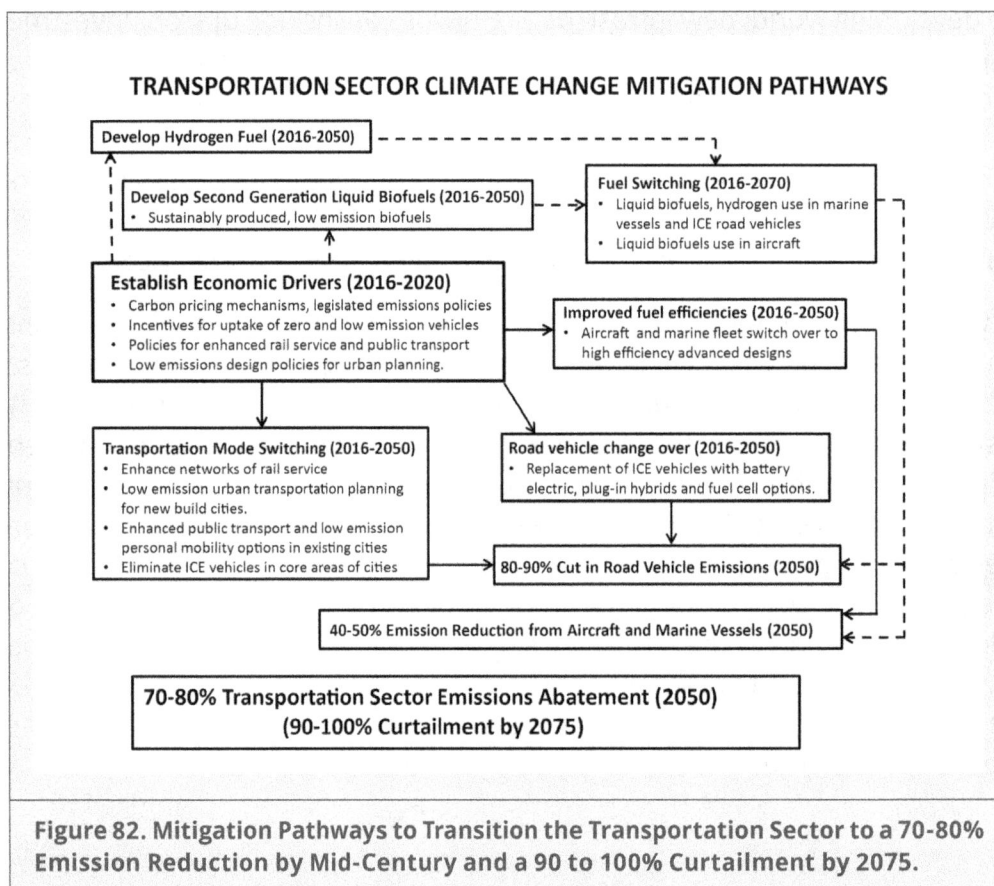

Figure 82. Mitigation Pathways to Transition the Transportation Sector to a 70-80% Emission Reduction by Mid-Century and a 90 to 100% Curtailment by 2075.

5.5 Emissions Reduction Pathways in the Buildings Sector

The IPCC defines the buildings sector as the total energy use and emissions from the heating, cooling, ventilation, and lighting of buildings, along with energy use and emissions from all activities occurring within buildings such as water heating, cooking, and the operation of appliances and electronic devices.[320] In 2010, the buildings sector was responsible for 18.4% of global GHG emissions.[320] Sixty-five percent of these emissions were indirect coming from the production of electricity used within buildings, and the remaining 35% were direct emissions primarily from the combustion of fossil fuels and biomass for space heating, water heating and cooking purposes.[320]

In 2010, residential buildings consumed 74% of total sector energy use with commercial buildings accounting for the remaining 26%.[320] On a global average basis, within residential buildings, space heating, cooking, and water heating were the major contributors to energy use follow by appliances, lighting, and building cooling.[320] In commercial buildings, the operation of electronic equipment accounted for 32% of total energy use.[320]

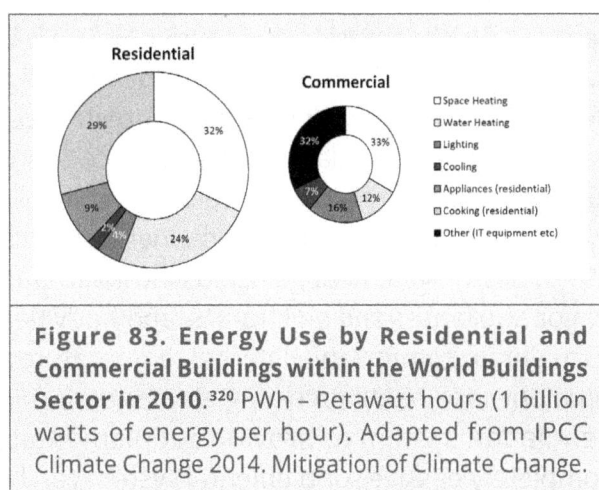

Figure 83. Energy Use by Residential and Commercial Buildings within the World Buildings Sector in 2010.[320] PWh – Petawatt hours (1 billion watts of energy per hour). Adapted from IPCC Climate Change 2014. Mitigation of Climate Change.

Successful mitigation of climate change requires near complete decarbonization of the electricity supply sub-sector by mid-century. Given that most of the emissions from the buildings sector are indirect and come from the production of electricity, one could conclude that improving energy efficiencies within the buildings sector is of less significance than achieving efficiency targets in other sectors. However, this conclusion is overly simplistic and discounts the importance of reducing the intensity of electricity use to minimize future power demands. Abating future emissions within the transportation and industry sectors requires a substantial switch in energy use from fossil fuels to electricity. This new demand must be offset by improved efficiencies across all economic sectors.

Over the next 30 years, a business as usual scenario would lead to a doubling or tripling of energy use and emissions from the world buildings sector, largely due to the increase in population and GDP per capita projected for developing nations.[320] This scenario is unsustainable and will lead to a catastrophic increase in world GHG emissions. The common theme of decoupling economic growth in the developing world from emissions applies to the maintenance of environments and the energy consuming activities within newly constructed residential and commercial buildings. While the challenges are significant, considerable opportunity exists for widespread adoption of advances in building technology, and in the use of high efficiency lighting, appliances and electronics. Implementation of cost-effective energy saving measures in the design and construction of new buildings can achieve a 2 to 10-fold reduction in energy use when compared to conventional buildings. Retrofits to existing buildings can result in a 2 to 4-fold cut in energy use.[320] Construction of new high-efficiency buildings and upgrades to existing structures in combination with the use of energy saving lighting, appliances and electronics will minimize emissions and limit the demand for electricity from the world buildings sector.

5.5.1 High Efficiency New Buildings. The Passive House (PH) standard is an international certification that prescribes specific maximum energy use for heating and cooling and the energy demands of domestic appliances per surface area of livable space.[321] The final design of a Passive House is based on computer models that account for the specifics of climate and the location of a house in orienting windows and other design features to either maximize or restrict passive solar gain. Design features include the use of super-insulation, and advanced windows that create an extremely air-tight high performance thermal envelope. A controlled mechanical ventilation heat recovery system is used to optimize atmospheric air exchanges with a heat recovery rate of over 80%. Passive solar heat gain, plus intrinsic heat sources from lighting, the operation of major appliances and electronics, and body heat from occupants is generally sufficient to maintain comfortable internal temperatures year-round. In more extreme climates, dual purpose heating/cooling elements or micro-heat pumps can be installed in the ventilation system to provide supplemental temperature control. Alternatively, a completely passive supplemental system can be installed whereby incoming air passes through underground pipes that act as an air-to-earth heat exchanger. The key feature of any PH design is complete elimination of conventional forced air heating and air conditioning systems. Energy use is confined to the electricity used by the mechanical ventilation heat exchanger and any supplemental heating/cooling system incorporated in the design. Regardless of climate, to achieve PH standard certification, the total energy use for heating and cooling must be less than 15 kWh/m2 of living surface/year.[321]

The International Passive House Association estimates that, worldwide, over 60,000 residential and non-residential passive design units have been constructed with 14,000 certifications in place.[322] Energy efficiency gains are greatest for home heating in cold climates and

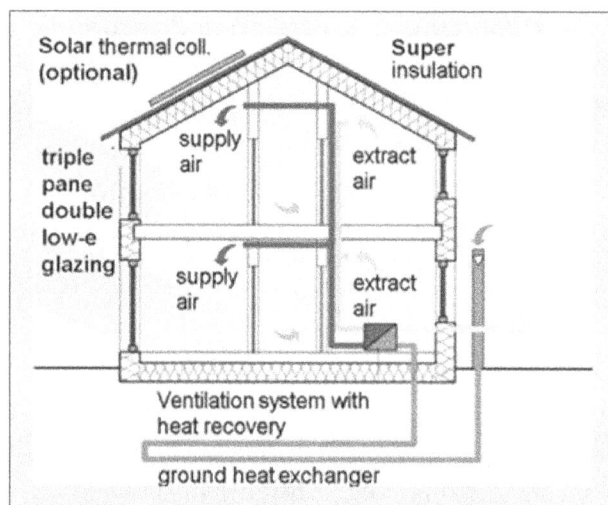

Figure 84. **Passive House Design with Optional In-Ground Heat Exchanger and Roof-Top Solar Collector.** The design is based on complete elimination of conventional heating, ventilation and air conditioning systems.

can range from 50% to over 90%.[323] Cooling applications in hot humid climates are less energy intensive than home heating in cold climates but, on a percentage basis, similar energy savings are achieved with PH designs. Curtailment of emissions will vary between countries depending upon conventional fuel sources and intensity of emissions per unit of electricity consumed. On average, across European countries, total direct and indirect emissions from a Passive House will be 50–60% less than a conventional new home.[323] Electricity is the only source of energy and thus full decarbonization of the electricity supply sector will result in zero net emissions from a Passive House.

Elimination of conventional HVAC systems largely offsets the additional costs of super-insulation and windows and other design features of a Passive House. The upfront costs of construction are currently about 5–10% higher than a conventional house. However, costs of design and construction will decline as the technology advances and building practices and materials become standardized. Currently, there are examples of schools and small office buildings that have been built to PH standards at no additional costs.[320] Passive House design principles can be used with new commercial buildings and will achieve similar efficiencies in space heating and cooling. The annual savings in energy costs to maintain a Passive House are such that any incremental upfront construction costs are quickly recovered within the first few years of occupancy.

Net zero-emissions buildings (NZEBs) refers to a balance of energy whereby on-site production of electricity from zero-emissions renewables and export to the grid matches energy use by the building.[320] Positive-energy buildings are an extension of this concept whereby production of energy from photovoltaic cells exceeds energy consumption.[320] Buildings with a large roof area per unit of living or working space are best suited to on-site power production. From a pure least-cost perspective, the most efficient integration of zero-emissions electrical energy supply and buildings sector demand is based on construction of highly efficient new buildings, retrofits of existing buildings, and a supply of electricity from larger scale zero-emissions power plants. However, broad-scale adoption of building codes set to net zero or nearly zero energy standards will extend the capacity of supply from distributed zero-emissions sources and will reduce demand for power from centralized plants.

The construction of high efficiency new buildings and the retro-fit of existing buildings are by far the most important emissions abatement opportunities within the buildings sector. The European Union has established the Energy Performance of Buildings Directive with the intent of achieving deep cuts in energy use and emissions from buildings. All member states are required to draw up plans to institute Nearly-Zero-Emissions Building standards that would apply to new public buildings as of December 31, 2018 and to all newly constructed buildings as of December 31, 2020.[324]

5.5.2 Retrofits of Existing Buildings to Improve Energy Efficiency and Reduce Emissions. Buildings typically have a life-span of 50 to 100 years and a substantial portion of the total inventory of current buildings will be in existence by mid-century.[320] While there is considerable opportunity to achieve high energy efficiency standards in new buildings, the sunk cost into existing buildings provides a barrier to achieving similar efficiencies across the entire sector.

Deep retrofits to existing buildings consist of replacing windows and insulation and sealing the structure to achieve comprehensive upgrades to the building envelope. By adding mechanical ventilation with heat recovery, home heating requirements for single family dwellings can be reduced by 50–75%, and by 50–90% for multi-family housing units.[320] In some cases, energy use following deep retrofits can approach PH standards. However, in the absence of potent incentive packages, deep energy efficiency retrofits to PH standards are often cost-prohibitive.

Modest upgrades to building envelopes and other energy saving retrofits can provide acceptable returns on investment. A study of 643 commercial buildings in the US found that a simple recommissioning of control systems resulted in an average saving of 16% in energy use with a pay pack of 1.1 years.[320] Other studies in the US have found that a 29–48% energy saving can be achieved by implementing near-cost-neutral energy retrofit packages for residential buildings.[320] In general, cost for retrofits increase in proportion to energy savings of up to 70–80%.[320] In France, 500,000 major energy retro-fits of existing buildings are targeted per year with tax credits to home owners used to incentivize renovations.[325]

5.5.3 Improved Efficiency of Energy Use by Lighting, Appliances and Electronics.

Solid state lighting using light-emitting diodes will become the most widely used lighting option in the near future. Currently, the best available LED lights are 7-fold more energy efficient that standard incandescent light bulbs and 1.6-fold more efficient than compact fluorescent bulbs and have the added advantage of resistance to shock and vibration.[320]

In recent years, the energy efficiency of major appliances has increased substantially with considerable potential for further improvement. In comparison to current best standards, energy savings of at least 40% are projected for refrigerators and freezers.[320] Continued advances in energy savings are expected for dishwashers, washing machines, electric ovens, microwaves, air conditioners, ceiling fans and all electricity consuming appliances. Heat pump clothes dryers eliminate external venting of hot moist air by condensing moisture from the air that exits the drying chamber and then recirculating the warm dry air within the appliance. These dryers operate on less than 50% of the energy used by conventional clothes dryers.[320] Heat pump water heaters extract energy from air in the room and can provide a 3 to 4-fold advantage in energy use relative to standard water heaters.[320]

The energy efficiency of electronics is projected to continually improve over time. Future energy savings of 60% are estimated for computers and monitors, while energy use by televisions could decrease by 30–50%.[320]

Population and economic growth in the developing world will increase worldwide consumer demand and thus partially offset the projected energy savings for future lighting, appliances and electronics. Overall, by 2030, consumer uptake of best available technologies should result in a net reduction in the global use of electricity within buildings.

5.5.4 Energy Use Within the Buildings Sector of the Developing World. High tech energy abatements such as heat pump clothes dryers and Passive House designs with mechanical ventilation and heat exchangers are well removed from the current reality for much of the developing world. The World Bank estimates that 70% of houses in the developing world were constructed informally without building standards.[326] In 2010, an estimated 1.1 billion people lacked access to electricity and 3 billion continued to burn wood, charcoal, coal, dung and crop residues in inefficient, highly polluting traditional solid fuel stoves.[305,327]

Over time, economic advances in the developing world will be characterized by population shifts to urban areas and a housing transition from informal structures to formal buildings. These apartment blocks and houses will be built on grids that provide electricity, water and sewage. Improved living standards will come with an increase in the use of domestic lighting, appliances and electronics that are taken for granted in advanced economies.

A transition to formal housing and improved standards of living in the developing world that is based on the high energy intensity typical of the housing sector in established economies is unsustainable and will lead to a catastrophic accumulation of GHGs during the current century. As discussed for other sectors, the challenge, and indeed the opportunity, for the developing world is to build quality housing and to reduce poverty based on low energy intensity pathways that are uncoupled from GHG emissions.

The opportunity exists for new builds in the developing world to place an early onset priority on energy saving solutions. Energy intensive central air conditioning and heating can be avoided in new structures while providing comfortable and safe internal environments. High standards of energy efficiency can be applied to lighting, appliances and electronics. This approach will reduce household energy costs and thus contribute to improved standards of livings. A future scenario of wide-scale adoption of low energy housing as a critical component of economic development will minimize power demands and thereby facilitate the construction and expansion of efficient electricity supply networks in the developing world. By reducing pressures on the electricity supply sector, efforts can be focused on meeting demand through sustainable zero or low emissions power generating options.

5.5.5 Summary and Overview of Buildings Sector Climate Change Mitigation Pathways. Long-term sunk costs and lock-in to existing buildings and infrastructure are considerable barriers to achieving deep cuts in direct GHG emissions from the buildings sector in advanced economies over the next 25 years. These barriers will likely delay transition of the world buildings sector to complete carbon neutral status until sometime after 2070.

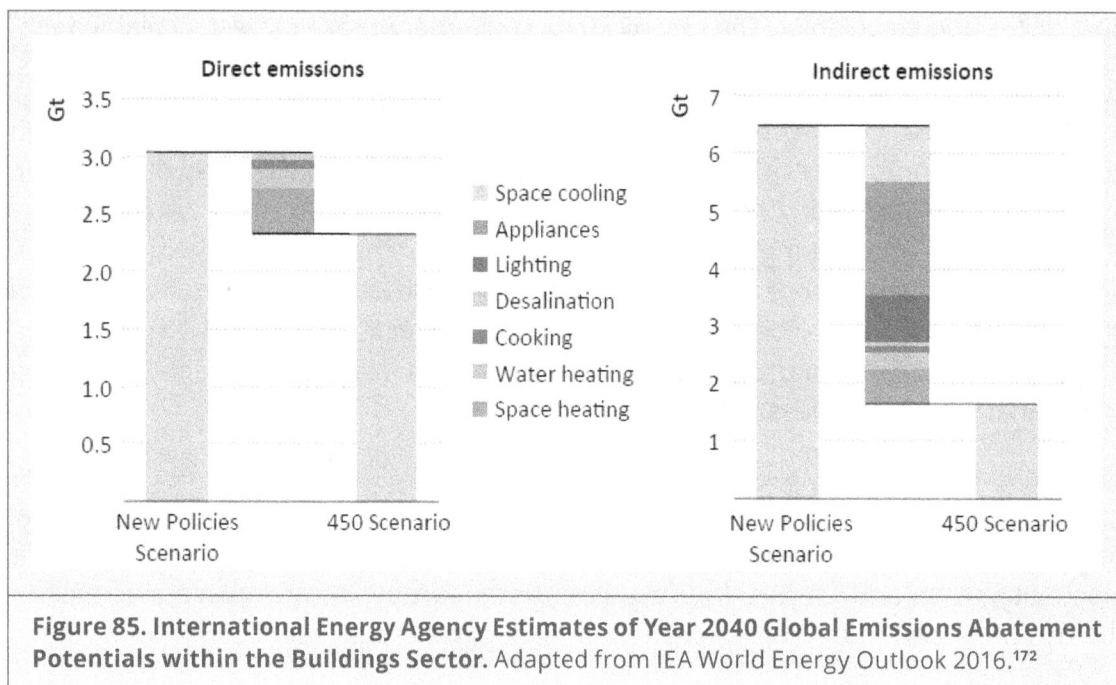

Figure 85. International Energy Agency Estimates of Year 2040 Global Emissions Abatement Potentials within the Buildings Sector. Adapted from IEA World Energy Outlook 2016.[172]

Cost-effective technologies such as Passive House designs can be applied to new build residential and commercial structures in economically advanced and developing nations to avoid continued long-term lock-in of high energy consumption and associated emissions. Carbon pricing mechanisms and incentive programs can be implemented, along with changes to building codes to achieve ambitious levels of energy efficiency and to eliminate the use of gas and oil as fuels in new buildings. These tools can also be used to incentivize retrofits of existing buildings to improve energy efficiency and reduce emissions.

Legislation to mandate construction of buildings to PH standards are in place many regions of Germany including the states of Bavaria, Saarland, Rhineland-Palatinate, Hesse, and Hamburg, along with the cities of Frankfurt, Cologne, Nuremberg and Heidelberg.[328] In Belgium, the city of Brussels and the province of Antwerp have enacted PH standards for new builds and complete renovations.[328] In Austria, the state of lower Austria, several municipalities within the state of Vorarlberg, and the city of Wels all require that new public buildings and retrofits must be constructed to PH standards.[328] Passive House standards are mandated for new buildings in Luxembourg and cities and municipalities in Ireland, Norway, and Spain.[328] On October 1, 2014, the mayor of New York City released a long-term plan to achieve a 35% cut in emissions from buildings by 2025 continuing to an 80% emissions reduction by mid-century.[329] To achieve this level of ambition, the city will look to "Passive House, carbon neutral or zero net energy" designs to develop standards.[329] Within buildings, mandated standards for lighting, appliances and electronics can ensure use of best available options for energy savings and promote the use of high efficiency heat pump clothes dryers and water heaters. The lifespan of appliances, lighting and electronics is such that continued advances in efficiencies can be quickly implemented. Improved efficiencies upon the switchover of devices in the buildings sector will be an important contributor to reducing the demand for electricity and thus cutting indirect emissions.

Fuel switching from wood, coal, dung and crop residues to electricity in the developing world and replacing gas and oil heating with electricity in advanced economies will reduce and eventually eliminate direct and indirect GHG emissions from the world buildings sector. This conclusion is based on an assumed transition of the electricity supply sector to near carbon neutral status by mid-century. Progressive reductions in the per capita energy intensity of the buildings sector will minimize demand and thus alleviate pressure on electricity supply networks. Efficiency in electricity use will facilitate the critical decarbonization of world power supply that is central to all climate change mitigation models aligned with achieving the core objectives of the Paris Agreement.

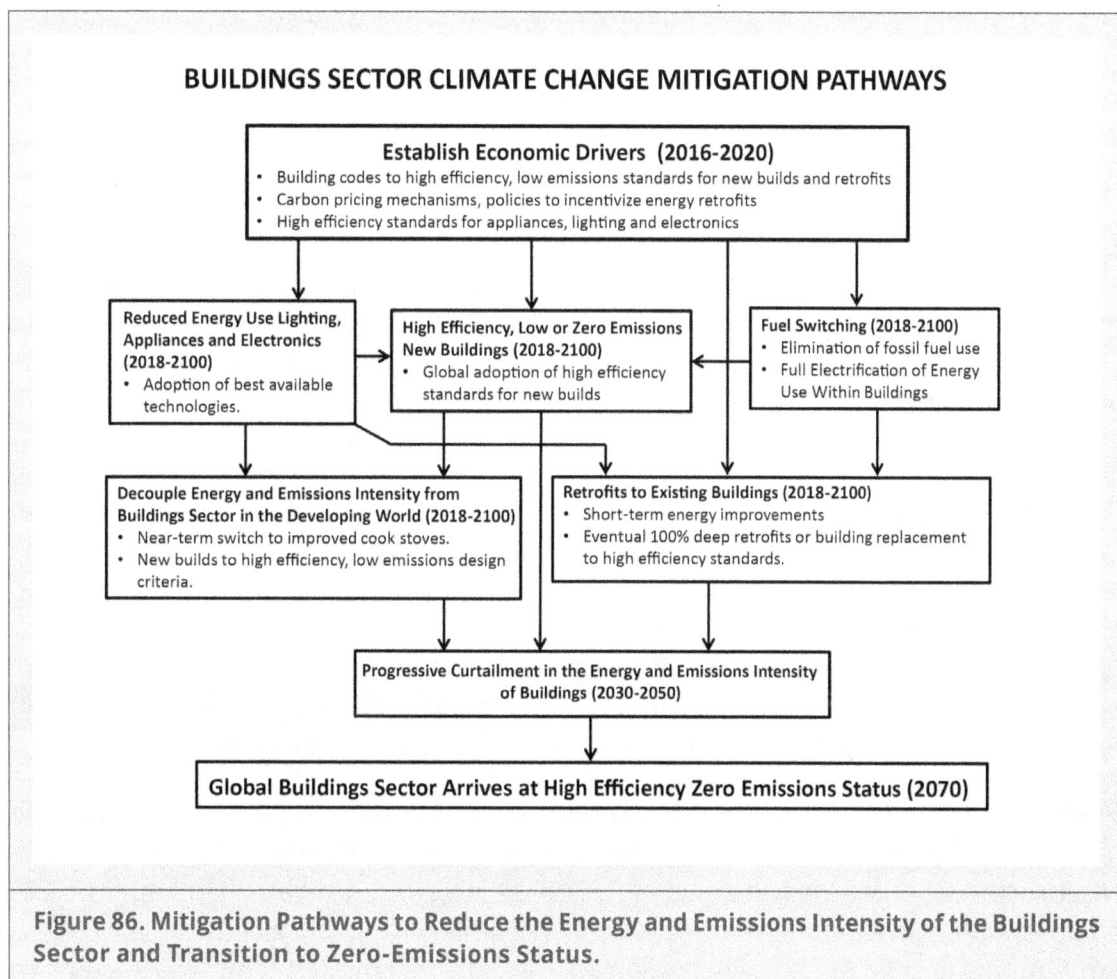

Figure 86. Mitigation Pathways to Reduce the Energy and Emissions Intensity of the Buildings Sector and Transition to Zero-Emissions Status.

5.6 Geo-Engineering and the Control of Surface Temperatures

Geo-engineering or climate engineering refers to planetary scale deliberate manipulation of the Earth's atmosphere and surface temperatures. By 1900, greenhouse gas emissions from fossil fuel combustion had begun to accumulate in the atmosphere and our species had unknowingly begun to alter the climate systems of the planet. Over the next 60 years, anthropogenic GHG emissions increased in correlation with economic progress and population growth with no clear understanding of effects on surface temperatures. By 1960, David Keeling's exacting measurements of well mixed atmospheric gases at the Mauna Loa observatory in Hawaii provided definitive proof of annual incremental increases in atmospheric carbon dioxide concentrations. Since the early 1960s, the science of climate change has advanced as the annual rate of GHG emissions has continually increased. The 1979, public statement from the United States Research Council as to a cause and effect relationship between continued accumulation of atmospheric carbon dioxide and future elevations in global surface temperatures can be considered as a transition point in human understanding of anthropogenic changes to the environment.[330] Since 1979, our species has knowingly geo-engineered the atmosphere of the planet and altered climate systems through combustion of fossil fuels, and other activities leading to incremental accumulations of greenhouse gases.

The concept of climate change mitigation through geo-engineering is to deliberately manipulate climate systems to compensate for and reverse the detrimental effects of the anthropogenic GHG accumulations. Broadly, geo-engineering is divided into engineering concepts for active carbon dioxide reduction and concepts to reduce the warming effect of incoming solar radiation.

5.6.1 Carbon Dioxide Removal (CDR). As discussed, reforestation and afforestation to increase the carbon-sequestering biomass on the planet is integral to all mitigation pathways arriving at 450 ppm CO_2 by the end of the century. Deliberate manipulation of global forested areas to achieve net removal of atmospheric carbon dioxide is an exercise in geo-engineering. There are no logical arguments against geo-engineered CDR through forest expansion. As discussed, CDR that is based purely on reforestation and afforestation is limited by available land mass and requires effective management of land use to balance the competing needs of agriculture and biofuel production.

Bioenergy with carbon capture and storage (BECCS) combines geo-engineered CDR by forests or other biomass and biomass harvesting for power production with CO_2 capture and storage. The net effect of BECCS is active withdrawal of atmospheric CO_2 and permanent isolation from the environment while producing electricity. Successful implementation of BECCS will allow for a higher efficiency of land use to remove atmospheric carbon dioxide than simpler programs based solely on an increase in forest biomass. As discussed BECCS, while technically feasible, is in the early stages of development with considerable uncertainties as to the cost and complexity of combining multiple technologies. However, the transformation pressures and costs of achieving deep and timely emissions cuts in the housing, transportation, industry, energy supply and AFOLU sectors are such that most least-cost aggressive climate change mitigation models incorporate BECCS despite the early stage of development and cost uncertainties of the overall process.

Direct air capture (DAC) of carbon dioxide uses sorbent materials to chemically capture CO_2 from the atmosphere. So-called artificial trees would function to sequester approximately 1,000 times the atmospheric CO_2 that would be removed by a similar-sized actual tree.[331] Captured carbon would then be isolated from the environment using the same permanent geological storage sites proposed for BECCS. DAC is an early phase concept with considerable uncertainties as to costs, energy requirements and practicality.

Marine based bioenergy with carbon capture (MBECS) would use low oxygen sub-tropical convergence zones in the world's oceans as vast three dimensional algal farms.[332] The algae would be grown with nutrients normally present in the water and would function as a carbon sink by sequestering dissolved CO_2. The marine algal biomass crop would be harvested for the production of biofuels and other useful products. Carbon capture and storage could be incorporated at one or more steps in the overall process. In theory, MBECS bypasses land use and other impediments that may limit the utility of terrestrial biomass and crops for use as feedstocks in the production for first and second-generation biofuels. Conceptually, energy products from marine algal farms can be sustainably produced in sufficient quantities to completely replace fossil fuels. MBECS is another early phase concept requiring systematic research and development to assess viability and commercial potential.

5.6.2 Solar Radiation Management. Stratospheric injection of sulphuric acid to reduce incoming solar radiation and limit surface warming can likely be accomplished without significant technical or cost barriers.[333] Aerosols in the upper atmosphere would create a reflective layer that would dissipate a small amount of solar radiation back to space and reduce surface heating by the sun. The process would mimic the aftermath of a large scale volcanic eruption such as the 1883 Krakatoa explosion and could be used to reduce surface temperatures in direct proportion to the quantities of sulphuric acid injected into the stratosphere.

Controlled aerosol injection into the upper atmosphere would counter the surface warming of elevated GHG concentrations. Aerosols have a limited life span in the atmosphere and would need to be replenished every few years to allow for continued control of surface temperatures.[333] Direct experience with volcanic eruptions allows for an understanding of the risks associated with anthropogenic injection of sulphur aerosols. Ozone layer depletion, acid rain, and uncertain changes to global precipitation and weather patterns are the primary risks.[333] In theory, the risks and damaging effects associated with controlled stratospheric aerosol injection are considerably less than the consequences of sustained elevated surface temperatures.

It is difficult to envision public acceptance of controlled programs of solar radiation management by injection of sulphuric acid into the stratosphere. However, failure to adequately curtail emissions over the next 35 years may bring solar radiation management to the forefront of remaining options to prevent catastrophic surface warming.

Geo-engineering to deliberately manipulate the Earth's climate should not be viewed as a magic bullet solution to global warming. At present, reforestation and afforestation are the only proven options for carbon dioxide removal. Solar radiation management should only be given consideration under dire circumstances of impending catastrophic surface temperature elevations. Efforts to combat climate change must focus on the primary objectives of reducing emissions and expanding forest carbon stocks. Research and development of geo-engineering technologies should continue but cannot detract from efforts to establish aggressive emissions reduction pathways in the energy, AFOLU, industry, transportation and housing sectors of world economies.

5.7 Summary and Integration of Technical Pathways to Mitigate Climate Change

Globally, the aggregate of aggressive climate change mitigation pathways defined within economic sectors are interrelated and co-dependent. Failure to achieve emissions reduction or energy efficiency targets in one economic sector will impact the capacity and practicality of achieving targets in other sectors and very likely result in an over expenditure of the remaining emissions budget.

Effective national climate change mitigation programs that are aligned with the core objective of the Paris Agreement are based on a mixture of short-term measures to curtail emissions from current practices and long-term action plans to progressively decarbonize economic sectors. Limiting average surface warming to well below 2°C above pre-industrial temperatures will require deep emissions cuts by 2050 followed by a complete transition to net zero GHG emissions in the second half of the century.

5.7.1 Near-Term Emissions Reduction from Fossil Fuel Extraction and Use. Over the next 5–15 years, measures to curtail fugitive methane emissions and eliminate routine flaring in the oil and gas sector have the potential to provide a significant early phase contribution to controlling global GHG emissions. Other initiatives such as fuel switching from coal to natural gas in the electricity supply sector will reduce net emissions but can lock-in continued fossil fuel use based on capital cost sunk into new build natural gas power plants.

5.7.2 Improved Efficiencies in the Industry, Transportation and Buildings Sectors. Overall, the energy intensity of the industry sector can be improved by 25% based on adoption of best available technologies and potentially up 45% with further innovation before reaching the technical limits to efficiencies of energy use. Within the building sector, Passive House and similar standards applied to new builds and deep retrofits can achieve an 80–90% savings in energy use and eliminate the use of fossil fuels for space heating and other applications. Adoption of high efficiency standards for lighting, appliances, and electronics can markedly reduce the intensity of energy use within buildings. In the transportation sector, energy inefficient road vehicle kilometres can be reduced through enhanced networks of passenger and freight rail service between centers. Within cities energy savings can be realized through urban planning that promotes and facilitates the use of public transit and zero or low emissions personal transportation options. Optimization of energy use within all economic sectors will control demand for electricity and thus facilitate a timely and cost-effective transition of the electricity supply sector to zero or very low emissions status.

5.7.3 Decarbonization of the Global Electricity Supply Sector. Phase out of emissions intensive coal and other fossil fuel power stations and replacement by zero and low emissions options for the production of electricity is a core requirement of all aggressive climate change mitigation models. Decarbonized power supply to electrical grids will come from a mixture of non-dispersible renewables such as wind and solar and dispersible zero and low emissions power sources including hydro, geothermal, biomass, waste to energy, and nuclear energy sources. Points of supply will vary from large power stations to smaller distributed sources including rooftop solar panels. For some electrical grids, sources of dispersible zero or low emissions energy may be insufficient or unacceptable to the public. To control costs, these grids may require flexible, minimal use of natural gas power production to balance supply and demand.

Transitioning the world supply of electricity to near zero-emissions status by mid-century is essential to mitigation pathways that are dependent upon eliminating indirect emissions from the production of electricity consumed within the buildings, transportation and industry sectors.

5.7.4 Electrification of Economic Sectors. Based on a supply of zero or very low emissions electricity, transitioning world steel manufacturing practices to electric arc furnaces will achieve sharp cuts in direct emissions. Road vehicle changeover from internal combustion engines to battery electric propulsion systems can eliminate 75% of emissions from the global transportation sector.

There are no serious technical or cost barriers to electrification of steel production and road vehicles. Transition to these low emissions options can be completed over the next 3 decades and will be an important contributor to achieving aggressive cuts in global GHG emissions.

Deep energy retrofits to existing buildings can replace fossil fuel consuming central heating systems, water heaters and appliances with energy saving alternatives that use electricity. The life-span of existing buildings and costs of deep energy retrofits will likely delay complete elimination of direct emissions from the buildings sector until the second half of the century. New buildings constructed to high efficiency standards will not use fossil fuels.

5.7.5 Fuel Switching to Zero or Low Emissions Options. Industrial requirements for thermal energy can only be replaced by electricity in a limited number of applications. However, fossil fuels can be switched out to low emissions alternatives such as solid and liquid biofuels, biogas, and district and waste heat from low emissions sources.

5.7.6 Optimization of Land Use for Agriculture, Forestry and Biofuel Production. Intensive Green Revolution agricultural practices must be adopted across the globe to feed an expanding population while optimizing the efficiency of land use by agriculture. Ruminant meat consumption and thus the size of the world herd of ruminant animals cannot exceed sustainable limits based on enteric methane emissions and land requirements to feed and raise these animals.

Programs of afforestation and reforestation to increase carbon-sequestering plant biomass are vital components of all aggressive climate change mitigation models. Forested areas must be expanded and protected across the planet such that forest biomass functions to actively withdraw carbon dioxide from the atmosphere.

Progressive decarbonization of world economies must arrive at a state of net zero-emissions during the second half of the century. Net zero-emissions will consist of a balanced state of atmospheric carbon flows whereby carbon dioxide reduction from forest biomass and potentially industrial bioenergy with carbon capture and storage equals residual anthropogenic GHG emissions. There is a practical limit to the carbon sequestration capacity of the world's forests. Forested areas cannot impact agriculture land requirements to feed a year 2100 world population of 11 billion. Land use management may be further complicated by potential requirements for the production of second generation liquid biofuels.

5.7.7 Carbon Capture and Storage Integrated to Power Production and Industrial Processes. Elimination of CO_2 emissions from chemical reactions within industrial processes such as cement manufacturing requires on-site carbon capture with subsequent permanent isolation from the environment in deep geological formations. Potentially, CCS can be applied to the generation of electricity from the combustion of biomass. The development of economically viable processes of CCS that contribute to reducing emissions must be independent of enhanced oil recovery.

Bioenergy with carbon capture and storage (BECCS) has considerable potential as an industrial process to reduce atmospheric carbon dioxide. Forest biomass or biomass grown on dedicated plantations could be sustainably harvested and used as fuel for power production in stations equipped with carbon capture technology. Captured carbon would be transported to permanent geological storage sites. Successful global scale implementation of BECCS will extend the efficiency of land use for the growth of carbon-sequestering biomass beyond the limitations of programs based on purely on an expansion of forested area.

5.7.8 Low Emissions Second Generation Liquid Biofuels and Hydrogen. Hydrogen burns with zero-emissions and can fuel road vehicles and many thermally intensive industrial processes. Potentially hydrogen could replace marine oil in next generation advanced ocean-going vessels. However, cost-effective, low emissions processes for commercial scale production of hydrogen do not exist. The development of these processes would provide an additional clean fuel option without land use implications.

The future of low emissions second generation liquid biofuels is uncertain. Options for road vehicle changeover to battery electric and hydrogen fuel cell propulsion systems may restrict the utility of liquid biofuels to niche applications such as replacement of jet fuels.

5.7.9 Climate Change Mitigation Pathways With and Without CCS. The lowest cost climate change mitigation pathways assume successful development and adoption of carbon capture and storage in the industry and electricity supply sectors. Conceptually, full-scale installation and operation of BECCS power plants around the globe can function to remove up to 10 Gt per year of atmospheric CO_2.[267] This level of CDR equals 25% of current peak emissions and thus could effectively balance residual emissions from other economic sectors. Carbon dioxide removal through full-scale application of BECCS technology could facilitate a controlled mid-century overshoot in atmospheric GHGs concentrations of up to 10% above the end-of-century target of about 450 ppm CO_2eq with minimal impacts on surface warming. This scenario would partially alleviate time constraints and thus reduce transition costs for decarbonization of the transportation, industry and buildings sectors. By the end of the century, world economies would function under net negative emissions with continued withdrawal of CO_2 from the atmosphere in excess of emissions until a safe equilibrium is attained.

As discussed, the barriers to full-scale implementation of BECCS are substantial, and the technology remains in the development stage. Operation of BECCS power plants will require potent economic drivers to cover the cost of sustainable biomass harvesting, the energy costs of carbon capture, and the costs and complexities of transport and permanent storage of captured carbon. Conceivably, these barriers may delay or even prevent the deployment of BECCS.

Climate change mitigations models without carbon capture and storage all require a global scale expansion of forest biomass to continually drawdown atmospheric CO_2. This expansion of forested area must occur in combination with a complete transition of the electricity supply sub-sector to zero-emissions status by mid-century, along with aggressive programs to progressively decarbonize other economic sectors. The full mitigation potential of forest biomass is uncertain and varies with assumptions as to future land requirements for agriculture and other uses. Estimates for the net carbon dioxide removal potential of the world forestry and land use sector range from 0.2 up to a theoretical maximum of 13.4 Gt of CO_2 annually.[267] A complete elimination of ruminant animal products would be required to fully maximize carbon-sequestering forest areas. This scenario will be unacceptable to many societies. As a result, the practical capacity for carbon dioxide removal by forest biomass will likely be less than 7 Gt of CO_2 per year. An effective expansion of forested areas on the planet will require optimization of land use by agriculture with sustainable limits placed on the world ruminant animal herd size and successful implementation of food waste reduction programs.

Long-term global and national action plans to decarbonize economies must strive to achieve aggressive targets for emissions reductions in all sectors and to enhance the size of the world's forests without assumptions as to the future availability of carbon capture and storage.

5.7.10 Comparison of 450, Well below 2°C and 1.5°C Scenarios. Prior to the Paris Agreement, the International Energy Agency had focused on models of transition in energy use and emissions reduction consistent with a 50% chance to limit surface warming to less than 2°C. The core objective of the Paris Agreement will require a higher level of ambition to control emissions than assumed in the IEA's 450 Scenario. About 75% of global GHG emissions come from the extraction, transformation and use of energy. In the World Energy Outlook 2016, the IEA has begun to assess pathways for reducing emissions and transforming the energy supply sector such that future surface warming will be limited to well below 2°C and ideally 1.5°C above pre-industrial temperatures.

In comparison to the 450 Scenario, the IEA has identified the road transportation, power production and buildings sectors as pillars to achieve further emissions reduction, and accelerate the process of decarbonizing world economies as required under the Well Below 2°C Scenario.[172]

The recent advances in zero-emission light-duty vehicle technology provides additional emissions abatement opportunities within the transportation sector. Under the 450 Scenario, the IEA projected that 710 million electric passenger vehicles, equal to one-third of the world fleet, would be on the road by 2040.[172] This transformation away from the internal combustion engine could be enhanced to further curtail emissions from fossil fuel combustion. By 2040, zero-emissions electric vehicles could be the norm accounting for 75% of the total of passenger vehicles in service around the world.[172]

Considerable opportunity exists to further increase efficiencies in the buildings sector beyond projections under the 450 Scenario. By further advancing high efficiency standards for residential buildings in industrialized countries and emerging economies, heating and cooling energy demands can be cut by over 40%.[172]

Relative to the 450 Scenario, further electrification of the transportation and other sectors of the world economy will increase the demand for low carbon electricity by about 10%.[172] Renewables and nuclear energy would largely make up this difference and would need to become increasingly prominent in the global energy mix. The IEA suggests that when compared to the 450 Scenario, an additional 20% reduction in the intensity of CO_2 emissions per unit of power produced will be required by 2040.[172] The contribution of low emissions sources to global power production would increase from 70% under the 450 Scenario to 80% for the Well Below 2°C Scenario.[172]

Currently, there is little available information on transformation pathways under a 1.5°C Scenario. There are uncertainties as to what remains in the carbon emissions budget at this level of ambition; however, the IPCC estimates the budget could be in the range of 400–450 Gt, which is less than half of what can be released under the Well Below 2°C Scenario. At current rates of emissions, this budget will be exhausted within 10 years. To avoid an over expenditure, all avail-

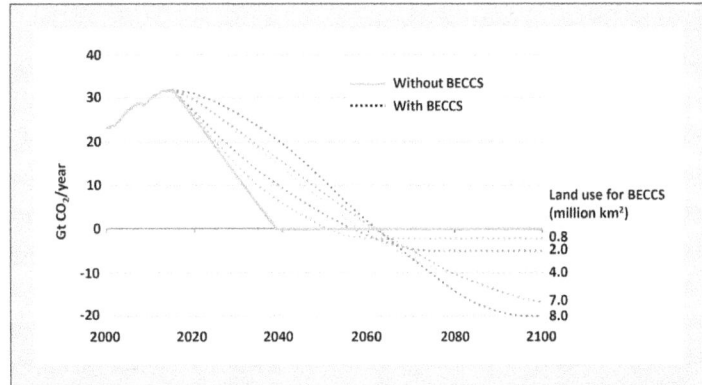

Figure 87. IEA Projections for Possible Energy Sector Emissions under a 1.5°C Global Warming Scenario.[172] Dashed lines indicate the potential of deployment of BECCS to extend emissions budgets. The carbon storage capacity of BECCS deployment is dependent upon land mass allocations to produce biomass feedstock for the power plants.

able technologies and options to maximize energy efficiencies and switch to low emissions options would need to be deployed immediately and economic sectors would be required to quickly and aggressively begin the process of deep decarbonization. By 2040, all passenger vehicles and light-duty trucks would be electric. Road transport of freight would either shift to rail or be carried by trucks propelled by electricity or advanced low emissions liquid fuels. By 2040, almost all buildings would need to be retrofitted to zero-emissions standards. A full 90% of power production would need to come from zero and low emissions sources.[172]

To limit future surface warming to 1.5°C in the absence of BECCS, the IEA concludes that world economies must arrive at net zero CO_2 emissions from the energy sector prior to 2040.[172] Potentially, deployment of BECCS would provide future capacity to withdraw atmospheric carbon dioxide, which would allow for a less abrupt transformation of world economies. As the collective size of the global installation of BECCS power plants increases, the time scale to zero-emissions status lengthens. The IEA estimates that 8 Gt/year of BECCS carbon storage capacity would allow about a 10-year extension to complete the transformation of the global energy sector to zero-emissions status.[172] This level of deployment of biomass power stations would require a total land mass equal to half the land area of India. In theory, under the 1.5°C Scenario, zero-emissions status for the energy supply sector could be delayed until 2060 if 20 Gt/year of BECCS carbon storage capacity was deployed. This level of BECCS would be extreme and double IPCC estimates of the practical limits of deployment. The land area required to produce the biomass feedstock for these power plants would be equal to the entire continent of Australia.

The reports published by the International Energy Agency and the Intergovernmental Panel on Climate Change are state of the art assessments of the world body of scientific and economic studies on the transformations required to limit future emissions and minimize the damage of climate change. Achieving the objective of limiting future surface warming to well below 2°C will be a challenge but can be accomplished by extending the emissions reduction pathways defined under the 450 Scenario and IPCC pathways to limit surface warming to 2°C. Limiting future surface warming to 1.5°C is another level of ambition and may well be impractical without massive deployment of BECCS and other technologies at the development stage.

If the intention of the world community is to seriously combat climate change, then it is incumbent upon all levels of governments, along with leaders of industry, to make full use of the technical expertise provided by agencies such as the IEA and the IPCC. The information exists to design cost-effective policies and measures to drive down emissions at a level of ambition that would fall within the globally agreed upon objective to limit future surface warming.

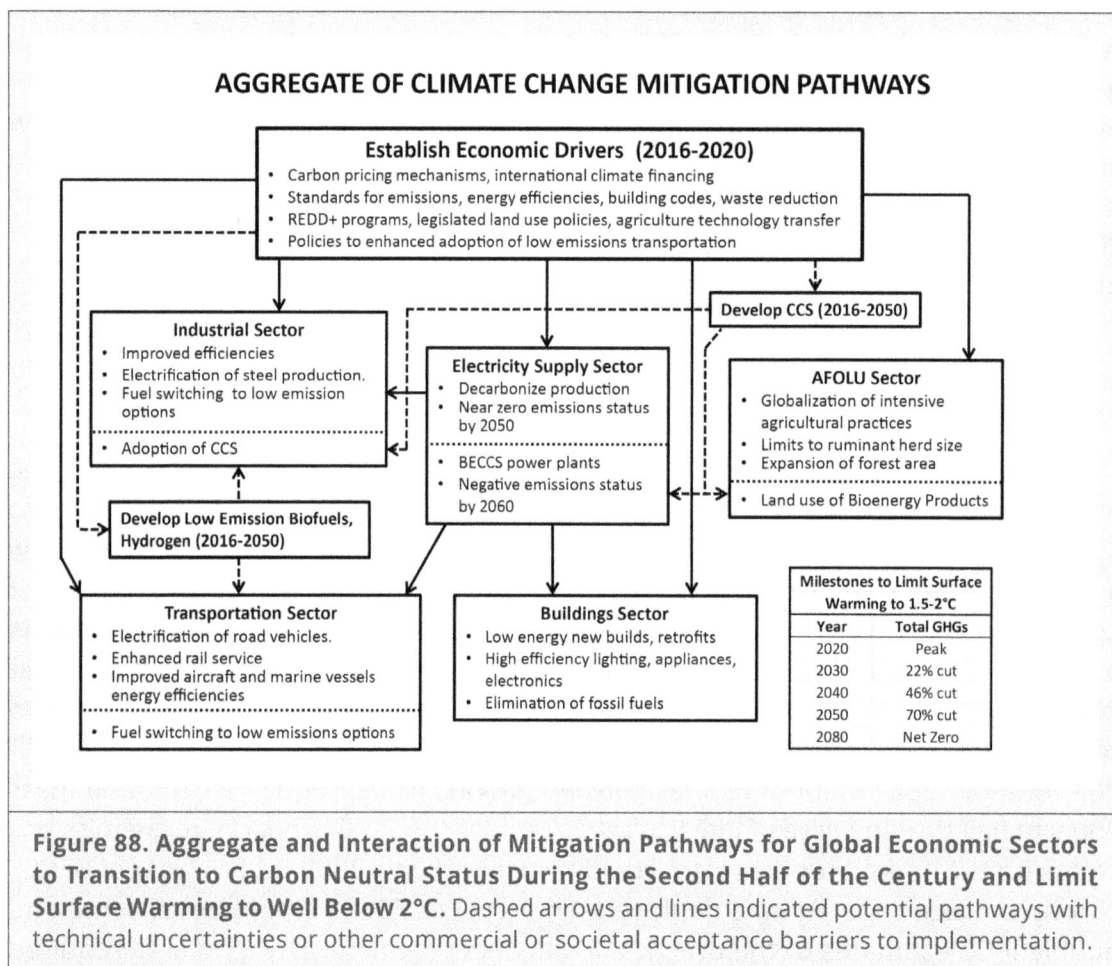

AGGREGATE OF CLIMATE CHANGE MITIGATION PATHWAYS

Figure 88. Aggregate and Interaction of Mitigation Pathways for Global Economic Sectors to Transition to Carbon Neutral Status During the Second Half of the Century and Limit Surface Warming to Well Below 2°C. Dashed arrows and lines indicated potential pathways with technical uncertainties or other commercial or societal acceptance barriers to implementation.

Chapter 6: THE PRICE OF CARBON

The nation that leads the clean energy economy will be the nation that leads the global economy.

Barack Obama, State of the Union address January 27, 2010.

The concept of global warming was created by and for the Chinese in order to make US manufacturing non-competitive.

Donald J. Trump, Twitter, November 6, 2012.

In economics, a "negative externality" or "spillover effect" is defined as a cost borne by a third party outside of a transaction. An example of a transaction with a negative externality would be the sales of chlorofluorocarbons, such as Freon, for use as refrigerants or aerosols. These compounds will accumulate in the stratosphere and, upon exposure to ultraviolet radiation, breakdown to release highly reactive chlorine. Chlorine, in turn, will react with ozone and disrupt the integrity of the ozone layer. Unrestricted passage of mutagenic ultraviolet radiation through a hole in the ozone layer will increase the incidence of cancer among terrestrial life forms. The health costs of unregulated sales of Freon were not part of transaction between the chemical manufacture and the customer, and this negative externality would be incurred by third parties exposed to carcinogenic UV radiation. As the science of chlorofluorocarbon reactions in the upper atmosphere advanced, the dangers of continued use of these compounds came into focus. In 1989, the Montreal protocol came into effect whereby world governments agreed to the implementation of an international treaty to phase out the use of chlorofluorocarbons.

An important role of government is to make full use of the expertise of the scientific community in assessing the risks and benefits of industrial and consumer practices. When appropriate, responsible governance requires the implementation of measures to protect citizens and ecosystems from the harmful effects of negative externalities from transactions that, in the absence of intervention, would continue unabated.

Excess emissions of GHGs over the next 20–30 years will result in a deferred negative externality in that the costs of surface warming will be largely incurred by subsequent generations during the second half the century.

The first 2 trillion tonnes of anthropogenic CO_2 released since the onset of the Industrial Revolution were low-cost emissions, resulting in a 1°C increase in average surface temperatures and relatively slight damage to the environment and human infrastructure. Going forward there remains less than 1 trillion tonnes of CO_2 in the budget of anthropogenic emissions under the Well Below 2°C Scenario. A significant over expenditure of this budget will result in the accumulation of atmospheric CO_2 at concentrations in excess of 500 ppm, and end-of-century surface warming will exceed 2°C above pre-industrial temperatures. The true cost of carbon emissions progressively increases as the planet warms, sea levels rises, oceans acidify and incidents and severity wildfires, floods, cyclones, droughts, heat waves, crop failures, famine, climate change migration, conflict and loss of life and livelihood increases. The high cost of GHG emissions must be recognized well in advance of potential catastrophic outcomes. Emissions must be priced to direct economies away from fossil fuel combustion and other emissions intensive practices. Effective carbon pricing mechanisms and other legislation to limit emissions will drive changes in industrial and consumer practices to more profitable, lower cost, low emissions options. Ultimately, a mixture of market-based mechanisms of carbon taxes and/or emissions trading systems, along with government policies, incentives, and specific restrictions on emissions, are required to adequately control atmospheric GHG concentrations.

6.1 Emissions Trading Systems

Emissions trading systems (ETSs) or cap and trade markets can be broad in scope covering emissions across multiple economic sectors or these systems can focus on emissions from defined sources within a nation or region. ETSs that apply specifically to emissions from heavy industry and the production of electricity tend to predominate although broad scope cap and trade systems are in place in California, and in the Canadian provinces of Quebec and Ontario. The advantage of an ETS is clarity and control over total annual emissions by defined sources within the jurisdiction. Total emissions are capped, and the cap progressively lowered with time. A limited number of emissions permits that total up to the cap are either issued for free or sold by auction to emitters and these permits are then bought and sold on the open market. The net effect is that the market functions to reduce emissions using least-cost pathways. The disadvantage of an ETS is the complexity and costs of monitoring, reporting and verifying emissions. A given system can be burdened by excessive administration and, for low level emitters; the costs per tonne of emissions can become unreasonable for the business. However, considerable experience in operating the European Union ETS has been gained since the program was launched in 2005. Well-designed cap and trade programs can incorporate learning from the EU ETS to optimize efficiencies and stream line the process to minimize costs and administrative complexity. As of this writing there are 18 emissions trading systems in force around the world. The size, percentage of total regional emissions covered, and the emissions reduction targets vary considerably between systems.

6.1.1 The European Union Emissions Trading System (EU ETS). The EU ETS came into effect in 2005, beginning with a 3-year pilot learning phase. Coverage was initially restricted to CO_2 released from power plants and energy-intensive industries, with permits issued for free. Emissions caps were set within each country of the EU.[334] Permits or allowances were then traded between companies operating within the spectrum of industries covered by the ETS on a free market basis. One unit of emissions allowance was defined as 1 metric tonne of CO_2 equivalents.[334] Compliance was defined by total emissions over the course of the reporting period relative to permits in-hand. Penalties of €40 per tonne of CO_2 emitted over in-hand permits were imposed on non-compliers.[334] All emissions permits expired with the conclusion of phase 1.

Implementation of phase 2 coincided with the onset of the first Kyoto reporting period beginning in 2008.[334] Iceland, Liechtenstein and Norway joined the ETS, and nitrous oxide emissions were added to increase GHG coverage.[334] Relative to phase 1, the total emissions cap was lowered by 6.5% and fines for non-compliance were increased to €100 per tonne.[334] Phase 2 allowed companies within the EU ETS to purchase international carbon credits up to a total of 1.4 Gt of CO_2eq.[334] International carbon credits could be sold by other Annex I countries that had ratified the Kyoto accord and were committed to binding mechanisms for specific emissions reductions. Based on the experience gained in the phase 1 pilot program, the process of monitoring and verifying emissions was considerably improved during the second phase of the EU ETS. The economic crisis of 2008 lead to a down-turn in industrial activity and a consequential drop in emissions from developed countries across the globe. This unexpected event led to a surplus of allowances and international carbon credits and a subsequent low price for carbon on the trading market. The effectiveness of the system as a mechanism to further reduce emissions was marginal under these conditions. However, by 2012, with the recovery of world economies, the system had rebounded and 7.9 billion allowances were traded at a total value of €56 billion.[334]

Phase 3 of the EU ETS began in 2013 with the onset of the second Kyoto reporting period and will extend to 2020.[335] The previous system of individual national emissions caps was replaced by a single EU-wide cap. Allowances were auctioned off as opposed to freely issued and perfluorocarbons from aluminum manufacturing were added to the GHGs covered by the ETS.[335] Coverage was expanded to include additional heavy industries and commercial aviation for flights within the EU and participating countries. Based on learnings from phase 1 and phase 2, some small emitters within economic sectors were excluded, provided that governments had implemented other measures or fiscal programs that resulted in equivalent curtailments. Over 11,000 power stations and energy intensive manufacturing installations are covered under phase 3, and approximately 45% of the total GHG emissions by EU and participant countries fall under the agreement.[335] The system is based on a 1.74% reduction in the cap for total emissions imposed on an annual basis.[335] Since the onset of phase 3, emissions from EU countries have declined by an additional 5% and are in line with achieving a 21% reduction by the year 2020 relative to emissions on record for the year 2005.[335]

Phase 4 of the EU ETS will begin in 2021, and the annual rate of emissions cap reduction will be increased to 2.2%.[336] The year 2030 objective is to achieve a 43% cut in the annual rate of GHG emissions covered by system relative to the 2005 baseline.[336] The European Commission has established a long-term mid-century objective of reducing total emissions by 85–90% with the EU ETS functioning as a core driver of the ongoing process of decarbonizing the economies of member states.[337]

The EU ETS is designed to facilitate the flow of funds and technology between companies operating within the emissions trading market and to international projects. Under the Kyoto Protocol, the Clean Development Mechanism (CDM) or Joint Implementation Mechanism (JIM) issues international carbon credits to specific projects that achieve real reductions in emissions. International carbon credits can be purchased by EU ETS participants to offset domestic emissions.

As the world's largest most experienced emissions permit trading system, the EU ETS can become a hub for an eventual international linking of carbon trading systems. Linking other compatible cap and trade systems with the EU ETS can potentially lead to an expanding region of capped emissions and international trading of permits.

6.1.2 Emissions Trading Systems in China. China has 7 large pilot scale ETSs operating in regions and cities in the country. These programs are designed as a learning phase to acquire the experience and expertise necessary to implement a nationwide ETS, which is scheduled to rollout in the second half of 2017. China's national ETS will be the largest emissions trading market in the world and will require the participation of companies operating within electricity and heat generation, petrochemicals, chemicals, iron and steel, non-ferrous metals, building production and materials, pulp and paper and aviation sectors.[338] The China national ETS will be twice the size of the EU ETS and cover about 33% of total national emissions.[338,339] Details as to the size of the emissions cap and annual cap reduction protocols have not been released. However, the fact that China has committed to establishing a national ETS is indicative of political will within the world's largest GHG emitter to bring into force carbon pricing mechanisms as required to effectively combat climate change.

6.1.3 Republic of Korea. In 2015, South Korea launched a national trading scheme covering emissions from 525 companies accounting for 68% of the country's total emissions.[339] Korea's ETS is an essential component of a climate action plan designed to achieve a year 2030 emissions reduction of 37% relative to a business as usual baseline.[339] This level of ambition is stated in the Nationally Determine Contribution submitted to the UN.

6.1.4 California, USA. In January of 2015, California expanded the scope of its ETS to cover 85% of total GHG emissions.[339] The program extends across all economic sectors including recycling, waste and agriculture and is integral to the objective of the state to achieve an 80% cut in emissions by mid-century relative to a year 1990 baseline.[339]

This level of ambition aligns with IPCC recommendations for advanced economies as required to limit future surface warming to less than 2°C.

6.1.5 Provinces of Quebec and Ontario, Canada. Along with California, Quebec and Ontario have the world's broadest ETSs covering 82–85% of total emissions and extending across all economic sectors.[339] These Canadian provinces have aligned future emissions abatement targets with those of the EU. Ontario has ambitions to cut emissions by 20% in the year 2020, 37% by 2030, and 80% by mid-century relative to the 1990 baseline rate.[339,340] The Quebec ETS is linked to the California system and the Ontario ETS is scheduled for linkage to the combined California-Quebec system in 2018.

6.1.6 Kazakhstan. Kazakhstan launched a broad spectrum ETS in 2013 that covers 50% of total emissions and is now in the third phase of implementation.[339]

6.1.7 US Regional Greenhouse Gas Initiative (RGGI). The states of Connecticut, Delaware, Maine, Maryland, Massachusetts, New Hampshire, New York, Rhode Island, and Vermont established a regional ETS that came into effect on January 1, 2014.[339] The ETS is focused on the production of electricity and covers 20% of the region's total emissions. The RGGI has established an ambitious year 2020 goal of achieving a 50% cut in emissions relative to emissions covered by the ETS that are on record for the year 2005.[339]

6.1.8 Australia. Australia's controversial ETS came into effect on July 1, 2016. The system covers 50% of the nation's total GHG emissions and includes 140 companies.[341] Emissions caps are relatively modest at the onset of the program and set to historical maximums between 2009–2010 and 2013–2014. However, the mechanism is now in place to progressively reduce the cap as part of any future national climate change action plan.

6.1.9 New Zealand. New Zealand has a broad spectrum ETS in force that includes all economic sectors and covers 50% of total emissions.[339] Relative to a 1990 baseline, New Zealand has established modest emissions reduction targets of 5% by 2020, 11% by 2030 and 50% by mid-century.[339]

6.1.10 Japan and Switzerland. Japan has two small regional ETSs in force that cover about 8% of overall emissions.[339] The Swiss ETS is limited in scope and covers 10% of total emissions.[339]

6.1.11 Other Emissions Trading Systems Scheduled for Rollout or Under Consideration. Mexico recently announced plans to implement a voluntary pilot scale emissions trading system with the intent of launching a full-scale ETS in 2018.[342] The Ukraine has plans to implement a national emissions trading system that would be compatible for linking to the EU ETS.[339] Within Canada, Manitoba has indicated intentions to roll out a provincial ETS designed for linkage to other North American trading systems.[339] With the 2017 rollout of the national ETS in China, a total of 6.8 Gt of GHGs or 16% of annual world emissions will be governed under cap and trade systems.[339]

As of this writing, Russia, Brazil, Chile, Japan, Mexico, Taiwan, Thailand, Turkey, Vietnam, and Washington State are considering the implementation of ETSs.[339]

6.1.12 Carbon Markets and the Paris Agreement. Article 6 of the Paris Agreement on Climate Change provides clarity and an apparatus for the development of international carbon trading networks and has been widely praised by climate change economists. In essence, article 6 provides the foundation for developing 2 pathways of international trading in emissions.

- Pathway 1. International trading between national and sub-nationally defined trading systems. Jurisdictions are free to design and implement their own carbon trading systems and to trade carbon credits internationally provided that international trading passes clear and robust accounting rules to be developed by the UNFCCC CMA*.[343,265]

- Pathway 2. The CMA shall create a platform to facilitate implementation of emissions reduction programs using Internationally Transferred Mitigation Outcomes (ITMOs). ITMO is a deliberately broad term that functions to include all foreseeable international agreements to facilitate emissions reduction programs. As written, ITMOs are not units of emissions but would include agreements between parties to trade in carbon offsets and other market and non-market emissions reduction mechanisms. Parties engaged in an exchange of ITMOs are required to follow the guidance of the CMA in applying robust monitoring, verification and accounting practices. Key to a successful outcome of the second pathway is a clearly stated provision that prevents double counting of emissions reductions. After completion of an international transaction designed to achieve a specific cut in emissions within the jurisdiction of a selling country, this country can no longer apply the emissions reduction toward its own carbon inventory.[343,265]

Systems of linked ETSs can potentially become overly complicated in situations where the level of ambition for emissions reduction varies between individual trading systems. A carbon credit issued within a jurisdiction with a high level of ambition will have a higher value than a carbon credit issued by a jurisdiction with a modest emissions cap. The price of carbon, then, will vary directly with the level of ambition or cap on future emissions within a given emissions trading system.

In the post-Paris Agreement world, countries and jurisdictions within countries are free to voluntarily implement carbon pricing mechanisms and other non-market drivers of emissions reductions as they choose. Market and non-market drivers will be part of national action plans to limit emissions in keeping with the objectives stated in each country's NDC. In the case of linked cap and trade systems, the CMA only becomes involved during international trading of emissions units and functions to provide the

---

* CMA refers to the post-Paris Agreement meetings of the Conference of Parties and acknowledges the transition from pre to post-Paris Agreement worlds.

guidance for development and implementation of systems to monitor, verify and report and thus certify the validity of the exchange. Parties are required to follow the guidance developed by the CMA. The incentive exists for jurisdictions to create emissions trading systems that are suitable for linkage to broader markets. Ideally, linkage between ETSs is based on compatibility of the systems. In North America, Quebec and Ontario designed their cap and trade systems to match the broad GHG coverage and emissions cap protocols of the California ETS. The end result will be compatible carbon pricing and a seamless exchange of carbon credits between companies in California, Quebec and Ontario. Given the blueprint for ETS design and linkage, other Canadian provinces, American states, and possibly Mexico may choose to create their own compatible ETSs and link these into an expanding North American carbon market.

The second pathway for international programs of emissions reductions is based on a central role of the CMA in providing guidance for ITMO agreements. Many developing nations filed an NDC with unconditional and conditional targets to control emissions. Conditional targets were based on a flow of funds to facilitate a higher level of ambition to control future emissions. Greenhouse gases are globally distributed and mixed in the atmosphere. It does not matter where or how emissions are reduced and least-cost pathways to reduce global emissions are based on executing programs with the lowest cost

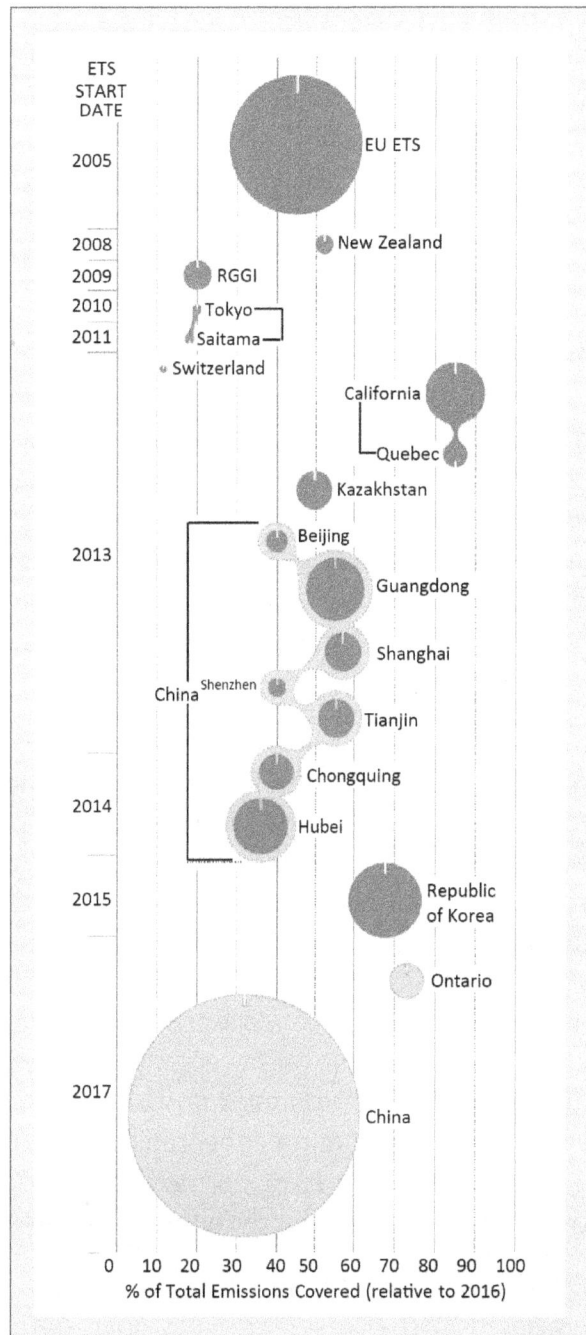

Figure 89. Emissions Trading Systems in Force (Dark Circles) and With Planned Future Rollout (Light Circles). The size of the circle indicates relative quantity of emissions covered. Notches indicate percentage of total emissions within the country or region covered by the ETS. Linked ETSs are indicated by dark lines. The Ontario ETS came into force in January of 2017 and will link with the California-Quebec ETS in 2018. Adopted from the International Carbon Action Partnership (ICAP) Status Report 2016.[339]

per unit of emissions reduction ahead of higher cost programs. As an example, the most cost-effective means to reduce global emissions for a wealthy nation or a large company may well be through participation in programs to reduce deforestation and forest degradation in a developing country. The seller of Internationally Transferred Mitigation Options would be motivated by the economic activity and flow of funds to support programs to reduce deforestation and degradation. The buyer of these ITMOs would receive credit for reducing emissions following the robust accounting rules developed by the CMA.

Over half of parties to the Paris Agreement indicate that they either plan to use international emissions reduction markets or are considering the use of markets as part of their action plan to meet emissions targets stated in their NDCs.[344] Most of these are developing countries on the supply side of the equation. These countries intend to sell ITMOs. Among advanced economies, the US and the EU indicate an intention to achieve emissions targets internally without the use of international carbon offsets.[344] Buyer nations would appear to be restricted to Canada, Japan, New Zealand, South Korea, Switzerland and Norway.[344]

The future of carbon trading and ITMOs in the post-Paris Agreement world is largely dependent upon political will and the specifics of climate change mitigation plans within national and sub-national jurisdictions. Setting a cap on emissions with a stated pathway toward overall future emissions reductions provides long-term clarity to industry. This clarity allows companies to develop business plans to change practices toward lower emissions options and to engage in national and international trading in emissions permits. As the economic reality of achieving targets aligned with the globally agreed upon objectives of the Paris Agreement comes into focus, the strategy of high GDP nations and companies will likely evolve toward an initial phase of implementing lowest cost solutions. Some of these options may be local and internal. However, lower cost international opportunities may foster a global expansion of linked emissions trading systems and may promote the trading of ITMOs. Ideally, article 6 of the Paris Agreement will facilitate the creation, harmonization and linkage of international carbon markets by establishing a robust monitoring system that is globally applied.

Emissions trading systems are well suited to achieving specific targets to curtail future emissions. Other market drivers, such as the direct application of carbon taxes, can change practices toward lower emissions options and contribute toward national and subnational climate change action programs.

6.2 Carbon Taxes

Generally, carbon taxes are based on the carbon content, and thus emissions potential, of a given fuel, and are applied as an indirect tax at a point of transaction. In theory, carbon taxes offer the advantage of simplicity of application and lower administration costs when compared with emissions trading systems. Carbon prices within an ETS will tend to fluctuate with economic downturns and booms and thus carbon taxes provide additional cost certainty for industry. The disadvantage of carbon taxation is the uncertainty as to the effect on total emissions from a given region or economic sector. Carbon taxes assume a consistent correlation between the carbon content of a given product and emissions. This assumption is accurate for the use of petroleum products in vehicles and for typical industrial and home heating applications but does not apply to the use of biomass as a fuel or to industrial processes that incorporate carbon capture and storage. Emissions taxes are less commonly used and are based upon monitoring, quantifying and taxing actual emissions from a power station or industrial installation.

A review of carbon taxes in force in national and subnational jurisdictions around the world reveals a wide dichotomy in coverage, pricing, and revenue distribution. Token taxation accomplishes little other than revenue generation. To be effective, the level of taxation must curtail consumption and direct industrial and consumer practices to lower cost, lower emissions options. The level of ambition of a given carbon tax program to achieve emissions reductions is based on the combination of coverage and tax applied per unit of CO_2 equivalents. Some governments use carbon taxes as an additional source of revenue for general accounts. However, carbon taxes can be designed as revenue neutral programs that are based on rebates or cuts in other taxes to balance carbon tax income. Alternatively, the funds generated from carbon taxes can be specifically used for climate change mitigation purposes.

Since 1990, 22 carbon tax programs have been initiated at national and subnational levels. Many of these programs are modest in ambition, based on GHG coverage and/or low levels of taxation per unit of emissions. The following list summarizes the more ambitious and significant carbon taxes in force around the world.

6.2.1 Finland. In 1990, Finland introduced the world's first carbon tax applied to the carbon content of fossil fuels. The tax has changed several times since inception but now sits at the equivalent of $66 USD per metric tonne of carbon dioxide equivalents ($/tCO_2$eq) for transportation fuels and $62 USD/tCO_2$eq for heating fuels.[345] The carbon tax in Finland is designed to be revenue neutral with increased government revenues balanced by reduced levels of other taxes. The primary purpose of the tax is to promote the use of lower emissions biomass and natural gas as heating fuels and to reduce coal consumption.

6.2.2 Sweden. The carbon tax in Sweden has evolved to become the highest in the world at the equivalent of $126 USD/tCO_2$. The tax is applied in full to consumer use of fossil fuel products for transportation and space heating.[345] Concerns over imposition of a competitive burden on industry led to the development of a two-tiered system whereby industry, agriculture, forestry and fisheries paid only 21% of the full carbon tax.[346] In addition, certain exemptions apply to installations that fall under the EU ETS. However, recent adjustments have reduced the gap between the full rate applied to consumers and the lower tax rates applied to industry. The Swedish government is considering alignment to a single carbon tax rate that would take effect in 2018.[347]

Sweden makes extensive use of district heating, and carbon taxes have been an important factor in fuel switching from coal and oil to biomass. Biomass fuel mostly consists of waste materials from forestry industries and is classified as carbon neutral. Fuel switching and improved energy efficiency across multiple economic sectors have led to a dramatic reduction in emissions since 1990. Sweden's per capita GHG emissions rate ranks among the lowest for developed nations and declined by 21.5% between 1990 and 2012.[124] When changes in land use and forestry are included in the calculation, 3.3 metric tonnes of CO_2 equivalents were released per person in Sweden during 2012.[124] This rate was 18% of the per capita GHG emissions rate for the US over the same period.[124]

In Sweden, the transition away from fossil fuels to lower emissions options has been driven primarily by ambitious carbon pricing mechanisms. The carbon tax goes directly to government revenue; however, energy taxes have been reduced since implementation. In 2015, Sweden ranked 11th among world nations in GDP/capita[213], and there is no evidence of negative impacts of high levels of carbon taxes on overall economic performance.[124]

6.2.3 Norway. Norway's long-standing carbon tax has evolved since inception to become a complex set of taxes that vary across economic sectors. Norway disperses revenues from carbon taxation to general government accounts. In 2015, approximately 55% of total CO_2 emissions were covered by taxes ranging from $3–52 USD/$tCO_2$eq dependent upon fuel type and usage.[345,348] The highest rates of taxation are applied to petroleum production and natural gas industries. Norway has an ETS that is linked to the EU ETS. Lower rates of carbon taxes or exemptions are in place to protect specified domestic industries and to reduce the burden of carbon pricing for industries covered by the emissions trading system. Overall, Norway has some of the highest rates of carbon taxation in the world; however, a system of variable taxation rates and multiple exemptions is less than ideal. The Norwegian government is considering reforming the system to a single taxation rate of $51 USD/$tCO_2$eq that would be applied across all economic sectors not covered by the ETS.[345]

6.2.4 Denmark. As of 2014, Denmark has applied a carbon tax equivalent to $25 USD/$tCO_2$eq for consumption of fossil fuels across all economic sectors with provisions for exemptions and refunds for some industries covered by the EU ETS.[348] Electricity production is taxed separately dependent upon fuel type with no tax applied to power production from low and zero-emissions renewables. Forty percent of the revenues from carbon taxation are used for environmental subsidies and 60% are returned to industry.[349]

6.2.5 Switzerland. In January of 2016, Switzerland increased the nation's carbon tax from the equivalent of $60 to $84 USD/t$CO_2$eq.[350] The tax is applied to all fossil fuels used for heating, industrial processes and transportation. Companies that partake in the Swiss ETS are exempt from the carbon tax. The Swiss carbon tax is revenue neutral with carbon tax revenues distributed to industry and the general public.

6.2.6 France. France introduced a national tax on domestic consumption of energy products not covered by the EU ETS. The tax is based on the carbon content of the energy source and was increased to €22/tCO_2eq in 2016 and extended to cover transportation fuels and heating oils.[348] On July 22, 2015, the government of France passed the Energy Transition for Green Growth Act that will increase the national carbon tax to €56/tonne in 2020 and €100 per tonne by 2030.[351] Progressive carbon taxation is a key component of market drivers designed to decarbonize the nation's economy by mid-century.

6.2.7 Ireland. In 2010, Ireland initiated a carbon tax at €15/tCO_2eq covering the use of liquid fossil fuels by homes, offices, vehicles and farms.[352] Coal and peat was added in 2013 and the tax was increased to €20/tCO_2eq.[352] Industrial sectors covered by the EU ETS are excluded from the carbon tax. Additional emissions taxes are applied to vehicle purchases and yearly registration with the level of taxation dependent upon the vehicle's emissions.[352] Revenues from carbon taxation are balanced by income tax cuts in Ireland.[352]

6.2.8 British Columbia and Alberta, Canada. In 2008, British Columbia enacted a broad spectrum, revenue neutral, carbon tax applied to the consumption of fossil fuels used for transportation, home heating, and the production of electricity.[352] Government revenue from carbon taxes are offset by other tax reductions.[352] The tax was expanded to include biodiesel fuels and increased to $30 Can/t$CO_2$eq in 2012.[352]

Recently, the government of Alberta initiated a $20 Can/t$CO_2$eq carbon tax applied to the purchase of fossil fuel products.[353] The rate of taxation will increase to $30 Can/t$CO_2$eq as of January 1, 2018.[353]

6.2.9 Slovenia. Slovenia implemented a carbon tax of $18 USD/t$CO_2$eq on April 1, 2016.[345] The tax applies to all fossil fuels.

6.3 Carbon Floor Pricing

A carbon floor or minimum price is a type of carbon tax that comes into play within an emissions trading system to prevent carbon prices from falling below a defined level. Fluctuating carbon prices can dissuade industry from investment in low emissions options. Implementing a carbon floor within an ETS provides additional certainty to the market as to the minimum price of carbon. The UK has established a carbon floor price equivalent to $15.75(USD)/tCO$_2$eq for carbon fuels used to produce electricity.[348] The government of France has proposed implementation of an aggressive national carbon floor price of €30/tCO$_2$eq that would apply to carbon fuels used to produce electricity.[354] This national initiative follows an earlier proposal by France to the European Commission to implement an EU wide carbon floor price that would function within the EU ETS.[354]

6.4 Carbon Leakage

Carbon leakage is defined as the process whereby emissions from a jurisdiction covered by effective programs of carbon pricing are transferred to regions without carbon pricing mechanisms or with relatively low ambition climate change mitigation policies. In theory, the process takes place when regions differ in the price of carbon, resulting in an economic driver to transfer emissions to lower cost markets. Potentially, carbon leakage can occur following physical relocation of an emissions intensive industry. Alternatively, the demand for carbon intensive fuels and commodities may decline in countries or regions with ambitious carbon pricing mechanisms, resulting in increased supply and lower costs in low ambition countries or regions. The increased use of carbon intensive fuels and commodities in countries without effective carbon pricing mechanisms can negate emissions reductions achieved in high ambition jurisdictions.

The Kyoto protocol divided the world into parties with binding agreements to limit emissions and parties without binding commitments. The lack of universal acceptance of emissions reduction protocols created the potential for carbon leakage during the Kyoto reporting periods extending to 2020. Economic models rated the potential for emissions relocation for various industries operating within the EU ETS. The most at-risk industries were characterized by high production volumes coupled with high intensities of emissions. In response, The European Commission created a carbon leakage list that included manufacturers of cement, iron and steel, aluminum, paper and paperboard, and refined products as the most significant sources of emissions exposed to risk of carbon leakage.[355] In phase 3 of the EU ETS (2013–2020) these industries were issued a higher share of free emissions allowance compared to other industries.[356] Within the California ETS, sectors with a high risk of carbon leakage will receive 100% of their emissions allocations for free up until at least 2020.[357] In countries that make use of carbon taxes, emissions intensive industries vulnerable to carbon leakage are often exempt from carbon taxes or taxed at a lower level in comparison to taxes applied to the general population for consumption of fossil fuels.

Since the inception of the EU ETS and implementation of other ETSs and national programs of carbon taxes, there is little evidence of asymmetrical carbon pricing policies, resulting in actual examples of carbon leakage. As discussed, the EU is on track to exceed Kyoto targets of a 20% cut in emissions by 2020. The success in achieving significant emissions reductions has occurred within the EU without economic consequences. Provisions within carbon pricing mechanisms designed to offset the incentive for international transfer of industrial operations are often described by critics as free handouts to industry. Policies to limit carbon leakage are far from perfect; however, these policies may have contributed to retaining industry within the EU and other Kyoto signatories with ambitious targets to limit emissions.

The Paris Agreement should, in theory, undermine and largely eliminate the significant asymmetry in climate change mitigation ambitions between nations and regions that is required to drive processes of carbon leakage. All nations have submitted NDCs with clearly stated future limits to emissions. While the mechanisms to limit emissions and details of national climate action plans are uncertain and will vary between countries, havens for high emissions industries and practices are unlikely to emerge. Individual nations will be unable to achieve emissions as targets stated in their NDCs if the economic and environmental policies of these nations facilitates international relocation and concentration of heavy emitting industries to their soils. As the reality of the post-Paris Agreement world begins to focus, the need for carbon leakage provisions in national climate change action plans should diminish. Beyond 2020, sufficient climate change mitigation symmetry should exist between countries and regions to facilitate the implementation of broader, simpler and more effective carbon pricing mechanisms that do not include special treatment and exemptions for heavy emitting industries. This conclusion assumes that all countries honour their commitment to the principles of the Paris Agreement.

By 2017, 40% of the nations on Earth will have put into action some mechanism of carbon pricing either as direct taxes or through the implementation of an emissions trading system.[345] By the end of 2017, carbon pricing mechanisms will cover 23% of global GHG emissions.[345] These mechanisms range from low ambition largely ineffectual carbon taxes to ambitious well-designed programs to achieve emissions reductions aligned with regional obligations to meet the objectives of the Paris Agreement. Overall, the current degree of global coverage of emissions by carbon pricing mechanisms is inadequate and will not provide the necessary economic drivers to sufficiently decarbonize world economies. However, in December of 2015, the nations of Earth reached an agreement to limit future global warming to less than 2°C and ideally 1.5°C and put into place the foundations for building national and international protocols to accomplish this goal. The world of Kyoto and non-Kyoto nations no longer exists. Logically, the Paris Agreement marks the beginning of a new era of symmetry of purpose to implement effective national climate change action plans. These plans must include carbon pricing mechanisms as economic drivers to limit emissions. A successful outcome of the Paris Agreement will see an expansion and linking of harmonized emissions trading systems, implementation of effective broad-spectrum carbon taxes at national and subnational levels, and the international transfer of mitigation options to facilitate emissions reduction programs and GDP growth in developing countries that is uncoupled from GHG emissions.

Carbon pricing mechanisms are essential components of programs to mitigate climate change. However, well designed national action plans incorporate a mixture of market drivers and non-market incentives and legislation to redirect practices within economic sectors toward lower and zero-emissions options.

6.5 Climate Funds

Article 4.3 of the 1992 United Nations Framework Convention on Climate Change consists of a single paragraph committing developed countries to meet the full incremental costs incurred by developing countries in combating climate change.[252] The agreement acknowledged that developing countries were not significant contributors to GHG emissions and had limited capacity to implement climate change mitigation programs. Essentially, all countries of the world agreed to the "Polluter Pays" principle whereby advanced, emissions-intensive, wealthy countries would cover the additional costs incurred by developing and emerging countries in following low-emissions climate-resilient pathways of economic growth.

The UNFCCC classifies developed countries as OECD industrialized nations plus several other economies in transition. All other countries are termed to be "developing," and this includes some middle-income countries. Brazil, China, Mexico and South Africa are notable middle GDP per capita nations considered to be developing countries by the UNFCCC.

The concept of climate funds is intended to capture all value transfer to support climate change mitigation, adaptation, and forest preservation and enhancement in developing countries. Funding and technology transfer in support of "clean coal" projects is not included in the UNFCCC assessment of eligible financial flows to climate funds.

The architecture of global climate financing is a complex web of agencies, banks, funds and initiatives. Private and public funds from advanced economies flow through implementing agencies to UNFCCC and non-UNFCCC financial mechanisms. These mechanisms consist of market-based trading of ITMOs, which can include international carbon credits, and an extensive series of climate funds and programs. Ultimately, funding is used to finance low emissions projects in developing countries.

In 2009, advanced economies committed to jointly mobilize $100 billion per year by 2020 to be used for climate change mitigation, adaptation, and programs to reduce emissions from deforestation in developing countries.[358]

The bulk of international climate change funding is allocated for mitigation programs. The Clean Technology Fund currently receives 71% of total mitigation funding with the remaining dollars flowing to the Green Energy Fund, the Scaling Up Renewable Energy Program, and the Global Energy and Renewable Energy Fund.[359] Funds are used to support a full spectrum of low emissions projects and generally to assist developing countries in the planning and building of low carbon economies.

As will be discussed, in general, the level of ambition to control future emissions, as evident in the NDCs and climate action plans and policies of developing countries, is such that future emissions are projected to be well below population based concepts of fair share contributions to limit future surface warming to less than 2°C. Access to climate funds by developing countries will be required to maximize the potential for economic growth that is uncoupled from GHG emissions.

6.6 National Climate Change Action Plans

When the nations of the EU are considered as single region, the EU plus the top 8 GHG emitting countries outside of the EU accounted for 67% of world GHG emissions in 2012.[124] The actions of these countries and the EU over the next 33 years will largely define the extent of global warming and thus the impact of climate change on ecosystems, human infrastructure, and the health and well-being of the global population. While the responsibility to curtail emissions is universal, the top emitters have the greatest share and must lead by example. Further, the cooperation required to create emissions trading zones and to facilitate the transfer of climate funds and ITMOs to smaller developing nations must come from the largest economies on Earth. This leadership will be essential if smaller countries and developing nations are to uncouple GDP growth from GHG emissions and to partake in the process of decarbonizing world economies.

As discussed in chapter 5, the UNEP concluded that global GHG emissions in the year 2030 should not exceed 42 Gt of CO_2 equivalents if future surface warming is to be held to less than 2°C.[122] Based on population projections, the year 2030 global ceiling on per capita emissions equates to 4.8 metric tonnes of CO_2 equivalents per person. Estimates of year 2030 populations provide a simple method for calculating a fair share emissions ceiling for individual countries and regions. NDC and policy-based projections for excess emissions above the fair share ceiling constitute a positive emissions gap. A negative emissions gap exists when projections are below the fair share ceiling for a country or region. The global net emissions gap equates to the sum of all positive emissions gaps less the sum of all negative emissions gaps. Year 2030 fair share emissions calculations are based on a 2°C surface warming limit and should be considered as a minimum acceptable level of ambition. Further cuts of about 7% are likely to be required to achieve the more ambitious target of limiting surface warming to 1.5°C.[122]

6.6.1 China. The NDC submitted by China targets a 60–65% reduction in the intensity of emissions per unit of GDP by 2030 relative to the emissions intensity on record for the year 2005.[254] China's NDC indicates an intention to reach peak emissions by 2030.[254] Based on projections for economic growth, achieving the NDC target specified by China would result in a 34% increase in emissions between 2012 and 2030.[254] A 3.7 Gt/year increase in emissions from China over the next 13 years cannot, realistically, be balanced by emissions cuts from other nations. China's projected 2030 emissions gap as calculated using the current NDC ceiling is a staggering 7.6 Gt/year.[254] This gap equals 63% of the net world emissions gap.[122] Based on NDC projections, the world's largest emitter would be responsible for an estimated 27% of global emissions in 2030.

In 2012, China's GDP per capita ranked well below world average, and the argument has been made that China is a developing nation and further economic development will be hindered by aggressive cuts to total emissions. This position lacks credibility when emissions are considered on a per capita basis. The per capita rate of emissions based on China's NDC target is twice the global fair share per capita emissions ceiling and thus China will have markedly exceeded any concept of a reasonable allowance for emissions to facilitate economic growth.

Clearly, China must increase the level of ambition to control future

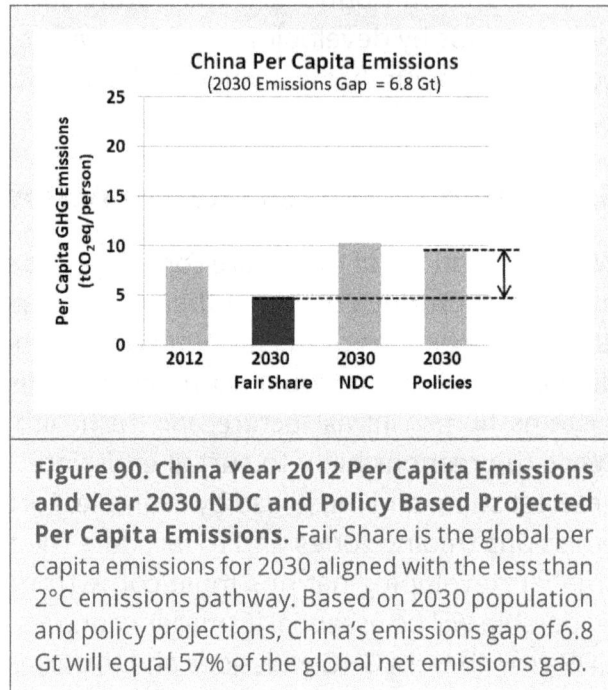

Figure 90. China Year 2012 Per Capita Emissions and Year 2030 NDC and Policy Based Projected Per Capita Emissions. Fair Share is the global per capita emissions for 2030 aligned with the less than 2°C emissions pathway. Based on 2030 population and policy projections, China's emissions gap of 6.8 Gt will equal 57% of the global net emissions gap.

emissions well beyond the current NDC target. However, a detailed assessment of progress within China to design and implement national climate change policies is consistent with China reaching peak emissions around 2025, followed by annual declines to arrive at a year 2030 emissions rate of about 13.6 Gt/year.[215] China would appear to be on a path to exceed the target indicated in the NDC, and the actual emissions gap would be cut to 6.8 Gt. The 2017 rollout of what will be the world's largest ETS is the cornerstone of China's climate action plan. China's willingness to implement carbon pricing mechanisms indicates a commitment toward transitioning to lower emissions practices. China's 13th five-year plan was released in 2016 and extends to 2020. The plan sets an upper limit on energy consumption with a shift in economic growth to service industries, along with a push toward innovation and improved efficiencies in manufacturing.[360] Details are sparse, but the plan calls for implementation of emissions reporting and verification systems for key industries, establishing a green finance system, and promotion of low-carbon industries and low emissions transportation. Commercial logging in natural forests will be banned and forested areas are to be expanded. Under the 5-year plan, non-fossil fuels use is to progressively increase to make up 15% of total energy consumption by 2020.[360] The plan indicates that China intends to be actively involved in global efforts to mitigate climate change and will continue to advance its own national contributions to curtail emissions.

Based on the combination of the NDC submitted by China, along with strong evidence for the development of climate change policies, CAT gives China a rating of "medium with inadequate carbon intensity targets."[215] This rating indicates that China's efforts are not consistent with limiting global warming to below 2°C unless other countries achieve much deeper levels of ambition to compensate for China's projected emissions. China's efforts to reduce emissions are, at present, not in line with the concept of a fair share contribution; however, recent indicators from the world's largest GHG emitter are consistent with an acknowledgement of responsibility and a willingness to implement carbon pricing mechanisms and other ambitious climate change mitigation policies. Given the magnitude of emissions from China, the level of ambition and success of these measures will be critical defining factors in global efforts to achieve the core objectives of the Paris Agreement.

6.6.2 United States. In 2012, the US emitted 5.8 Gt of CO_2eq and accounted for 12.2% of world GHG emissions.[124] The US is among the most intense emitters of GHGs with a per capita emissions rate that is double that of the European Union, and 2.7-fold greater than the world average.[124] Under the obligations of the Paris Agreement, the US has considerable responsibility to dramatically reduce emissions over the next 30 years.

In June 2013, then President Obama directed the Environmental Protection Agency (EPA) to develop an action plan to reduce emissions from power facilities across the US.[361] The Clean Power Plan (CPP) was announced in August 2015 and was designed to achieve a nation-wide 32% reduction in carbon emissions from the production of electricity by 2030 relative to 2005 base line levels.[361] The CPP was the centerpiece of the Obama Administration's climate change agenda. Under the CPP, states were required to submit plans to reduce emissions from power stations, and these plans were to comply with EPA standards. Failure to submit would result in the imposition of plans developed by the EPA. Potentially, the CPP could have fostered an expansion of the Regional Greenhouse Gas Initiative. By linking to the RGGI ETS, other states could have followed market-based least-cost pathways to satisfy the requirements of the CPP.[362] However, 24 states, along with support from the coal industry and some power companies, filed lawsuits challenging the legal authority of the EPA to impose the CPP.[363] In February 2016, the Supreme Court ordered the EPA to halt enforcement of the plan, pending decisions by lower courts.[361]

Other significant initiatives contained in President Obama's Climate Action Plan included an increase in permits issued for clean energy power production and further investment in clean energy research and development.[364] Fuel economy standards for heavy-duty vehicles were introduced for the first time, and national standards for passenger vehicles were to be toughened. New minimum energy efficiency standards extending to 2030 were introduced for household appliances, along with initiatives to improve the energy efficiency of buildings. The plan also described initiatives to reduce methane and hydrofluorocarbon emissions and to conserve forests.

On January 20, 2017, Donald J. Trump was inaugurated as the 45th president of the United States. Mr. Trump had previously declared climate change to be a hoax and had stated that the EPA was engaged in "totalitarian tactics." On June 1, President Trump followed up on a campaign promise to withdraw from the Paris Agreement.[365] The US joined Syria and Nicaragua as the only non-participant countries in what would otherwise be a global agreement to combat climate change.[365]

Legally, US withdrawal from the Paris Agreement would require a full four years.[366] Possibly, the US could accelerate this process through an executive order from the president to cancel the original UNFCCC agreement. By cancelling the original agreement, the US would invalidate ratification of the Paris Agreement.[366] The Paris Agreement does not contain provisions for punitive action against developed countries that fail to achieve emissions reduction targets or to meet obligations to contribute to international climate funds. Conceivably, the US could bypass formal proceedings to withdraw from the agreement by ignoring obligations contained in the NDC.[366] Mr. Trump's simple declaration of withdrawal implies that the current administration has no intention of following through with the implementation of polices designed to reduce emissions as specified under the US NDC.

The Trump administration has shelved the Obama Climate Action Plan, including the Clean Power Plan. At the federal level, the US has withdrawn from international obligations to curtail future emissions. However, policies at the highest level of government do not equate to complete inaction on the part of the US to combat climate change. President Obama's Climate Action Plan was based on legislation and non-market policy initiatives, and the cost-effectiveness and capacity of this approach to progressively decarbonize economic sectors is questionable. To a limited extent, ambitious climate action plans, including carbon pricing mechanisms, have been implemented in the US at the sub-national level. These programs provide some indication of the potential for the US to curb emissions and contribute to the global effort on a regional and state by state basis.

On September 27, 2006, then governor of California Arnold Schwarzenegger signed into law the Global Warming Solutions Act of 2006.[367] The act established future targets for the state to reduce GHG emissions to 1990 levels by 2020. The bill included all greenhouse gases as defined in the Kyoto Protocol and gave authority to implement the measure to the California Air Resource Board. The Board implemented mandatory reporting and verification of GHG emissions from the largest emitters. On January 1, 2013, the California ETS came into force as part of compliance to the Global Warming Solutions Act.[357] Cap emissions are absolute and progressively decline each year. In the current form, the ETS covers 85% of total GHG emissions across all economic sectors with emissions caps defined until 2020. The California ETS is the cornerstone of the state's climate action plan and is by far the most comprehensive market-based carbon pricing system currently in force. In April 2015, Governor Jerry Brown issued an executive order to extend the level of ambition of California's climate change law.[368]

The revised objective is to achieve a 40% reduction in GHG emissions by 2030 and an 80% reduction by 2050 relative to 1990 baseline levels. This level of ambition can be summarized as "40 in 30; 80 in 50" and matches EU targets. California's climate strategy[369] is among the most ambitious on Earth. Specific year 2030 objectives within the overall targets to curtail emissions can be summarized as follows:

- Increase electricity production from renewable sources to 50%.
- Reduce petroleum use in vehicles by 50%.
- Double the energy efficiency of existing buildings. As of 2020, all new homes are to be constructed to net zero-emissions standards.
- Reduce GHG emissions from agriculture, natural and working lands through the introduction of effective land management and conservation practices.
- Reduce emissions of short-lived climate pollutants (black carbon, fluorinated gases, and methane).
- Safeguard the state from risks of climate change.

California's annual emissions rate is second to that of Texas and, in 2014, accounted for 6.6% of total CO_2 released by the US to the atmosphere.[370] However, as the union's most populous state, California's intensity of emissions on a per capita basis was about half the national rate and 35% of the per capita emissions rate of Texas.[370] Assuming that carbon pricing and other climate change mitigation measures are maintained, California's future emissions will progressively decline at a rate that approximates the region's obligations to meet the objectives of the Paris Agreement.

Texas is by far the largest emitter of GHGs in the US, and, in 2014, accounted for 13% of total CO_2 emissions. The intensity of emissions from Texas on a per capita basis is well above the national average.[370] Texas is highly vulnerable to climate change-dependent increased frequencies of heat waves, floods, wild fires, and the damaging effects of storms and sea level rise. However, Texas lacks a climate change action plan and was part of the coalition to file suit against the EPA over the CPP.

There is no evidence of consistency between states in comparing environmental policies and positions on climate change. Eighteen states accounting for 43% of total emissions essentially have no climate change action plans, and many of these states are actively opposed to federal imposition of measures to reduce GHG emissions.[371] Outside of the California ETS and the RGGI ETS, there are no other carbon pricing mechanisms in force in the US and climate action plans, where they exist, are based on legislation and non-market emissions reduction policies. Many states have compiled extensive reports containing detailed policy recommendations to progressively reduce emissions. The degree of implementation of policy recommendations varies between states. No state other than California has adopted a comprehensive climate change action plan. New York State participates in the RGGI and has an executive order in place that mandates the development of policies to achieve an 80% emissions reduction by 2050.[372] However, these policies have yet to be fully articulated and measures that are currently in force will not achieve such an ambitious cut in emissions by mid-century. Similar situations of disconnect between policy implementation and policy recommendations are found in other states. In 2008, Florida developed a detailed set of 50 policy recommendations designed to achieve a 34% reduction in emissions by 2025 relative to 1990 baseline emissions.[373] These recommendations included establishing an economy wide emissions trading system that could be linked to other projected ETSs. The proposed Florida ETS and other aggressive climate change policy recommendations have yet to come into practice.

At the municipal level, there are examples of ambitious climate change action plans. New York City is the largest metropolis on Earth to commit to an 80% reduction in GHG emissions by 2050.[329] NYC released a long-range plan to cut emissions primarily through retrofitting of existing buildings to high efficiency standards. The 2014 US Conference of Mayors produced the US Mayors Climate Protection Agreement.[374] The agreement urges Congress to enact a range of climate change mitigation policies and programs and to adopt ambitious post 2020 emissions reduction targets at the national level. The agreement provides an extensive list of local actions that can be implemented at the municipal level to increase energy efficiency in buildings, to promote efficient low emissions modes of transportation, and to reduce waste and increase recycling.

On June 1, 2017, hours after the announcement by President Trump of his intention to withdraw from the Paris Agreement, the governors of California, New York, and Washington jointly announced the formation of the US Climate Alliance.[375] The alliance is committed to achieving the original goal for the US, as specified in the NDC submitted under the Paris Agreement, to reduce year 2025 GHG emissions by 26-28% relative to emissions on record for 2005.[376] Since inception, membership of the US Climate Alliance has grown to include 12 states and Puerto Rico. These states and territories represent over 36% of the US GDP and 18% of total GHG emissions. The governors of another 10 states and the District of Columbia, accounting for another 22% of US emissions, have pledged support for the Paris Agreement but have not joined the US Climate Alliance.[376] In addition, over 200 US cities, as members of the Mayors National Climate Action Agenda, are committed to the Paris Agreement.[376] The 10 most populous cities in the country have indicated support for the US Climate Alliance.[376]

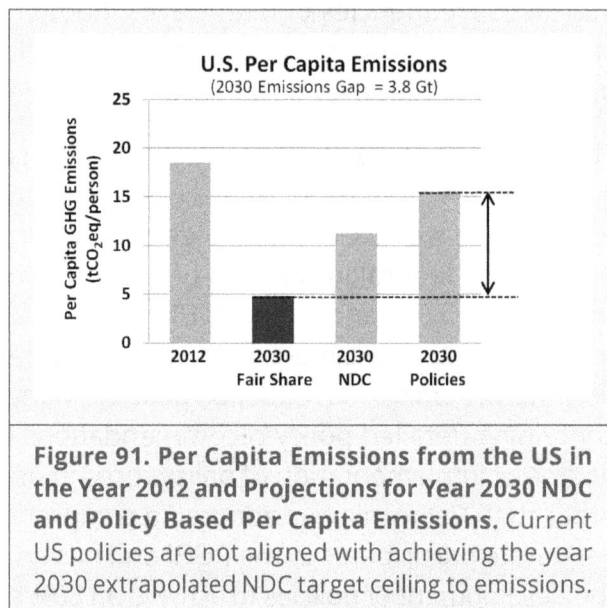

Figure 91. Per Capita Emissions from the US in the Year 2012 and Projections for Year 2030 NDC and Policy Based Per Capita Emissions. Current US policies are not aligned with achieving the year 2030 extrapolated NDC target ceiling to emissions.

The level of ambition specified in the US NDC is insufficient as a fair share contribution to adequately curtail global emissions. Assuming a linear extrapolation of the NDC target to the year 2030, emissions from the United States would be 2.4-fold in excess of the fair share ceiling.[254] With cancellation of the CPP and the rollback of other climate action polices at the federal level, the 2030 emissions gap extents to 3.8 Gt or 3.2 fold in excess of the fair share ceiling. This emissions gap is estimated on the basis of climate change policies currently in place at all levels of government.[215] The combined emissions gaps for the US and China equal 88% of the net global emissions gap.[122]

Recently, Climate Action Tracker downgraded the US rating on climate action from "medium" to the lowest category of "inadequate,"[215] reflecting the rollback of previous federal initiatives to curtail emissions and the Trump administration's decision to withdraw from the Paris Agreement. If other countries were to follow America's example, end-of-century average surface temperature would increase by at least 3–4°C. The actions and inactions of the current US president and Congress on climate change are detrimental to global efforts to limit future surface warming. If year 2030 emissions from the US were to approximate estimates based on current policies, it is unlikely that compensatory cuts by other countries could balance this massive excess in emissions. Potentially, a change in government could re-establish a meaningful climate change agenda at the federal level. Alternatively, the actions and influence of the US Climate Alliance could compensate for inaction at the federal level and drive down national emissions through regional and state initiatives. Going forward to mid-century, the path taken by the US to cut emissions and decarbonize economic sectors will be a critical component of global efforts to minimize surface warming.

6.6.3 The European Union.

Between 1990 and 2012, when land use and forestry were included in the calculation, the total emissions from the 28 member states of the European Union declined by nearly 20%, and the region will easily exceed year 2020 Kyoto targets for emissions reduction.[124,215] In 2012, the EU accounted for 8.8% of world GHG emissions at a per capita rate that was 21% above the world average.[124] The NDC submitted by the EU targets a 40% reduction in GHG emissions by 2030 relative to a 1990 baseline.[254] A net-net approach

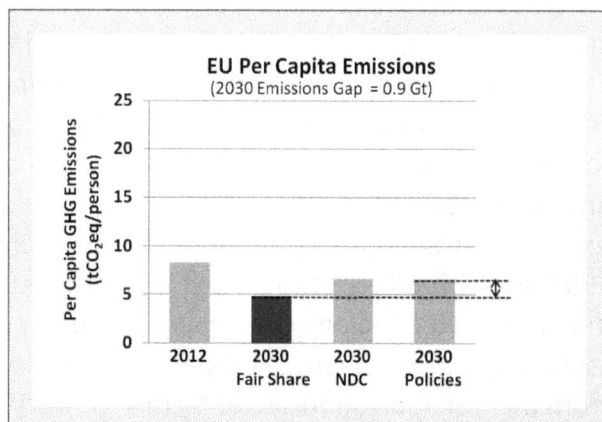

Figure 92. EU Year 2012 Emissions and Year 2030 NDC and Policy Based Projections. Based on EU policies this group of advanced economies will be a minor contributor to the global emissions gap in 2030.

whereby net emissions or withdrawal from forestry and land use are used to calculate baseline and future targets is assumed although the EU's NDC is not completely clear as to the inclusion of the forestry and land use sector in these calculations.

In 2009, the EU Heads of State formally adopted an objective to reduce total GHG emissions by 80–95% by 2050.[377] Subsequently, the EU developed a low carbon roadmap with interim targets to achieve a 40% cut in emissions by 2030, and a 60% curtailment by 2040, relative to a year 1990 baseline.[378]

The level ambition in the EU low carbon roadmap follows IPCC guidelines for advanced economies as required to limit future surface warming to less than 2°C above pre-industrial temperatures. Emissions reductions are to be accomplished internally without reliance upon international credits. Progressive decarbonization will occur across all economic sectors. By mid-century, zero and very low emissions technologies are projected to account for nearly 100% of power

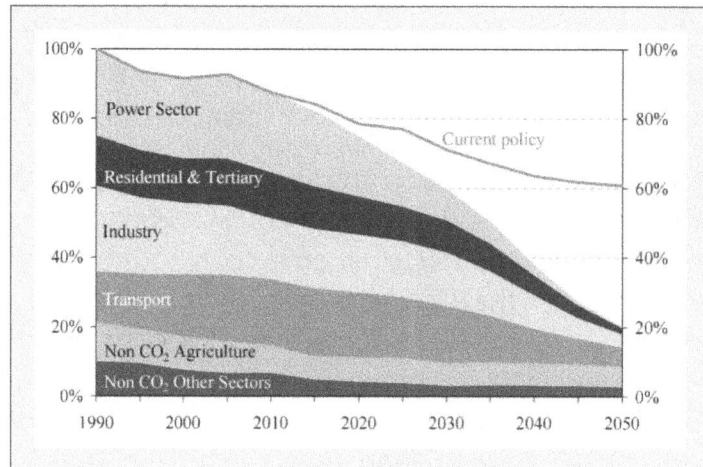

Figure 93. Emissions Reduction Pathways within EU Economic Sectors. Projections are based on the implementation of policies contained in the EU Roadmap for Moving to a Competitive Low Carbon Economy by 2050. The red line indicates projections based on 2011 EU policies.[378]

production. Fuel switching to electricity in the transportation, industry and buildings sectors will increase demand; however, this will be largely offset by increased energy efficiencies across economic sectors. A 90% emissions reduction from the buildings sector by mid-century would follow an update of building codes to Passive House standards, high efficiency retro-fits to existing buildings, and fuel switching to electricity and renewables for building heating, cooling and cooking. Industry is targeted for an 83–87% reduction in emissions based on improvements in energy efficiencies and increased use of electricity and renewables in place of fossil fuels. Non-CO_2 GHG emissions from industry are projected to decrease by 80% by mid-century relative to 1990 levels. Total emissions from industry could be further reduced if carbon capture and storage technology can be applied to in-process CO_2 release from cement manufacturing and other processes. The roadmap targets a 60% reduction in emissions from the transportation sector by mid-century. Improved fuel efficiency and light-duty vehicle fleet changeover to plug-in electric hybrid, fully electric, and fuel cell automobiles will dramatically reduce emissions. Increased use of low life cycle emissions biofuels in aviation and heavy-duty road vehicles will contribute to emissions reduction from the transportation sector. Emissions from agriculture, forestry and other land use are targeted for a 42–49% reduction through improved efficiencies of food production, waste reduction and increased forested areas. Requirements for food production will limit the potential for emissions reduction from the agriculture sector. By 2050, non-CO_2 emissions from agriculture consisting primarily of methane release are projected to become the single largest source of residual GHG emissions.

During the current Kyoto reporting period from 2013 to 2020, the EU set binding annual GHG emissions targets for each of the member states.[379] These targets applied to GHGs not covered by the EU ETS. The Effort Sharing Decision established annual emissions allocations based on the GDP per capita of each country. Wealthier nations were required to achieve a higher ambition to limit non-ETS covered emissions. Year 2020 targets ranged from a 20% decrease in emissions for Germany to an allowance for a 20% increase in emissions for Bulgaria. The system has proven to be effective and when combined with the EU ETS provides full coverage of emissions.

Going forward from 2021 to 2030, emissions reduction in the EU will be driven by the same combination of the EU ETS and binding national emissions allocations as used during the second Kyoto reporting period. However, the EU ETS will be strengthened to achieve a 43% reduction in emissions from power production and heavy industry by the year 2030 relative to baseline emissions on record for 2005.[380] Year 2030 emissions not covered by the EU ETS are to be reduced by 30% relative to 2005 using binding national annual emissions allowances.[379] In combination, the overall objective of these programs is to cover all GHG emissions and to achieve the EU NDC target of a 40% reduction in total emissions by 2030 relative to the 1990 baseline.[381]

The EU ETS is the main economic driver for decarbonization of electricity production and heavy industry. In addition to the cap and trade system, the EU has established year 2030 binding targets to increase the share of renewables in energy consumption by at least 27%, and to decrease energy use by at least 27% through improved effi-ciencies.[381] The Energy Performance of Buildings Directives specifies that, as of 2021, all new buildings are to be constructed to near zero-emissions standards.[382] In the transportation sector, EU wide CO_2 emissions standards are in place for cars and vans and these standards will toughen in 2021 to achieve a 40% reduction in emissions for new light-duty vehicles relative to emissions from the 2007 fleet.[383]

Germany is the largest economy in the EU and, in 2012, Germany accounted for 19% of the total EU GHG emissions.[124] Between 1990 and 2012, Germany reduced emissions by 27% when changes in land use and forestry were factored into the calculation.[124] However, on a per capita basis, Germany's emissions were 20% greater than that of the entirety of the EU.[124] In 2012, 40% of the total emissions from Germany came from the production of electricity and heat.[124] In recent years, Germany has been a leading nation in the adoption of wind and solar options for power production and in the implementation of high efficiency practices to reduce energy consumption across economic sectors. By 2020, an estimated 40% of the nation's electricity will come from renewable zero-emissions sources consisting primarily of wind and solar.[384] However, Germany has committed to a complete shutdown of nuclear power plants by 2022[384], and this policy will likely require continued interim use of coal and natural gas to provide dispatchable power to balance the high penetration by wind and solar. Based on current energy policies and available technologies, further reductions in emissions from power production may reach a practical limit.

The European Commission has proposed a year 2030 national binding emissions allowance for Germany that would target a 38% reduction in GHG emissions not covered by the EU ETS.[379] Consequently, Germany is obligated to develop climate change mitigation policies of sufficient ambition to meet this target.

Recently, Germany published a revised draft of Climate Action Plan 2050.[385] The plan does not include provisions for additional carbon pricing mechanisms beyond participation in the EU ETS, and thus non-market policies will be used to meet Germany's fair share obligations to reduce emissions not covered by the EU ETS. The plan states that Germany will become "largely greenhouse gas emissions neutral by mid-century." However, no specific emissions reduction targets for economic sectors are provided. The document deliberately avoids rigid provisions and instead provides stakeholders with guiding principles. As an example, future production of electricity is described as "almost entirely devoid of emissions" by mid-century, but no timeline for phase out of coal-fired power plants is provided. Fuel switching from fossil fuels to electricity and renewables is stated to be an important contributor to long-term decarbonization of the transportation, industry, and buildings sectors. An earlier draft of the document indicated that most newly registered cars were to be propelled by electric motors or by renewable synthetic fuels by 2030. However, this section was dropped and replaced by less specific language indicating that the "government aims to significantly lower car emissions by 2030." Emissions from agriculture are expected to be half of current levels, but no specific measures to achieve this target are provided. Germany's Climate Action Plan 2050 has been criticized for lacking details. In response, the German government has indicated that Germany remains strongly committed to emissions reduction but will not commit to overly rigid climate change mitigation pathways. According to the German government, specific policies and emissions reduction targets within economic sectors will be developed over time as future options are fully assessed.

On June 23, 2016, a slim majority voted in favour of the United Kingdom leaving the European Union.[386] The future status of Britain's participation in the EU ETS and commitment to other EU wide climate change mitigation policies is uncertain. The UK would appear to be committed to an ambitious climate change action plan. A review of options would strongly suggest that continued participation in the broader efforts of the EU to combat climate change would be in the best interest of the UK. For the purpose of this discussion, Britain is assumed to continue to partake in the EU ETS and to be covered by other EU-wide climate change mitigation policies.

In 2012, the UK accounted for 13% of the total GHG emissions from the European Union, and the annual rate of emissions had dropped by 26% relative to 1990.[124] For the year 2030, the European Commission has proposed a binding national emissions allocation for the UK that would target a 37% cut in GHG emissions not covered by the EU ETS.[379]

In 2010, the UK government released "Beyond Copenhagen" as a detailed climate change action plan follow-up to the Copenhagen Accord.[387] The plan is designed to meet regional obligations to maintain future surface warming to a maximum of 2°C above pre-industrial temperatures. Beyond Copenhagen established a short-term target to reduce emissions by 34% by 2020, along with a long-term goal of progressive decarbonization of the economy leading to at least an 80% cut in emissions by 2050 relative to a 1990 baseline.[387] At the time of release, these goals exceeded the EU level of ambition for abatement of future GHG emissions. The plan called for an expansion of nuclear power, increased contribution from renewables up to 15% of total energy consumption, and substantial investment in developing carbon capture and storage technology.[387] Additional highlights included 2020 targets to reduce household emissions by 29% through energy efficiency initiatives, the announcement of an extensive network of electric vehicle charging stations, and incentives for the purchase of electric vehicles. Aside from the initiatives contained in the climate action plan, the UK introduced a carbon floor price applied to power production as an additional economic driver to supplement and strengthen the effectiveness of the EU ETS.[348]

In 2012, GHG emissions in France equaled 6.6 metric tonnes of CO_2 equivalents per person, and this per capita intensity of emissions was among the lowest rates for developed nations.[124] For the year 2030, the European Commission has proposed a binding target for France to achieve a 37% cut in emissions not covered by the EU ETS.[379] France has chosen to implement a progressive national carbon tax as a market driver to reduce non-EU ETS covered emissions.[351] France's Energy Transition for Green Growth Act includes a long-term target to achieve a 50% reduction in energy use by mid-century relative to 2012.[388] Incentives for energy retrofits to existing buildings, along with high efficiency standards for new builds are key policies designed to reduce energy consumption. The act stipulates that 7 million electric vehicle charging stations will be in place by 2030. As of 2015, citizens can receive a bonus of up to €10,000 for the purchase of an electric vehicle, along with the scrapping of emitting vehicles. This combination of carbon pricing mechanisms and non-market policies is targeted to achieve a minimum of a 40% reduction in total GHG emissions by 2030 relative to the 1990 reference year. If this this can be achieved, year 2030 per capita emissions in France would be about 27% below the projected rate for the EU, and would be well below the "fair share" emissions ceiling for France that aligns with the core objective of the Paris Agreement.

Italy has the 4th largest economy in the EU, and in 2012 accounted for 10% of total EU GHG emissions.[124] For the year 2012, the intensity of emissions on a per capita basis in Italy was 88% of that of the EU.[124] Going forward from 2021 to 2030, the European Commission has proposed a binding target for Italy to reduce non-EU ETS GHG emissions by 33% relative to 2005 levels.[379] Italy does not have any additional carbon pricing mechanisms outside of participation in the EU ETS, and the focus of the government has been on policies to improve energy efficiencies. Italy makes use of energy certificates (called white certificates) within an energy efficiency trading system to promote reduced energy consumption in the industry sector.[389] In this system, industry is required to achieve defined efficiencies of energy use and failure to meet targets results in the imposition of fines. White certificates are issued by the government to participants covered by the program after a target of energy savings is achieved in their operations. These certificates can be bought and sold and finally used as offsets to avoid fines for parties that do not meet efficiency standards. The system is analogous to an ETS and in theory will allow companies to follow market driven least-cost pathways toward improving energy efficiencies. Italy has the fourth lowest intensity of energy use in the EU and policies designed to reduce energy consumption have contributed to reducing emissions.[389] In the transportation sector, Italy provides significant incentives for the purchase of zero and low emissions vehicles.[389]

Since 2005, the complementary mechanisms of an emissions trading system covering power production and heavy industry coupled with national "fair share" binding targets to reduce emissions of GHGs not covered by the ETS has proven to be effective in achieving the relatively ambitious emissions reduction targets of the EU. A comparison of current intensity of emissions between the EU and the US provides proof of the success of EU efforts to curtail emissions since the onset of enforcement of the Kyoto protocol. The EU ETS plus the system of binding national emissions reduction targets will continue past 2021 and will be instrumental in the efforts of the region to achieve future emissions abatements aligned with regional obligations to limit future surface warming to less than 2°C.

Climate Action Tracker rates the combination of the EU NDC and climate change mitigation policies within the region as "medium," and not consistent with fair share contributions to limiting surface warming to less than 2°C.[215] CAT acknowledged the leadership role of the EU in global efforts to combat climate change but concluded that the current set of policies within the EU lacked sufficient ambition to achieve NDC targets.[215] CAT expressed concern over the slight slow down in the projected rate of emissions reduction over the period from 2020-2030 relative to cuts over the entirety of the Kyoto period extending to 2020. Further, CAT was critical of the apparent lack of certainty as to inclusion of land use and forestry in the EU's emissions accounting methodology. Climate Action Tracker's assessment indicates that the EU is discussing a more comprehensive package of climate policies. These policies would be designed to counter the recent slow down in emissions reduction and arrive at a 2030 rate of emissions below the target ceiling specified in the EU NDC.

A modest emissions gap of 0.9 Gt/year remains between the EU NDC projected rate of emissions and the region's fair share emissions ceiling for the year 2030. This gap is the equivalent of only 7.5% of the net global emissions gap. The EU is clearly the world leader among advanced economies in developing and implementing effective climate change action policies. This group of nations has begun the process of progressive and cost-effective decarbonization of economic sectors, and is well positioned to meet regional obligations as a contribution to achieving the core objective of the Paris Agreement.

6.6.4 India. In 2012, GHG emissions from India accounted for only 6.1% of the world total, and on a per capita basis, the intensity of emissions was one third of the global figure.[124] This low rate of emissions was directly related to India's state of economic development. In 2012, 17.8% of the world's population resided in India; however, this nation accounted for only 2.5% of world GDP. India's GDP per capita was 13.8% of the world average.[106,216,213]

India has the world's fastest growing economy and, since 2002, India's GPD has increased by an average of 7.4% annually.[205] Over the next 5 years, the population of India will surpass that of China.[210] India has put forward a position that, as a developing nation, actual cuts in the current low rate of emissions going forward to 2030 will restrict economic growth and contribute to a continued state of impoverishment for large sectors of the population. India's position is that economic growth can be accomplished with responsible control over emissions.

Climate change mitigation policies within India are based on a concept of share fair allowance within a world budget of continued emissions. The NDC submitted by India targets a 33–35% decrease in the intensity of emissions per unit of GDP.[254] Based on India's NDC and an assumed annual GDP growth of 7.2%, Climate Action Tracker calculates that, between 2012 and 2030, India's annual rate of GHG emissions would double and surpass the emissions rates of the EU and the US.[215] By 2030, India would account for 11% of global emissions. However, India's

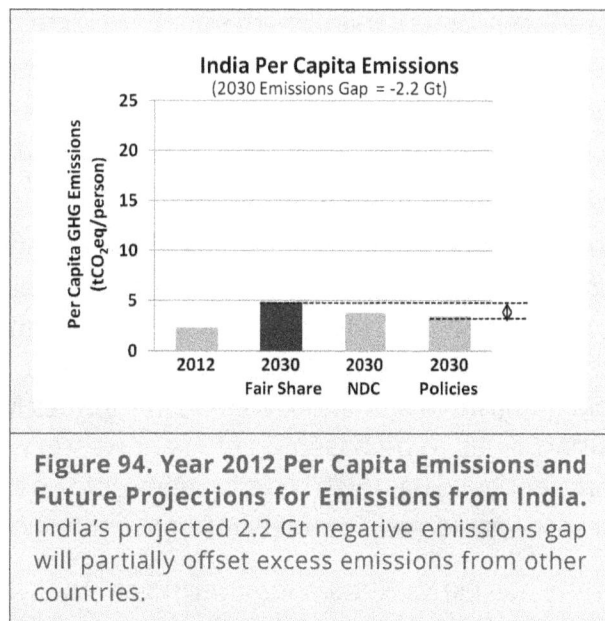

Figure 94. Year 2012 Per Capita Emissions and Future Projections for Emissions from India. India's projected 2.2 Gt negative emissions gap will partially offset excess emissions from other countries.

per capita emissions would remain well below the world average and thus would follow the concept of a reasonable allowance to increase emissions with economic development.

In 2008, India released a National Action Plan on Climate Change.[390] The plan has been updated and revised over the years and currently contains 8 National Missions. By 2030, India will have substantially increased power production capabilities, with 40% of the total installed capacity coming from renewable energy sources.[254] India's National Solar Mission set out ambitious targets to install a rooftop solar capacity of 40 GW, along with 60 GW of capacity from ground mounted solar projects (solar parks) by 2021–2022.[391] This magnitude of solar installations equals 38% of the worldwide installed solar capacity as of 2015 and would increase India's solar power production capacity by 20-fold. Achieving this level of ambition would make India a world leader in solar power. The National Wind Energy Mission has set year 2022 as the target for 50–60 GW of total installed capacity from wind installations.[392] India's NDC also states that efforts will be made to achieve 63 GW of installed nuclear power by 2032 and to increase hydro installations, in light of India's considerable potential for hydro power production.[254]

As part of the National Mission on Enhanced Energy Efficiency, India launched "Perform Achieve Trade" (PAT), which came into force in 2012 as a market-based energy efficiency trading system.[393] The system covers power production and heavy industry totaling 36% of India's energy consumption[394] and over 60% of fossil fuel use.[293] Targets are set for facilities to achieve energy efficiencies relative to baselines, and Energy Savings Certificates are issued when targets are achieved. These certificates can be bought and sold, and used as offsets by facilities that do not meet efficiency targets.

In addition to PAT, the Indian government issues certificates to renewable energy producers as a market-based system to increase penetration of the power production sub-sector.[394] Renewable producers can either sell power at an advantageous feed-in tariff rate or sell the electricity and the renewable energy certificates separately. Certificates can then be used as offsets for power distribution companies to meet state imposed obligations to purchase energy from renewable providers.

India's National Mission on Sustainable Habitat includes an initiative to mandate energy conservation through higher standards in building codes and to support recycling, waste management and high efficiency urban planning.[395] By 2020, India is expected to be the world's third largest construction market. To provide direction and improve urban planning, India has launched the Smart Cities Mission with the goal of establishing 100 smart cities to be based on efficient waste, water and energy management, energy efficient buildings, high rates of renewable energy use, and efficient multi-modal urban transportation and mobility infrastructure.[396]

India has established a long-term goal to increase forested and tree covered areas from 23.4% to 33% of the geographical area, and this objective is part of India's NDC.[254] Climate Action Tracker estimates that achieving this goal would create a biomass carbon sink of 0.17–0.2 Gt CO_2/year which equals about 6% of year 2012 emissions from India.[215,124] In 2015, the government passed the Compensatory Afforestation Fund Bill that established funding equivalent to $6.2 billion USD as a contribution to the massive afforestation effort.[397]

Based on India's NDC, projected year 2030 emissions are 1.6 Gt of CO_2eq below fair share requirements to limit future surface warming to 2°C.[254] The level of ambition contained within India's climate mitigation policies markedly exceeds that contained in the NDC submitted to the UN. Successful implementation of government policies could increase India's "negative emissions gap" to 2.1–2.3 Gt, which would compensate for the projected positive emissions gap of Russia.

Climate Action Tracker rates the combination of India's NDC and climate mitigation policies as "medium" and not consistent with fair share contributions to limiting surface warming to below 2°C unless other countries achieve compensatory deeper reduction. Specifically, CAT was critical of the apparent disconnect between the level of ambition contained in India's NDC and climate action policies. India's NDC can be revised to reflect the actual level of ambition of this densely populated rapidly growing economy to combat change. A revision of the international commitment to control future emissions would further advance India's leadership role among developing countries in establishing pathways of economic growth that are largely uncoupled from GHG emissions.

The NDC submitted by India provides some general indications of a gap between domestic resources and the financial requirements to implement India's climate action plans. Details have yet to be determined, but India will likely need to access international climate financing such as the Clean Technology Fund to achieve the full the potential to uncouple economic growth from GHG emissions. Potentially, some of the cost differentials between conventional and low emissions options for economic growth could be bridged with international climate change financial mechanisms as originally required under the 1992 UNFCCC agreement. The pathway that the developed world takes as contributors to the efforts of India and other developing nations to decouple economic growth from GHG emissions will be a significant factor in defining the rate of progress in decarbonizing world economies going forward to mid-century and beyond.

6.6.5 Russia. On December 26, 1991, the Soviet Union was officially dissolved leading to full autonomy for allied states and for states within the former union.[398] Dissolution of the Soviet Union led to an economic downturn in Russia and GHG emissions declined by 40% between 1990 and 1999.[215] Since 2000, emissions have steadily increased and, by 2012, when land use change and forestry was included in the calculation, Russian emissions accounted for 4.8% of the world total with a per capita intensity that was 2.3-fold greater than the world average.[124]

The NDC submitted by Russia targets a 25–30% cut in emissions by 2030 relative to 1990.[254] By using a Soviet era baseline, the year 2030 NDC target ceiling actually exceeds the rate of emissions on record for 2012. If year 2030 emissions were to approximate the NDC ceiling, Russia's per capita emissions would approach 25 tonnes of CO_2 equivalents per person. This intensity of emissions would exceed all other major economies, and Russia would

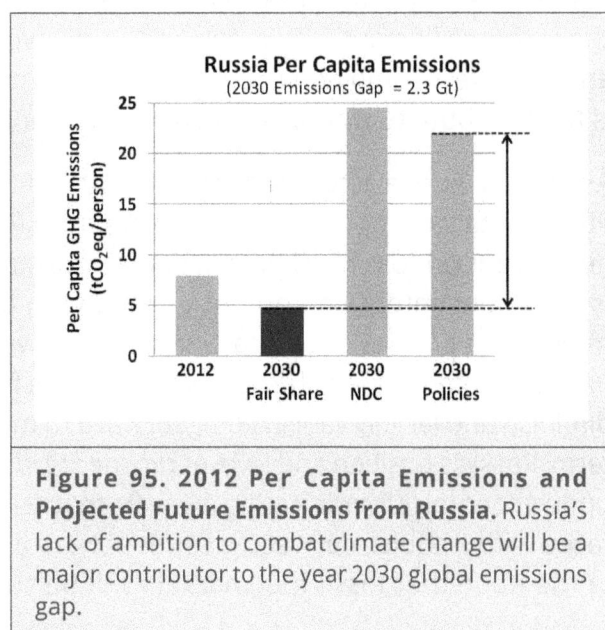

Figure 95. 2012 Per Capita Emissions and Projected Future Emissions from Russia. Russia's lack of ambition to combat climate change will be a major contributor to the year 2030 global emissions gap.

be a substantive contributor to the global emissions gap between actual emissions and the rate of emissions aligned with limiting future surface warming to less than 2°C.

Russia's economy is heavily dependent on fossil fuel resources. This type of economy is the classic example of a petro state. In 2012, crude oil, petroleum products and natural gas accounted for 70% of Russia's total exports and 20% of the nation's GDP came from the oil and gas sector.[399] In 2012, energy production accounted for 40% of Russia's total GHG emissions. Release of fugitive gases from oil and gas activities was the second largest contributor to emissions accounting for 18% of the total.[400] Natural gas power plants generate 49% of Russia's electricity with the remaining power production divided equally between nuclear, large hydro installations and coal-fired power plants.[401]

As of this writing, there is no evidence that the Russian government has intentions to implement significant measures to curtail emissions. Russia's climate change mitigation policies are contained within "Energy Strategy 2030"[401], the "Energy Efficiency Federal Law,"[402] and other decrees and resolutions. Government policies are focused on cuts in primary energy consumption. Russia has established a target to reduce the energy intensity per unit of GDP by 44% between 2005 and 2030.[399,400] There are some indications of policy initiatives to achieve a slight increase in the use of renewables. The government has set an official target of 4.5% of electricity coming from renewable energy sources by the year 2020; however, more recent documents indicate this target has been reduced to 2.5%.[215] Renewables are defined as wind, solar, geothermal, and small-scale hydro, and do not include large scale hydro and nuclear power production.

Climate Action Tracker rates the combination of Russia's NDC and climate change mitigations policies as "inadequate" and not aligned with a fair share approach to limit surface warming to below 2°C.[215] Based on an assessment of current government policies, CAT estimates that actual emissions from Russia will be about 10% less than the target ceiling outlined in the NDC. However, this difference is of no consequence, and the level of ambition to curtail emissions must be substantially increased if Russia is to contribute to global efforts to meet the objectives of the Paris Agreement.

As of 2013, Russia accounted for about 10% of world energy production. Within Russia, 90% of total energy consumption came from fossil fuels.[403] Russia's economy is highly dependent on fossil fuel production, export, and internal use. This type of economy is emissions intensive and not aligned with a future of aggressive climate change mitigations policies and decarbonization of world economies. If the objectives of the Paris Agreement are to be achieved, markets for the export of fossil fuels will steadily diminish and largely vanish going forward to mid-century and beyond. Over the longer term, Russia's position and level of prosperity in a low emissions world will be dependent upon economic diversification away from fossil fuel production and use. Russia is now faced with a choice between a continuation of current, petro state, economic policies in the hope that fossil fuel markets will be sustained over the next 33 years or the implementation of effective measures to reduce emissions and progressively diversify and decarbonize the economy.

6.6.6 Indonesia. As of 2011, approximately 13 Gt of CO_2 were sequestered in the forested areas of the archipelago nation of Indonesia.[404] This quantity of sequestered carbon exceeds the entire emissions of China in the year 2012.[124] Between 1990 and 2010, Indonesia's forested land area declined by 20%.[404] In 2012, emissions from deforestation and degradation of forests and peat land in Indonesia accounted for 61% of total national GHG emissions and equaled 2.6% of the world total.[124] Between 30 to 40% of total emissions from global deforestation come from Indonesia.[215] Indonesia is a significant contributor to global GHG emissions, and on a per capita basis the rate of GHG emissions from Indonesia exceeds the world average. However, Indonesia is a developing nation and emissions are not correlated with affluence and industrial activity. Excluding land use and forestry, Indonesia's per capita emissions were only 3.1 Gt of CO_2eq per person and less than half of the world average.[124]

The NDC submitted by Indonesia uses the net-net approach and thus includes emissions from land use and forestry in calculating baseline and year 2030 targets. Indonesia has committed to achieving an unconditional target of a 26% reduction in total GHG emissions by the year 2030 relative to a "business as usual" (BAU) scenario.[254] By achieving the unconditional NDC target, year 2030 emissions per capita would be about 75% of the world average.[124,215] Indonesia's NDC includes a second, more ambitious, target to reduce emissions by 41% relative to the BAU

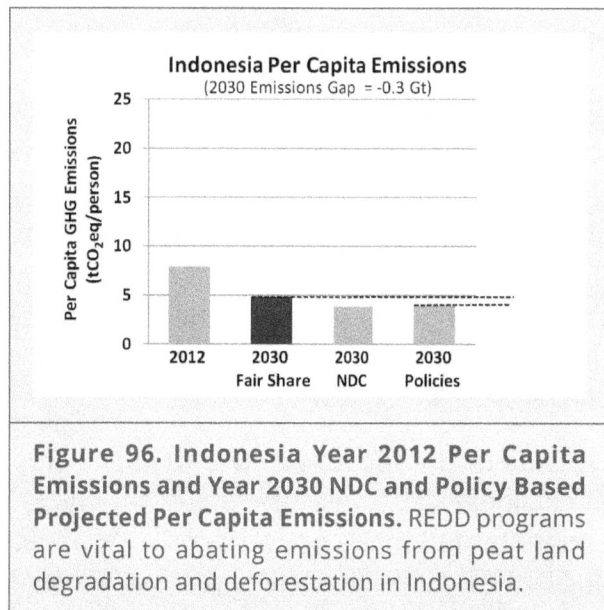

Figure 96. Indonesia Year 2012 Per Capita Emissions and Year 2030 NDC and Policy Based Projected Per Capita Emissions. REDD programs are vital to abating emissions from peat land degradation and deforestation in Indonesia.

baseline. This target is conditional to accessing international climate funds, along with technology transfers and cooperation, capacity building, and payment for performance in cutting emissions. Based on Indonesia achieving the conditional NDC target, projected emissions would be cut by an additional 19%.

In 2011, Indonesia's National Action Plan to Reduce Greenhouse Gas Emissions was created by presidential decree.[405] Climate change mitigation policies are broadly defined by a series of objectives to improve land management, reduce deforestation and land degradation, develop carbon sequestration capacity through forestry and agriculture, promote energy efficiency, increase use of renewables, reduce waste, and transition to low emissions transportation options.

Reducing emissions from land use and forestry in Indonesia is both a challenge and a substantial opportunity. Forests cover 52% of total land area and consist of 50% natural primary forests, 46% naturally regenerated forests and 4% planted forests.[405] Forest industries account for 3.5% of Indonesia's GDP, and the contribution of these industries to the nation's economy must be considered.[405] However, the status quo of a continual 1% annual rate of net deforestation, along with emissions from peat fires and land degradation is unsustainable. In 2010, Norway and Indonesia signed a letter of intent that would make available $1 billion USD over a 7-year period contingent upon success in developing and implementing REDD programs in Indonesia.[405] The agreement led to a moratorium on the issuance of new licenses for forestry and peat land developments.[405] The moratorium excluded existing licenses and only applied to primary forests.[405] As a follow-up to the agreement, a taskforce was assembled to develop a national strategy to reduce GHG emissions from deforestation and forest degradation.[406]

REDD programs must function to create an effective set of regulations to preserve and eventually enhance forested areas. These programs will require an independent agency for monitoring, verification and reporting. Potentially, UN monitored Internationally Transferred Mitigation Options could flow between Indonesia and countries such as Norway seeking to achieve cost-effective international emissions reductions. Given the long history of continuous deforestation, considerable land area is available in Indonesia for reforestation. The potential exists to transition Indonesia's land use and forestry sector from a serious source of emissions to net withdrawal of atmospheric CO_2. International financial mechanisms could see a significant flow funds to Indonesia upon verification of success in the preservation and enhancement of forest biomass carbon storage capacity.

Outside of the forestry and land use sector, Indonesia is faced with the same set of challenges as other developing nations to uncouple economic growth from GHG emissions. Indonesia has targeted a 1% annual reduction in energy intensity per unit of GDP.[405] By 2025, Indonesia plans to achieve a 41% reduction in total energy supply compared to BAU projections.[405] Details are few, but the government indicates that energy efficiencies and energy conservation measures will be implemented to achieve these goals.

Indonesia's National Energy Policy has set a year 2025 target to achieve at least a 23% inclusion level for renewables in the total primary energy supply.[407] In 2014, Indonesia introduced a system of feed in tariffs to provide economic incentives for the generation of electricity from renewables.[407]

Climate Action Tracker rates Indonesia's NDC and government policies to control emissions as "medium" and not consistent with a fair share contribution toward limiting surface warming to well below 2°C.[215] CAT is highly critical of government data claiming relatively stable emissions from deforestation and peatland degradation over the past 10 years. Independent scientific assessments concluded that high rates of deforestation continued over the period from 2004 to 2014.[408] This means that the temporary moratorium on new forestry and peat land activities that has been in place since 2010 appears to have had little effect. While Indonesia has made some progress in policy development to protect forested areas, these policies have yet to produce tangible results. CAT indicates that an earlier draft of Indonesia's NDC contained specific targets for forest preservation, but these targets were dropped from the final version of the document. CAT acknowledged Indonesia's ambitious target to increase the use of renewable energy sources but was critical of plans to build new coal-fired power plants to meet the increased demand for electricity as the economy develops.

Given the enormous potential for forest biomass to withdraw and sequester atmospheric CO_2, Indonesia has submitted an NDC with a low level of ambition to control future emissions. Effective REDD+ programs in Indonesia could function to largely eradicate deforestation, minimize peatland degradation, and improve management of land use. These programs could eliminate emissions from the forestry and land use sector. Further ambitions to reforest available lands could put Indonesia on a pathway toward net zero-emissions or even negative emissions by mid-century.

There is little evidence for a high level of ambition in Indonesia's NDC or other national action plans to combat climate change. This lack of ambition and inaction is contrary to the substantial potential for cost-effective control over emissions through implementation of REDD+ programs. In a post-Paris Agreement world, least-cost pathways may quickly evolve that would facilitate the international flow of funds to nations with the natural resources to preserve and enhance carbon-sequestering forest biomass. Economic drivers could motivate an increase in Indonesia's level ambition to reduce emissions. The pathway taken by Indonesia and the international community to preserve and enhance Indonesia's forested areas will be a contributing factor to the success or failure of global efforts to limit surface warming to less than 2°C above pre-industrial temperatures.

6.6.7 Brazil. During the period of peak deforestation of the Amazon, Brazil's GDP per capita was half of the world's average, yet the per capita rate of emissions was 60% greater than the global figure. In 2004, land use and forestry accounted for 59% of total GHG emissions in Brazil.[124]

Between 2004 and 2012, Brazil's economy underwent a period of rapid growth with a nearly 4-fold increase in GDP per capita.[216] This period of economic growth coincided with an 82% decrease in deforestation rates in the Brazilian Amazon, and, as such, the growth in emissions associated with economic activity was balanced by reduced emissions from deforestation. By 2012, emissions from land use and forestry in Brazil had declined by 33%, and total GHG emissions were 10% lower than the data on record for 2004.[124]

As described in chapter 4, efforts to halt deforestation in the Amazon are ongoing, and emissions from the forestry and land use sectors in Brazil and other countries with land in the Amazon basin remain as significant contributors to global GHG accumulation.

Brazil's NDC uses the IPCC recommended Global Temperature Potential (GTP) as a metric to calculate fair share future emissions.[409] The GTP-100 calculates an emissions ceiling that is consistent with limiting surface temperature elevations to less than 2°C above pre-industrial

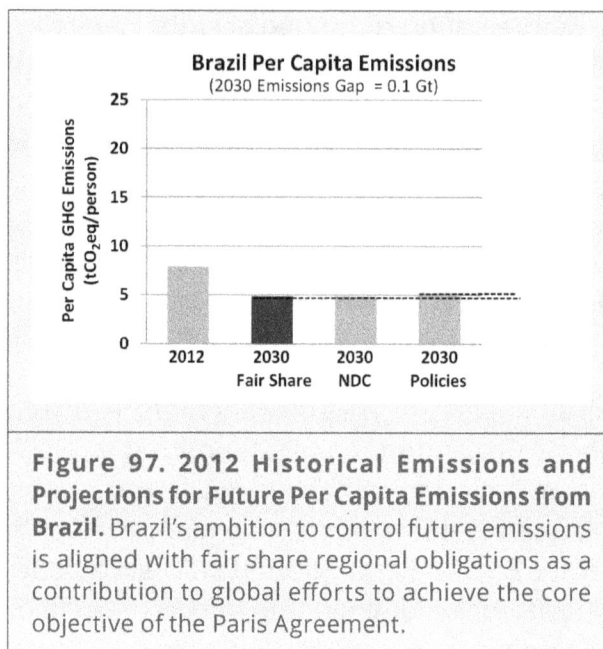

Brazil Per Capita Emissions
(2030 Emissions Gap = 0.1 Gt)

Figure 97. 2012 Historical Emissions and Projections for Future Per Capita Emissions from Brazil. Brazil's ambition to control future emissions is aligned with fair share regional obligations as a contribution to global efforts to achieve the core objective of the Paris Agreement.

temperatures over the next 100 years.[409] Brazil is one of the few countries to use a science-based metric to calculate regional obligations to reduce emissions.

Brazil's NDC uses the net-net approach and targets cuts in GHG emissions of 37% by 2025 and 43% by 2030 relative to emissions on record for 2005.[409] Achieving these targets would result in the release of 1.1 Gt of CO_2eq during 2030 at a per capita emissions rate of 4.8 tCO_2eq/person.[124,409] This rate of emissions is consistent with a population-based calculation of Brazil's fair share contribution to limit future surface warming to less than 2°C.

REDD+ programs are the cornerstone of climate change mitigation policies in Brazil. Policies and measures to control emissions from the land use and forestry sector are designed to achieve over half of the target for total emissions reduction. Brazil's NDC includes year 2030 objectives to eliminate illegal deforestation and to fully compensate for emissions from legal "suppression of vegetation."[409] Twelve million hectares are targeted for restoration and reforestation by 2030.[409] At the state level, considerable progress has been made toward achieving readiness for implementation of REDD+ programs. Financial mechanisms for REDD+ programs have yet to be fully developed but likely will involve issuance of tradable credits as emissions reduction units (U-REDD+).[410] A recent proposal would see a state-based decentralized system of regulation and management of REDD+ programs.[410] National and international compensation including access to the Amazon Fund would see a flow of funds to Amazonian states within Brazil based on verifying emissions reduction achieved through REDD+ programs.

By 2012, Brazil's emissions profile by economic sector had shifted from land use and forestry such that agriculture accounted for 24% of total GHG emissions.[124] Brazil is a net exporter of agriculture commodities with a large herd of emissions intensive ruminant animals. In 2009, Brazil enacted the Low Carbon Agriculture Programme, "ABC

Plan", designed to promote sustainable agricultural practices with a lower intensity of emissions.[411] Under the plan, food production and the efficiency of food production are to increase through implementation of higher intensity agricultural practices.[411] An important initiative of the program will see 5 million hectares allocated to a low carbon integrated system of livestock-crop-forest land use.[409] By 2030, 15 million hectares of agricultural land are targeted for restoration through pasture recovery.[87] These and other programs have the potential to reduce annual emissions from agriculture in Brazil by 33–37% relative to 2012.[412]

Continued long-term expansion of the ruminant animal herd in Brazil is unsustainable based on intensity of emissions and land use requirements. However, supply side suppression of the livestock industry is unlikely and thus national and international market demands will be the economic driver that defines future herd size. Ultimately, carbon pricing mechanisms applied to food products will likely be required to control the demand for ruminant meat.

In 2012, Brazil had one of the world's lowest intensity of emissions per unit of electricity produced. The electricity supply sub-sector in Brazil accounts for only 4.5% of total GHG emissions.[124] Large zero-emission hydro installations provide 80% of domestic electricity and provide flexibility for increased penetration by renewables.[413]

Brazil's National Energy Plan 2030 sets out to provide strategic direction in expanding the energy supply to meet the demands of continued economic growth.[414] The plan calls for a doubling in power production from hydro and renewables. Under the plan, zero and very low emissions sources would provide 92% of total capacity to produce electricity by 2030.[414]

In 2012, the transportation and manufacturing sectors accounted for 17.5% of Brazil's total GHG emissions.[124] Brazil's National Energy Efficiency Action Plan calls for improvements in energy efficiency across economic sectors and a 10% reduction in electricity consumption relative to a reference scenario.[415] Road vehicles account for 90% of emissions from the transportation sector.[415] Brazil continues to mandate high inclusion rates for sugarcane derived ethanol and other biofuels for use in road vehicles.[416] However, actual emissions savings from first generation biofuels is marginal and land use management is complicated by requirements for biofuel production. Achieving a balanced mix of transportation modes with a higher dependence on rail and waterways is expected to reduce future emissions.[415] However, the various action plans covering Brazil's transportation and manufacturing sectors provide few policy details, and the pathway toward controlling future emissions in these sectors is uncertain.

Climate Action Tracker rates the combination of Brazil's NDC and national climate change mitigation policies as "medium" and not consistent with a fair share contribution to limit surface warming unless other countries achieve deeper compensatory cuts in emissions.[215] CAT acknowledges the success of government initiatives to curtail deforestation of the Brazilian Amazon. However, CAT is critical of recent government decisions to open electricity supply auctions to coal and gas-fired utilities. This decision contrasts national plans for a 92% penetration of the electricity supply grid by hydro, nuclear and renewable energy sources by 2030. Construction of new fossil fuel power plants may lock in a higher percentage of electricity supply from emissions intensive sources than is stated in the National Energy Plan 2030. Over the last few years, the Brazilian government has discussed the possibility of a national emissions trading system, but these discussions have yet to evolve toward implementation.[417] ETS planning appears to be further advanced at the subnational level in Brazil, but no systems are in place at the time of this writing.[417] Other than emissions reduction certificates in the forestry sector, market-based mechanisms to reduce emissions are not part of Brazil's current climate change action plans.

Brazil's climate, geography, and natural vegetation are unique. Brazil has made considerable progress in implementing land use management practices to optimize and balance land requirements for food and fuel production with the global need to maintain and enhance carbon-sequestering forest biomass. The technical capabilities to monitor and police deforestation provide an advanced state of readiness for implementation of REDD+ programs. Brazil is well positioned to provide a leadership role in global efforts among developing countries to curtail deforestation and implement programs of reforestation and afforestation. Brazil has a long history of innovation in agriculture. Programs of integrated livestock-crop-forest land use and pasture land restoration are in place to optimize land use and reduce emissions from the agriculture sector. Brazil makes good use of abundant hydro power resources, and the potential exists to meet increasing demands for electricity through expanded use of zero and very low emissions energy sources. Potentially, Brazil could follow the lead of Europe and implement measures to improve the energy efficiencies in buildings and in manufacturing practices. Given the very low rate of emissions from the power supply sector, programs designed to transition road vehicles to electric propulsion options and, where possible, to switch industrial practices from fossil fuels to electricity would be effective in reducing emissions from the transportation and manufacturing sectors.

Brazil's NDC is ambitious and aligned with regional obligations to reduce emissions to limit future surface warming to less than 2°C. The potential exits for Brazil to introduce carbon pricing mechanisms and to strengthen climate change mitigation policies for economic sectors outside of forestry and agriculture. These measures would add to the natural advantages of abundant forest biomass, vast areas of agriculturally productive lands and considerable resources and options for zero and low emissions power supply. By increasing the level of ambition to curtail emissions across all economic sectors, Brazil would become a model nation in global efforts to decarbonize economies and combat climate change.

6.6.8 Japan. Between 1990 and 2009, net GHG emissions from Japan had declined by 4.4%[124] and Japan was well positioned to exceed the emissions reduction target for the first Kyoto reporting period ending in 2012.[418] In 2009, coal, natural gas and oil-fueled power plants produced 61% of Japan's electricity.[419] Nuclear energy was an important contributor, accounting for 29.2% of electricity production, while renewables provided the remaining 10%.[419] The electricity supply sector was responsible for 45% of Japan's total emissions.[124] Japan has almost no domestic fossil fuel resources and is entirely dependent upon imports.

In 2009, at the 15th meeting of the Conference of Parties to the original UNFCCC agreement, Japan pledged to achieve a 25% reduction in GHG emissions by the year 2020 relative to the rate of emissions on record for 1990.[418] This second Kyoto reporting period target exceeded the level of ambition of all other parties that had ratified the accord. One year later, Japan released the Basic Energy Plan 2010.[420] The plan called for construction of an additional 14 nuclear power stations to be online by 2030.[420] Nuclear power was targeted to supply 52% of the nation's electricity, with renewables providing another 19%.[421] Japan was to become a leader in the development and commercial implementation of next generation nuclear power plants and to contribute to global efforts to cut emissions through international deployment of advanced nuclear technologies.

On March 11, 2011, a magnitude 9.0 undersea earthquake occurred with the epicenter located 70 kilometres from the coast of Japan.[159] The quake was the fourth most powerful since modern recording began in 1900, and produced a massive tsunami that devastated Japan. The disaster completely destroyed 127,290 buildings and damaged another 1,020,777.[159] Over 18,000 people were either confirmed dead or classified as missing and presumed dead.[159] The World Bank estimated the total cost of the disaster at $235 billion.[159]

As described in chapter 3, a 13 metre tall tsunami wave easily breached the 10 metre seawall designed to protect Japan's Fukushima Daiichi second generation, 40-year-old, nuclear power plant.[153] The tsunami caused the second level 7 major accident in the history of the nuclear power industry.[153] Release of radioactive material following the partial core meltdown of 2 reactors was managed such that there were no fatalities or prospects for future fatalities.[422]

The Fukushima accident eroded public confidence in the safety of nuclear power. After the accident, all nuclear power plants in Japan were shut down because of safety concerns. To restart, a power plant must meet stringent post-Fukushima safety standards and must overcome court challenges. A limited number of plants have been restarted, and, as of December 2015, nuclear power accounted for only 2.2% of the supply of electricity in Japan.[423]

The Fukushima nuclear accident has reshaped Japan's energy usage and aspirations to reduce greenhouse gas emissions. In the immediate period following the disaster, electricity production from fossil fuels filled the gap left by the shutdown of nuclear plants.[419] The abrupt replacement of nuclear power with liquid natural gas and oil resulted in a 13% increase in GHG emissions.[124] At the 2013 COP meeting in Warsaw, Japan announced a revision to year 2020 emissions targets.[418] The new objective was to reduce emissions by 3.8% relative to 2005 levels.[418] Using emissions on record for 1990

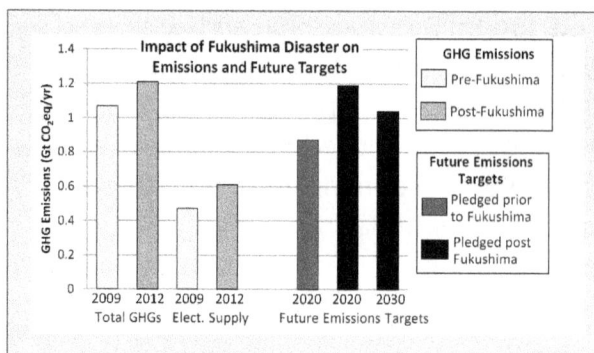

Figure 98. Impact of the Fukushima Nuclear Disaster on GHG Emissions from Japan and on the Ambition to Control Future Emissions. Total and electricity supply sector emissions on record for 2009 (pre-Fukushima) and 2012 (post-Fukushima). Year 2020 targets for emissions pledged pre-Fukushima, and then modified post-Fukushima. Year 2030 NDC target submitted in 2015.

as a baseline, Japan's Warsaw target would result in a 6.6% increase in emissions and thus was a complete rollback of the original Copenhagen ambition to achieve a substantial cut in emissions by 2020.[124,418]

Japan's NDC as submitted in advance of the Paris 2015 climate change meeting is based on a year 2030 target to reduce net GHG emissions by 25.4% relative to emissions on record for 2005.[424] Achieving this target would result in 1.1 Gt of CO_2eq released in 2030 at a projected per capita rate of 9.2 tCO_2eq/person.[424] The document states that Japan remains committed to a long-term goal of achieving a greater than 80% reduction in emissions by mid-century.[424] However, the projected rate of emissions that are based on Japan's NDC is nearly double the population-based calculation of a "fair-share" ceiling in emissions that aligns with regional obligations to achieve the core objective of the Paris Agreement.[215]

The NDC submitted by Japan provides a breakdown of emissions reduction targets by sector and greenhouse gas. The document effectively summarizes domestic policies developed by the Ministry of Economy, Trade and Industry, the Ministry of the Environment and other ministries and government agencies charged with developing

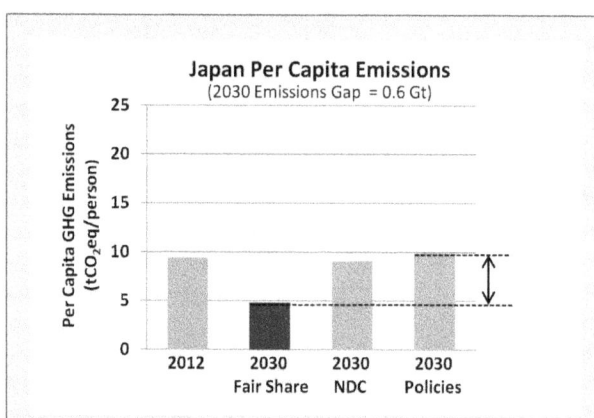

Figure 99. Japan's 2012 Per Capita Emissions and Year 2030 NDC and Policy Based Projected Per Capita Emissions. The Fukushima nuclear power plant disaster has set back Japan's level of ambition to combat climate change.

climate change mitigation policies and action plans. Approximately 10% of the total reduction in net emissions is to come from improvements in cropland and grazing land management plus re-vegetation.[424] Specific targets to reduce CO_2 emissions from each economic sector are provided, along with targets to reduce emissions of methane, nitrous oxide and fluorinated gases. A year 2030 target to limit overall energy consumption is provided in the NDC.[424] Renewables, consisting of hydro, solar, wind, geothermal and biomass are to produce 22–24% of the total electricity.[424] Nuclear power will be re-established as an important zero-emissions contributor to electricity production; however, year 2030 grid penetration by nuclear power will be less than half of what was planned pre-Fukushima.[420,424] The NDC indicates that fossil fuels will continue as important contributors to power production going forward to the year 2030. Overall, fossil fuels will account for 56% of electricity production.[424] This level of power production from high emissions sources is nearly double what had been planned for the year 2030 in the Basic Energy Plan of 2010.[420,424]

Japan's NDC provides an extensive list of bullet point policy objectives to reduce emissions within economic sectors. However, the list does not provide details as to specific measures to achieve these objectives. As an example, under the transportation sector, the policy objectives are simply listed as "improvement in fuel efficiency" and "promotion of next generation automobiles" without an indication of incentives or other mechanisms to transition the transportation sector to these lower emissions options.

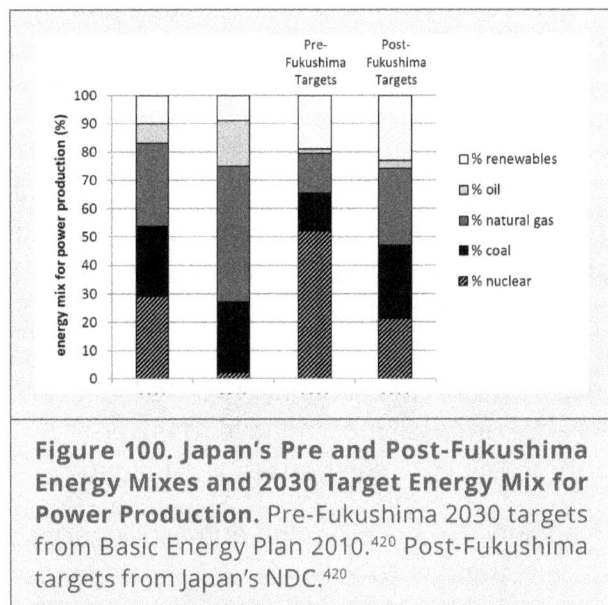

Figure 100. Japan's Pre and Post-Fukushima Energy Mixes and 2030 Target Energy Mix for Power Production. Pre-Fukushima 2030 targets from Basic Energy Plan 2010.[420] Post-Fukushima targets from Japan's NDC.[420]

Since the onset of the first Kyoto reporting period, Japan has utilized international credits as contributions toward achieving national emissions reduction targets. Japan did not participate in the second Kyoto reporting period but continued to partner with 16 nations on a voluntary basis to disseminate technology, and otherwise contribute to efforts to limit emissions in developing countries.[425] Japan indicates an intention to partake in supposed low carbon international transfers with developing countries going forward to the year 2030.[425] Presumably, these programs will be classified as Internationally Transferred Mitigation Outcomes and will be monitored by the UN under article 6 of the Paris Agreement with fair allocation of emissions credits where applicable.

In 2015, the Ministry of Economy Trade and Industry released Japan's Energy Plan extending to the year 2030.[426] The plan calls for a 17% reduction in electricity demand by 2030 through improved energy efficiencies in the buildings and industry sectors. In addition, the plan calls for a 20–30% reduction in emissions from fossil fuel power stations by through the implementation of advanced technologies.

In 2012, Japan enacted a system of feed-in tariffs to increase the use of renewable energy sources for power production.[427] The system provides cost-based compensation to renewable energy producers and provides long-term price certainty for the purchase of power generated from renewable sources. The system was revised in 2016 and will be a key component in transitioning Japan's electricity production mix toward increased use of renewables.[427]

Japan has two, relatively small, linked emissions trading systems that operate at the subnational level. The Tokyo ETS[428] was launched in 2010 followed by the Saitama ETS[429] in 2011. These systems cover about 9% of Japan's total emissions and by 2014 had contributed to achieving a 22–25% reduction in regional emissions. Longer term targets for the Tokyo and Saitama regions are to reduce GHG emissions by 25% and 30% by the years 2020 and 2030 respectively. There is no evidence of government intention to implement a national emissions trading system or to impose carbon taxes on a national basis.

Climate Action Tracker rates Japan's NDC and domestic policies and measures to reduce emissions as "inadequate."[215] CAT stated that if all countries were to adopt Japan's level of ambition to mitigate climate change, average surface temperatures by the end of the century would likely be in excess of 3–4°C above pre-industrial temperatures.[215] CAT is highly critical of Japan's proposed 2030 energy mix for the production of electricity. The mixture of energy sources proposed in Japan's NDC accomplishes little to reduce the reliance on fossil fuels relative to the energy mix that was used to produce electricity prior to the 2011 earthquake and tsunami.

In addition to concerns over domestic energy policies, CAT was highly critical of Japan's stated intention to assist developing nations with the adoption of "clean coal" technologies. Japan's NDC implies the use of credits for reducing the intensity of emissions from future international coal power installations that would employ advanced lower emissions technologies. Potentially, these technologies could reduce emissions by 30% relative to a conventional coal power plant; however, "clean coal" plants would still operate at a high intensity of emissions. In theory, the cleanest possible coal technology requires carbon-capture and permanent geological storage that is independent of enhanced oil recovery. This integration of processes has yet to be developed, and when compared to nuclear and renewables, it will be a costly and relatively inefficient method of power production. Further, residual CO_2 emissions from theoretical CCS equipped coal-fired power plants will be well in excess of nuclear and renewable energy options to produce electricity. Continued investment into coal and fossil fuel power plants and technologies is highly questionable. Japan's intention to support implementation of clean coal technology in the developing world may well exacerbate the current situation of "locked in" new build emissions-intensive fossil fuel power plants. While emissions may be 30% lower than conventional coal plants, economic growth in developing countries would still be correlated with GHG emissions.

Japan has clearly stated a long-term objective to achieve at least an 80% reduction in emissions by mid-century and to contribute to international efforts to reduce emissions. However, current policies and the level of ambition contained in Japan's NDC are inconsistent with deep decarbonization of the economy by 2050.

As previously discussed, an aversion to the continued use of nuclear power to produce electricity is irrational and, ultimately, will be far more damaging given the fossil fuel alternatives combined with practical limitations in the use of non-dispatchable renewables to generate electricity. Public education in combination with strong political leadership within Japan and in the broader international community is required to redirect efforts away from continued reliance on coal, natural gas and oil for the production of electricity.

By re-establishing pre-Fukushima targets to displace fossil fuels using a combination of nuclear energy and renewables to produce electricity, and by implementing carbon pricing mechanisms and other emissions reduction policies, Japan could align its level of ambition with regional obligations to achieve the objectives of the Paris Agreement. Further, through continued development and implementation of nuclear and other very low and zero-emissions energy options, Japan would reduce its future dependency on fossil fuel imports. By reverting to pre-Fukushima ambitions and energy policies, Japan could become a world leader in combating climate change and an important contributor to economic growth in developing countries that is uncoupled from GHG emissions.

6.6.9 Canada. In 2002, the Government of Canada ratified the Kyoto Protocol and committed to a 6% cut in GHG emissions by 2012, relative to a year 1990 baseline.[430,119] Canada's level of ambition was second only to the group of nations including the EU that committed to an 8% cut in emissions.[119] However, over the next ten years, Canadian governments at the national and provincial levels were largely inactive in developing and implementing meaningful climate change mitigation policies. The disconnect between actual emissions and the Kyoto target was such that, in December of 2011, Canada became

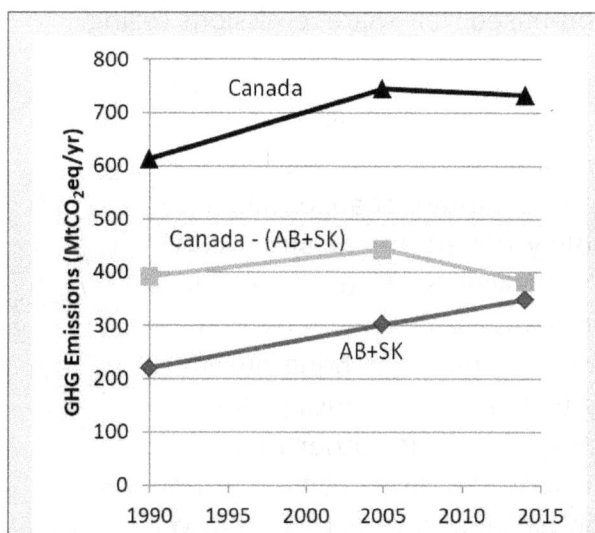

Figure 101. Greenhouse Gas Emissions from Canada. In 2015, half of total emissions came from the oil and gas rich provinces of Alberta (AB) and Saskatchewan (SK).

the only nation to have ratified and then withdrawn from the protocol.[119] By 2012, the rate of GHG emissions from Canada had increased by 26% relative to 1990.[124] Canada's per capita emissions were 3.6-fold above the world average and well in excess of the rate of per capita emissions from China, the US and the other nations and regions ranked among the top emitters.

Canada is a net exporter of fossil fuels with extensive coal, oil, and natural gas deposits that are concentrated in the western provinces of Saskatchewan and Alberta. In 2014, the oil and gas sector accounted for 26.3% of total national emissions and was the largest contributor to net emissions among economic sectors in Canada.[431] GHG emissions from the fossil fuel rich provinces have steadily increased since 1990 while emissions from the remaining regions of the country have declined slightly.[432] By 2014, only 15% of Canada's population lived in Saskatchewan and Alberta; however, this region of the country accounted for 48% of total emissions.[432,433] On a per capita basis, during 2014, 67 tonnes of CO_2 equivalents were released per person residing in Saskatchewan and Alberta.[432,433] This rate of emissions was over 5-fold greater than per capita emissions from the remaining regions of the country. Oil and gas sector activities in Saskatchewan and Alberta are major contributors to Canada's excessive rate of per capita GHG emissions.

In May of 2015, Canada submitted its NDC to the UNFCCC and pledged to reduce net GHG emissions by 30% by the year 2030 when compared to emissions on record for 2005.[254] This target is based on a net-net calculation and would equal an 8% increase in emissions relative to the 1990 baseline used by many other industrialized countries in developing their NDCs.[124,254] Based on NDC projections, Canada's per capita emissions in the year 2030 would be double the world average and 3-fold in excess of a population-based "fair share" emissions ceiling.

Recently, Canada has begun to deviate from the political inaction and indifference that have characterized climate change mitigation policy development since 2002. The provinces of Quebec, Ontario, British Columbia and Alberta have implemented or will bring into effect aggressive carbon pricing mechanisms, along with other measures designed to direct consumer and industrial practices toward lower emissions options. With rollout of the Ontario ETS in 2017 and future linkage to the California-Quebec systems, nearly 30% of Canada's

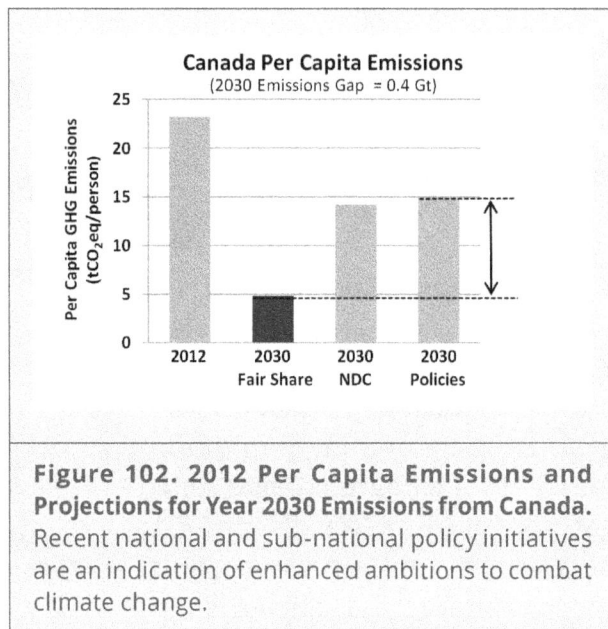

Figure 102. 2012 Per Capita Emissions and Projections for Year 2030 Emissions from Canada. Recent national and sub-national policy initiatives are an indication of enhanced ambitions to combat climate change.

total emissions will be covered by a large multinational broad-spectrum cap and trade program.[432]

In June 2016, Ontario released a detailed 5-year climate action plan designed to set the province on a pathway toward achieving a 37% reduction in GHG emissions by 2030 and an 80% reduction by mid-century relative to 1990.[434] The Ontario ETS provides the foundation to the plan; however, an extensive series of additional measures will be enacted beginning in 2017. All told, the Ontario government has allocated $7 billion toward the climate change plan. Highlights of the plan include an incentive package of up $14,000 for the purchase of an electric vehicle. Within the transportation sector, programs will be implemented to build a network of public and home based charging stations and to promote zero and low emissions buses and trucks. In addition, building codes are to be enhanced to high efficiency standards and fossil fuel use for building heating will be gradually phased out.

Prior to the release of the climate action plan, Ontario had shut down all coal-fired power plants. As of 2016, 84% of the electricity generated in Ontario came from a mixture of zero-emissions nuclear, hydro, and renewable sources.[435] Nearly 100% of the electricity produced in Quebec, British Columbia and Manitoba comes from zero-emissions hydro power stations.[436]

British Columbia's carbon tax has been in effect since 2008 and is a core component of provincial plans to achieve a 33% emissions reduction by 2020 and an 80% cut in emissions by mid-century relative to emissions on record for 2007.[437] The carbon tax rate in B.C. is currently set at $30/tonne.[352]

In 2014, 79% of the oil produced in Canada and 68% of total natural gas output came from Alberta.[438] This heavy concentration of oil and gas sector activities was such that Alberta was responsible for 37% of Canada's total GHG emissions.[432] The intensity of emissions in Alberta and Saskatchewan is extreme and is a direct result of oil and gas sector activities including the massive, Athabasca oil sands project to produce oil from bitumen. Thermal energy requirements for the production of fossil fuels are high, and emissions from production exceed emissions from fugitive gas leakage and routine flaring from mining and extraction activities.[439] In essence, there are practical limits as to what can be accomplished on a cost-effective basis to reduce emissions from oil and gas activities.

In November of 2015, the Alberta government announced the "Climate Leadership Plan."[440] A provincial carbon tax of $20/tonne came into effect in January of 2017 and will increase to $30/tonne in 2018. The tax will drive consumer and industrial practices, including oil and gas sector activities, toward lower cost, lower emissions options. The plan places an emissions cap on oil sands operations and introduces an ambitious target of a 45% reduction in methane emissions from the oil and gas sector by 2025. In addition, all coal-fired power facilities in the province will be phased out by 2030. The methane reduction target contained in the Alberta climate action plan is similar to the targets established by the US-Canada agreement to cut methane emissions from the oil and gas sector.

With rollout of the Ontario ETS and startup of the Alberta carbon tax, 86% of the population in Canada will live in jurisdictions with some form of carbon pricing mechanism in force. On October 3, 2016, the Government of Canada announced that, beginning in 2018, all provinces must have a carbon pricing mechanism (ETS or carbon tax) in place.[441] Mandatory minimum carbon pricing starts at $10 per tonne of CO_2 and must increase by at least $10/tonne annually, arriving at a minimum price of $50/tonne by 2022.[441] Taxes are to remain within each province, and the provinces are free to establish their own priorities for the allocation of carbon tax funds. Failure to enact a suitable provincial carbon pricing mechanism will result in the imposition of a carbon tax administered by the federal government.

Climate Action Tracker rates the combination of Canada's NDC and climate change mitigation policies as "inadequate" and inconsistent with various interpretations of an equitable approach to limit surface warming to less than 2°C.[215] This assessment applied to policies in place as of November 2016. CAT concluded that this set of policies were inconsistent with Canada achieving its NDC target, and that emissions would increase by 14-29% in 2030 when referenced to a 1990 baseline.[215] CAT acknowledged the recent announcements of carbon pricing and other initiatives to advance Canada's level of ambition to combat climate change and stated that if these measures were to be fully implemented Canada would be able to achieve the targets specified in the NDC submitted under the Paris Agreement.

As a net fossil fuel energy exporter, Canada intends to maintain international market share while demand lasts. This intention is sound economic policy and does not invalidate a national plan to combat climate change. If Canada were to pull back on fossil fuel production, demand would be filled by other suppliers with no net effect on global GHG emissions.

Based on recent activities at the federal and provincial level, Canada has embarked on an aggressive set of climate change action plans. In the short-term, Alberta's carbon tax, cap on oil sands emissions, and the initiatives to limit methane release will be important measures to control emissions from oil and gas activities. Moving forward to 2030 and beyond, as economies progressively decarbonize, domestic and export markets for oil and gas will diminish and eventually vanish. Declining demand will be the long-term driver of deep emissions cuts from Canada's fossil fuel sector.

Arguments have been made that whatever Canada does or does not accomplish to reduce emissions will have no bearing on global efforts to combat climate change. In 2012, Canada accounted for only 1.7% of global emissions and, based purely on this statistic, any unilateral actions of Canada will not impact future surface warming.[124] However, 35.3% of total global emissions come from countries with emissions rates that are comparable to or lower than that of Canada.[124] Canada, and all nations, must contribute to global efforts to combat climate change, based on a fair share concept of emissions reduction. There is no foundation for absolving national and subnational responsibility based on the magnitude of absolute emissions.

Opponents of carbon pricing mechanisms and emissions reduction policies applied to the oil and gas sector often cite the potential for carbon leakage to other lower cost, fossil fuel rich jurisdictions. These arguments inevitably lead to proposals for exemptions and thus a weakening of emissions reduction policies. In the post-Paris Agreement world, there is little justification for carbon leakage exemptions. Given the magnitude of emissions from the oil and gas sector in Canada, opposition to effective measures to limit these emissions is an opposition to reducing total national GHG emissions and, through inaction, a form of climate change denial.

As of this writing, the majority of political leaders in Canada who are in positions of power are in favour of effective carbon pricing mechanisms and other market and non-market-based measures to combat climate change. The current status of Canada as an oil and gas exporter may limit what can be accomplished in the short-term relative to other developed nations whose economies are less dependent upon fossil fuel exports. However, the province of Alberta as Canada's largest source of GHG emissions has developed a responsible climate action plan designed to provide a reasonable degree of control over the intensity of oil and gas sector emissions. In addition, carbon pricing mechanisms and other policies developed at the provincial level are either in place will come into force to drive down emissions from economic sectors outside of oil and gas. The provinces of Ontario and Quebec are home to 62% of Canadians. The level of ambition of these two provinces to combat climate change matches IPCC recommendations for developed countries to cut emissions by 40% by 2030 and 80% by 2050 relative to baseline emissions from the year 1990.

6.6.10 Remaining Countries. In 2012, China, the US, the EU, India, Russia, Indonesia, Brazil, Japan and Canada accounted for 67% of world GHG emissions.[124] Remaining emissions came from the other 158 countries. These lower emitting nations vary considerably by population, economic productivity, natural resources, and intensity of GHG emissions. All of these countries share a common characteristic of a national GHG emissions rate that is inconsequential when compared to the global total. However, the aggregate of emissions from these countries approximates 16 billion tonnes of CO_2 equivalents annually and, as a collective, these countries are a major contributor to the atmospheric accumulation of greenhouse gases.[124]

All nations share in regional obligations to combat climate change, regardless of population, status of economic development, and magnitude of total emissions. Over the next 13 years, if future surface warming is to be held to less than 2°C above pre-industrial temperatures, emissions from individual nations must begin to converge toward a common "fair share" per capita rate. By 2030, annual global emissions must be less than 4.8 tonnes of CO_2 equivalents per person. To achieve this target, aggressive cuts are required from countries with a high intensity of emissions, and emerging and developing nations must grow their economies while controlling emissions per unit of economic growth. The overall process will not be a simple convergence of per capita emissions rates between rich and poor nations. The pathways taken and the time frame to arrive at targets for limiting emissions will be dependent upon resources, circumstance, and the level of national and international ambition to combat the realities of climate change.

A full review of the climate action plans of the remaining 158 countries is beyond the scope of this narrative. Rather, a brief summary of NDCs and national climate action plans of Mexico, as an emerging economy, Nigeria, as a heavily populated developing country, and Australia as high GDP per capita nation are presented.

6.6.10.1 Mexico. Mexico's GDP per capita is close to the world average, and Mexico is currently ranked in 74th place on the UN Human Development Index.[203] Mexico, as such, is generally classified as an emerging economy. In 2012, Mexico had the highest rate of emissions among the remaining 158 countries, and the intensity of emissions on a per capita and on a per unit of GDP basis were close to the world average.[124]

Mexico's NDC is based on an unconditional 25% reduction in GHG emissions by 2030 relative to a business as usual scenario.[442] A second more ambitious target of a 40% cut in emissions is conditional to international cooperation and technology transfer, along with access to international climate financing.[442] The potential of a future global agreement on international carbon pricing is specifically mentioned in the NDC as a conditional contributor toward increasing Mexico's capacity to limit future emissions. Based on achieving the unconditional NDC target, year 2030 per capita emissions in Mexico would approximate 5.6 tCO_2eq/person.[442] Emissions would be reduced by another 20% if the conditional target was attained. The level of ambition contained in Mexico's conditional NDC aligns with year 2030 regional obligations to limit future surface warming to less than 2°C.

Mexico's NDC is an international pledge that follows a recent set of ambitious climate change policies introduced by the government. Mexico's General Law on Climate Change[443] and the National Climate Change Strategy 10-20-40 Vision[444] call for a progressive transition to the production of electricity from clean energy sources. Clean energy is to produce 35% of total electricity within 10 years, 40% within 20 years and 50% within 40 years. All economic sectors are targeted for cuts in emissions with the goal of achieving a 50% reduction in total GHG emissions within 40 years. On August 15, 2016, a cooperative agreement was signed by the Mexican stock exchange and the government of Mexico to implement a voluntary pilot scale emissions trading system.[342] The system will be operated over a 12-month period leading to the launch of a full-scale national ETS in 2018.[342] Potentially Mexico's ETS could be linked to an international system such as the California-Quebec-Ontario ETS.

Climate Action Tracker rates the combination of Mexico's NDC and national policies to combat climate change as "medium" and not yet consistent with limiting warming to below 2°C unless other countries make compensatory deeper cuts.[215] CAT described the level of ambition stated in Mexico's climate change policies as remarkable and ground breaking for an emerging country. However, CAT noted that while ambitious targets to cut emissions within economic sectors have been set, in most cases actual measures to meet these targets need to be further articulated. The assessment was completed prior to Mexico's announcement of intention to implement an emissions trading system. Mexico has the potential to be become a model country and a leader among emerging and developing nations in efforts uncouple economic growth from GHG emissions. To reach this goal, Mexico must continue to develop and enact specific measures such as carbon pricing mechanisms that are designed to cost-effectively direct practices toward lower emissions options. As is the case for other emerging and developing countries, international cooperation and access to low-cost carbon funds will facilitate a higher level of ambition to control emissions.

6.6.10.2 Nigeria. Nigeria is by far the most populous country in Sub-Saharan Africa, and the economies of Nigeria and South Africa are the largest in the region.[216] GDP per capita in Nigeria has increased steadily over the past decade but, as of 2015, was only 26% of the world average.[213] Nigeria remains an impoverished nation with over half the population living below the World Bank poverty line on less than $1.90 USD per day.[236,445] Agriculture accounts for 20.8% of national GDP and about 50% of employment.[446] Nigeria's food production has not kept pace with demands, and the country is a net importer of food despite the prominence of agriculture in the society. Agriculture remains highly inefficient and reliant upon low intensity antiquated practices. The country has vast reserves of oil and gas yet has struggled to leverage the potential of these resources to increase standards of living. In 2014, life expectancy was 52.8 years, and Nigeria ranked in the Low Human Development category of the UN Human Development Index.[203]

In 2012, 37.4% of Nigeria's total GHG emissions came from forestry and other land use.[124] Annual deforestation rates exceed 3% and are driven by the use of wood for cooking and heating, along with logging and subsistence level agricultural practices.[447] Between 1977 and 2010, forests and other wooded lands declined from 39.4% to 16.4% of the total land area of Nigeria.[447] Rampant uncontrolled deforestation, unsustainable agricultural practices and increased climate variability have depleted the soils such that 64% of the total area of Nigeria is impinged by desertification.[448]

Per capita emissions in Nigeria, including emissions from deforestation and land degradation, were less than half of the world figure in 2012 and reflected the state of economic development of the country. Transportation, manufacturing, construction, and electricity production sectors were relatively minor contributors to emissions. In 2012, 88% of total emissions came from land use and forestry, waste, agriculture, oil and gas sector activities and other fuel combustion.[124]

In the context of Nigeria's reality, concepts such as decarbonization of economic sectors and electrification of transportation and industry come across as abstract first world problems. In the absence of extensive infrastructure and industrialization, emissions from the poorest countries do not factor into equations for atmospheric GHG accumulations. However, the goal of Nigeria and all developing countries is to stimulate economic growth and improve standards of living. One billion people live in Sub-Saharan Africa, and this region will account for most of world population growth over the current century. Over time, the pathways of economic growth taken by low income African nations will become important components in the success or failure of global efforts to combat climate change. Given the current low rate of emissions, there is some short-term leeway for controlled responsible enhanced use of fossil fuels in the developing world. However, if a repeat of the emissions intensive industrialization of China were to occur among the nations of Sub-Saharan Africa, the consequence would be a future of catastrophic damage from uncontrolled climate change.

Nigeria's NDC presents a set of objectives for economic growth and social development combined with emissions reduction targets relative to a business as usual (BAU) reference scenario. Calculation of the emissions baseline is based on projections for sustained strong economic growth and a year 2030 BAU projected emissions rate of 0.9 Gt of CO_2 equivalents.[449] Nigeria has pledged to unconditionally reduce year 2030 emissions by 25% relative to the BAU baseline.[449] A second target of a 45% emissions reduction is conditionally dependent upon international cooperation, technology transfer and access to carbon funds.[449] Success in achieving Nigeria's conditional NDC target would result in a per capita emissions rate of about 1.9 tCO_2eq/person.[449] This rate is well below the fair share population-based concept of an emissions ceiling for Nigeria to meet regional obligations to limit future surface warming to less than 2°C.

The NDC submitted by Nigeria contains an assessment of climate impacts and future vulnerability. Under a business as usual scenario, the document predicts a 10–25% decline in national agriculture productivity with up to an 80% decline in northern regions exposed to extremes of desertification and drought.[449] Water security, soil erosion and sea level rise will have devastating consequences in Nigeria under a business as usual scenario. The southern region of the country is susceptible to increased incidence and severity of floods capable of destroying years of development. The NDC references the 2014 World Climate Change Vulnerability Index that classifies Nigeria as one of the 10 most vulnerable countries in the world.

Nigeria's NDC clearly articulates a position that failure to adequately curtail global emissions will be catastrophic to the country. The NDC pledges ambitious unconditional and conditional targets to limit future emissions as Nigeria's contribution to combat climate change. Among the key measures listed, is a program of climate smart agriculture including integration of livestock, crop production and reforestation.[449] Additional measures include the complete elimination of gas flaring by 2030, a 2% annual increase in energy efficiencies and the building of an extensive capacity for off-grid localized solar voltaic production of electricity.[449]

The positions set forth in Nigeria's NDC are encouraging and indicative of an acceptance of the realities of climate change and the need for sustainable economic development to be largely uncoupled from a future of intensive GHG emissions. However, Nigeria remains encumbered by conflict, corruption, widespread poverty, deforestation, and embedded, low productivity, land use practices. Actual implementation of effective measures to adapt to and mitigate climate change must overcome existing barriers and the limits of domestic financial resources. International cooperation, technology transfer, and access to carbon funding mechanisms can contribute to the efforts of Nigeria to embark on a low emissions pathway toward alleviation of poverty and sustainable economic growth.

6.6.10.3 Australia. Australia and Nigeria are at the extremes of economic productivity and social development. The latest World Bank indicators show a 22-fold difference in GDP per capita between Australia and Nigeria.[213] Australia ranks second among all countries on the UN's Human Development Index.[203]

By 2012, net annual per capita GHG emissions from Australia, including land use and forestry equaled 30.6 tonnes of CO_2 equivalents per person and exceeded all other high GDP, emissions intensive countries.[124] Fossil fuel combustion produced 85% of Australia's electricity in 2012, with coal-fired plants accounting for nearly 61% of total power production.[450] Agriculture is an important sector of Australia's economy and accounted for 28% of total emissions.[124] The extreme intensity of emissions in Australia is a result of high consumption of fossil fuels across economic sectors and the prominent role of agriculture in the national economy.

Australia ratified the Kyoto protocol and, despite a lack of action to curtail emissions, was able to meet targets for the first reporting period ending in 2012.[451] The level of ambition contained in Australia's Kyoto pledges can be accurately described as weak and close to non-existent. In 1990, Australia's net emissions were exceptionally high due to extensive land clearing programs. Australia demanded that forestry and land use be included in establishing year 1990 baseline emissions, and as such, Australia's Kyoto targets referenced a year of peak emissions. Australia then pledged to limit increases in future emissions during the first Kyoto reporting period ending in 2012 to less than 8% above the 1990 baseline.[451] While Australia led all other developed countries in per capita rates of GHG emissions, this country easily met first reporting period Kyoto targets. Under the rules of the protocol, Australia's surplus in emissions allowances were transferred to the second reporting period extending to 2020. Now, Australia is positioned to meet what are weak second reporting period Kyoto targets while continuing to operate an emissions intensive economy.

Australia's NDC, as submitted under the Paris Agreement, pledged a year 2030 target of an economy wide reduction in total GHG emissions by 26–28% relative to emissions on record for the year 2005.[452] Achieving this target would result in a per capita emissions rate of about 15 $tCO_2eq/person/year$.[452] This rate of emissions grossly exceeds any concept of a fair share contribution to global efforts to combat climate change.

Climate Action Tracker is highly critical of Australia's NDC and of current policies and measures to control emissions. Based on current polices, CAT projects a 39% increase in emissions by the year 2030 relative to a 1990 baseline that excludes forestry and other land use.[451] This rate of emissions will be almost double the target specified in Australia's NDC, and per capita emissions would be in the range of 23–24 $tCO_2eq/$ person.[451] Specifically, CAT is critical of federal government actions and inactions leading to uncertainty among stakeholders. In 2011, the labour government in Australia introduced a carbon tax of $23 per tonne that covered 60% of emissions.[453] Carbon pricing was in effect from 2012–2014 but was revoked following a change in government in 2014.[453] Recently, Australia placed an emissions cap (stated to be a safeguard mechanism) on the voluntary Emissions Reduction Fund trading system.[454] In essence, Australia has created a broad spectrum ETS with the potential to cover 50% of total emissions. However, the system is based on voluntary, incentive-driven, participation and initial caps on emissions have been set to high historical rates. The current structure of Australia's ETS will not result in a market price for carbon sufficient to drive down emissions.

Australia's Clean Energy Act states a long-term goal to reduce year 2050 emissions in Australia by 80% below year 2000 levels and to contribute to global efforts to limit future surface warming to below 2°C.[455] However, current climate change policies and measures are lacking in the potency required to achieve deep decarbonization of the economy by mid-century. There is potential for Australia to strengthen the national ETS and use carbon pricing as a market-based driver of transition to lower emissions practices. Based on Australia's NDC and current set of domestic policies and measures, CAT rates Australia's climate change mitigation efforts as inadequate and, if other countries were to follow Australia's approach, global warming would exceed 3–4°C.

Australia's history of participation in the Kyoto Protocol has been to promise little and then to deliver on these promises. With universal acceptance of the Paris Agreement, governments are beginning to acknowledge responsibility and to take meaningful action to reduce emissions. Australia's inaction and lack of ambition is an exception to the trend and, in essence, a form of climate change denial. Beyond 2030 toward mid-century, the aggregate of global economies must progressively decarbonize to minimize climate change damage. Countries that persist in operating emissions intensive economies will become outliers. These economies will be costly, isolated, and out of sync with international low carbon trading practices.

6.7 International Cooperative Initiatives (ICIs)

An International Cooperative Initiative refers to any international organization of states, municipalities or companies with shared interests in mitigating climate change. Examples of intergovernmental ICIs are The Climate Group States and Regions Alliance, The C40 Cities Climate Leadership Group, The Carbonn® Climate Registry, and the

Compact of Mayors.[122] Some of the more prominent international groups of companies that can be classified as ICIs include the Business Environmental Leadership Council, the Cement Sustainability Initiative, Ultra-Low CO_2 Steelmaking and the Global Fuel Economy Initiative.[122]

Calculating the potential contribution of an ICI to global mitigation efforts is complex since emissions reductions can fall under national programs and NDC targets. Often, ICIs function as facilitators and contributors to achieving national emissions reduction targets. This ground level role in the implementation of national climate change action plans is important, and the international transfer of information and technology within an ICI can contribute to the implementation of cost-effective solutions within countries. In some circumstance, participation in an ICI can not only facilitate but extend national and regional ambitions to curtail emissions.

The C40 Cities Climate Leadership Group is a network of over 80 megacities and smaller innovator cities representing 9% of world population and about 25% of global GDP.[456] C40 Cities recently published Deadline 2020 as an analysis of the contribution that member cities can make toward limiting future surface warming to 1.5°C.[457] The analysis uses a global carbon emissions budget of 375 Gt that can be released between 2016 and 2100. This figure is close to IPCC estimates for the 1.5°C Scenario and is only 37% of the budget that aligns with the Well Below 2°C Scenario. The C40 Cities model assumes that only 6% of this global emissions budget is available to members and set a year 2030 target for each city to converge to a common per capita emissions rate of 2.9–3.0 tCO_2eq/person. This level of ambition exceeds the global year 2030 per capita emissions rate of 4.8 tCO_2 that aligns with the 2°C scenario. Going forward to 2030, high emitting cities are required to substantially cut emissions while, in some circumstance, low emitting cities in developing countries have some leeway to increase emissions. The report goes on to define specific opportunities to control emissions through urban planning, transit, energy, buildings, and waste management pathways. Over $1 trillion in investment would be required over the interval to 2030 and $375 billion would be needed prior to 2020 to kick start climate change action plans. The Deadline 2020 report concludes that global deployment of C40 pathways has the potential for cities to deliver 40% of the total emissions curtailment required to limit surface warming under the 1.5°C Scenario.

There are no binding mechanisms to compel C40 cities to achieve emissions reduction targets. However, organizations such as C40 Cities have an important role in bringing together international cities, regions and industries with similar vested interests and commonalities in combating climate change. An interesting example of the potential for an ICI is Houston, Texas. The state of Texas does not have a climate action plan; however, the largest city in the state is a member of C40 Cities and has pledged to achieve an 80% cut in emissions by 2050 relative to a 2005 baseline.[458] This type of bottom-up initiative can, to some extent, compensate of inaction at higher levels of government.

6.8 The Price of Carbon

Surface warming in excess of 3–4°C above pre-industrial temperatures will have devastating consequences to future generations. An acceptance of the science and reality of climate change leads to a further acceptance of intergenerational responsibility to effectively curtail emissions over the short and medium term. The follow-up questions are economic and can be summarized as follows:

1. What will be the actual costs incurred by future generations if surface warming were to exceed 3°C above pre-industrial temperatures?

2. What are the immediate costs that will be incurred by the current generation to curtail emissions and limit future surface warming?

3. Is there a cost-benefit to curtailing emissions and limiting future surface warming to well below 2°C, and ideally 1.5°C, and can the current generation afford this cost?

Assigning a true cost to current emissions of GHGs requires an estimate of the costs of climate change damage that will be incurred by future generations and a conversion to present day values. This modelling exercise is extremely complex and compromised by uncertainties and ethical dilemmas. However, estimates of the true cost of carbon do provide a rough indication to policy makers as to the potential savings to future generations that can be achieved by cutting emissions.

6.8.1 The Social Cost of Carbon. On August 8, 2016, the 7th Circuit for the United States Court of Appeals rejected arguments put forward by a group of refrigerator manufacturers against the Department of Energy using the "Social Cost of Carbon" as a factor in establishing refrigerator efficiency ratings.[459] In Zero-Zone vs. the Department of Energy, the complaintive referred to the US government derived Social Cost of Carbon as "arbitrary and capricious." The court disagreed and, in a unanimous decision, dismissed the claim stating that the "DOE's engineering analysis, including its use of analytical models was neither arbitrary nor capricious."

Zero-Zone vs. the Department of Energy applies only to refrigerator regulations and covers the states of Illinois, Wisconsin and Indiana. However, the ruling provides a legal precedent for the use of the Social Cost of Carbon in developing regulations governing emissions related practices in the United States.

In 1993, by executive order, then President Clinton put in motion a prolonged process that would eventually lead to a requirement that government agencies assess and establish a value for the "Social Cost of Carbon" or the "Social Cost of Carbon Dioxide" (SC-CO_2).[460] Eventually, during the Obama administration, an Interagency Working Group (IWG) was established.[461] This group of experts came up with a distribution of the SC-CO_2 by year. The National Academies of Sciences, Engineering and Medicine reviewed the IWG models and agreed with the pricing mechanisms. In 2016, a revised technical support document was issued with recommendations for a year 2015 SC-CO_2

of $36 per metric tonne of carbon dioxide released to the atmosphere with future increases to $50/t by 2030 and $69/t by 2050.[461]

The US derived $SC-CO_2$ is based on a comprehensive estimate of long-term future climate change damages and includes changes in net agricultural productivity, human health, property damages and other impacts of climate change. The models used by the US and other countries to calculate a global $SC-CO_2$ assumes that climate change will not affect rates of economic growth. A recent Stanford University study introduced additional parameters of projected effects of climate change on economic growth rates in developed and developing countries, along with costs of climate change adaptations. When these factors where included, the model generated a revised $SC-CO_2$ that was 6-fold greater than the current US estimates developed by the Interagency Working Group.[462] The US $SC-CO_2$ is one of many estimates of the Social Cost of Carbon. The UK currently uses a range equivalent to $17–56 USD per tonne of CO_2 for the year 2020 Social Cost of Carbon.

Any attempt to establish a global $SC-CO_2$ can be criticized, based on differences in regional vulnerabilities to damage. While GHGs are evenly distributed, the damage incurred within highly vulnerable regions will be disproportionally greater. A metric tonne of CO_2 emitted by a coal burning power plant in Western Canada following a business as usual scenario will be evenly distributed in the atmosphere but, in 50 years, will contribute to extensive per capita damage and human misery in Nigeria and other highly vulnerable countries while having much less of an impact in Canada. Modelling the Social Cost of Carbon is beset by a range of other ethical issues such as applying a dollar value to water and food insecurity, and climate change induced human conflict and mass migration.

Given the complexities of attempting to place a cost on current CO_2 emissions for future climate change damage, one could simply conclude that the costs will be extreme but cannot be boiled down to an average global value due to differences in regional vulnerability and other ethical considerations. However, this conclusion ignores the utility of assigning a monetary value to the Social Cost of Carbon. A global $SC-CO_2$ becomes an external deferred cost of emissions that must either be avoided by elimination of the emissions or paid for by future generations in the form of climate damage. The recent US federal court decision in Zero-Zone vs. the Department of Energy is fundamentally an acceptance of the science and economic reality of anthropogenic climate change. The court concluded that there were no grounds to prevent the DOE from using the Social Costs of Carbon in developing regulations for refrigerators in Illinois, Wisconsin, and Indiana.

Ultimately, a legal acceptance of the concept of a Social Cost of Carbon can provide the foundation for the implementation of effective carbon pricing mechanisms. High Court rulings on the validity of the Social Cost of Carbon, in essence, support, and, indeed, may require governments to take action to implement regulations that apply costs to limit current emissions rather than have these costs deferred and imposed at much higher levels on future generations in the form of climate change damage.

The Social Cost of Carbon should not be considered to be a method for defining a carbon price through taxation or cap and trade systems. Rather, an acceptance of the concept of a Social Cost of Carbon should compel governments to implement carbon pricing mechanisms and other policies and measures designed to reduce emissions following the science-based timelines recommended by the IPCC and the International Energy Agency to limit future surface warming to less than 2°C.

6.8.2 Marginal Abatement Costs (MAC). Marginal abatement costs are estimates of the costs entailed to avoid emissions of 1 metric tonne of CO_2 equivalents using a specified technology or change in practice.[463] MAC assessments can also include the total annual abatement potential (tonnage of CO_2 release that can be avoided) through widespread adoption of a given process. A detailed MAC evaluation will include side benefits such as lower costs of energy through implementation of high efficiency technologies or practices. However, in many cases, the cost savings from side benefits are difficult to estimate and are not included in MAC estimations. Improvements in air quality following fuel switching away from coal combustion can lead to savings in health care costs; however, this secondary benefit is often not captured in the abatement cost estimates. MAC valuations are highly dependent upon the quality of the data that goes into estimates and, as such, are generally accurate for mature processes. Cost estimates are far less certain for new technologies for which there is often limited or no commercial experience. The costing process does not consider non-monetary barriers to implementation such as public preference and social acceptance of some low emissions options.

Some technologies and changes in practices come with a negative MAC value. As an example, energy saving through installation of high efficiency lighting will result in a cost savings when compared to continued use of low efficiency lighting. The emissions savings realized by reducing the demand for power production comes at a net negative cost. The same analysis generates negative MAC values for the implementation of other measures to improve energy efficiencies such as building management systems and smart metres that enable customer optimization of power use.

The majority of emissions abatement measures come at a cost when compared to business as usual scenarios. MAC values will vary considerably between countries and regions and economic sectors. In the US, emissions cuts based on improved crop rotation, enhanced soil sequestration of carbon, forestry management, low and mid-cost afforestation, energy efficiency retrofits to commercial and residential buildings and wind power installations come with a relatively low-cost per tonne of CO_2 avoided.[464,465] In comparison, solar thermal, residential solar photovoltaic, carbon capture and storage applications, and coal to gas fuel switching have a higher cost per unit of CO_2 abatement.

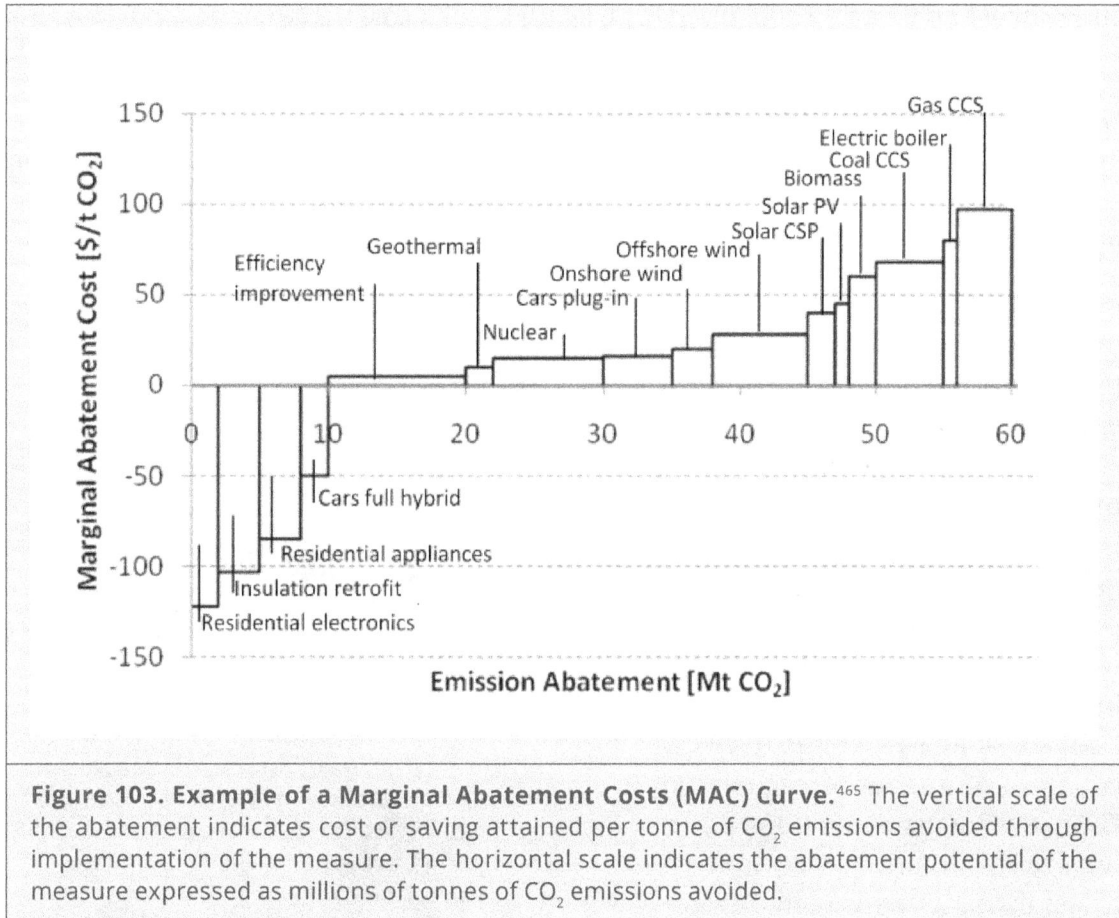

Figure 103. Example of a Marginal Abatement Costs (MAC) Curve.[465] The vertical scale of the abatement indicates cost or saving attained per tonne of CO_2 emissions avoided through implementation of the measure. The horizontal scale indicates the abatement potential of the measure expressed as millions of tonnes of CO_2 emissions avoided.

MAC estimates can provide governments and stakeholders with an indication of costs as emissions are progressively cut over time. This information is of greater significance in designing policies to cut emissions outside of carbon pricing mechanisms. An ETS will function to automatically arrive at lowest cost mitigations options that are based on free market trading of emissions allowances. Initially when emissions caps are high, the price of carbon will be low, and low-cost technologies and changes in practice will predominate. As the cap drops, the price of carbon adjusts to higher levels and brings into play higher cost mitigations. Progressive programs of carbon taxation will have a similar effect.

6.8.3 Costs of Mitigation. Since the year 2000, over 20 countries have reduced net emissions.[466] European countries have been the leaders among advanced economies in combating climate change, and there is now a significant database on emissions reduction and GDP change during the early phase of decarbonization of advanced economies. Germany, France, and the UK are the largest economies that have introduced effective climate change policies. Between 2000 and 2014, GHG emissions were reduced by 12–20% in these countries without evidence of a negative impact on economic growth.[466] There may be some indication of a slight decline in industry share of GDP; however, this trend was also evident in the US where emissions were cut by only 6%.[466] Since 2000, there is no evidence of a correlation between emissions reduction and declining economic performance.

Going forward toward deeper levels of decarbonization, costs will be incurred to implement effective climate action plans. These costs will vary considerably between countries based on resources, public perceptions, acceptance of low emissions options, and the cost-efficiency of emissions reduction policies and measures.

An idealized scenario would consist of unhindered access to available technologies and practices and a worldwide ETS covering all emissions. A study by the Organization of Economic Co-operation and Development concluded that mitigations costs expressed as a percentage of future GDP are negligible under such an idealized scenario.[467] In models based on a less than ideal scenario of an ETS limited to OECD countries or a series of fragmented ETSs, emissions reductions pathways to limit future surface warming to less than 2°C are projected to cost OECD countries about 0.5% of GDP in 2030 and 1.7% in 2050.[467]

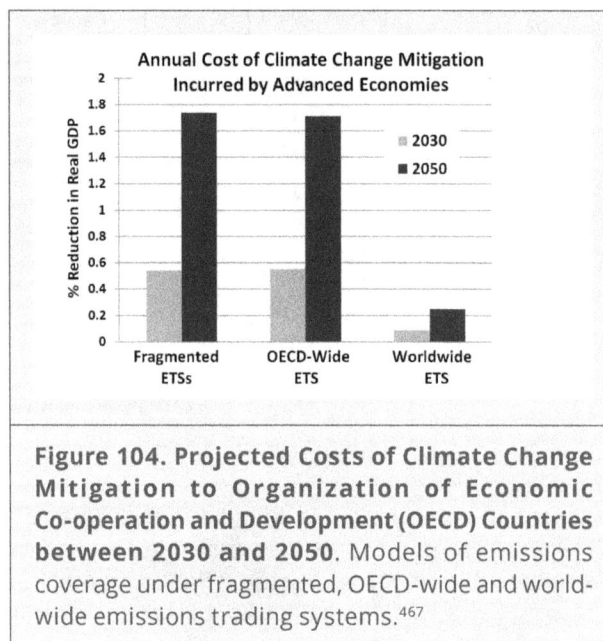

Figure 104. Projected Costs of Climate Change Mitigation to Organization of Economic Co-operation and Development (OECD) Countries between 2030 and 2050. Models of emissions coverage under fragmented, OECD-wide and worldwide emissions trading systems.[467]

Economic models that assess employment patterns during transition to decarbonized economies are dependent upon the pathway and efficiency of the transition process and other factors such as labour rigidity to reallocation and distribution of funds collected by carbon pricing mechanisms. The OECD predicts some degree of job reallocation but no net effect on total employment by the year 2030.[467] Basically, jobs will be lost in the traditional fossil fuel-based energy sectors while job creation will occur in other energy and non-energy sectors that will benefit from transitions to low emissions practices.

The European Commission estimates that 1.5% of annual GDP (€270 billion) will need to be invested to achieve deep decarbonization of economies and to cut mid-century emissions by over 80% relative to the 1990 baseline.[378] However, the commission indicates that this investment will be largely recovered or even overcompensated by annual fuel savings estimated at €175–320 billion. Improvements in air quality are projected to result in additional savings in health care costs plus savings in the costs of air pollution control measures that could total €88 billion annually by mid-century.

Recently, the Government of Canada published a detailed assessment of the overall cost to society by achieving NDC targets.[468] Between 2014 and 2030, under a business as usual scenario, Canada's GDP per capita is projected to increase by 11.5%. Achieving NDC targets would reduce year 2030 GPD per capita by 1–3% when compared to the BAU scenario. The model arrived at a range of costs given uncertainties as to uniformity in the application of carbon prices across the country, along with other uncertainties and complexities in assessing cash flows.

An idealized scenario of a global carbon price is unlikely to evolve over the short-term leading to 2030. Rather, the immediate post-Paris Agreement world will likely consist of a patchwork of subnational, national, and regional carbon pricing mechanisms with some degree of linkage. Many countries have yet to indicate an interest in carbon pricing or energy efficiency trading mechanisms. If these countries are to meet regional obligations to curtail emissions this will be dependent upon higher cost, non-market driven policies and measures.

The IPCC estimates that total global policy costs to limit end-of-century surface warming to less than 2°C will be in the range of 1–1.5% of global GDP over the period from 2010 to 2050.[267] About 80% of this cost applies to GHG emissions reduction. The remaining 20% will cover additional costs to optimize co-benefits of energy security and improved air quality that can be achieved, along with cuts to GHG emissions.

The Stern Review on the Economics of Climate Change concluded that the costs of mitigations would approximate 1% of global GDP.[469] The report was authored by Nicholas Stern, Chair of the Grantham Research Institute on Climate Change and the Environment at the London School of Economics, and published in 2006. Cost estimates were based on stabilizing end-of-century atmospheric CO_2 concentration between 500–550 ppm. Since the time of publication, the author has revised his estimates to a cost of effective climate change action approximating 2% of global GDP going forward to mid-century.

Over the long-term, extending to mid-century and beyond, carbon capture and storage will likely become a significant contributor to the package of technologies required to arrive at net zero GHG emissions. At a minimum, CCS will be needed in the latter stages of complete decarbonization of economies to limit emissions from cement manufacture and industrial processes where avoidance of CO_2 release is not possible or extremely impractical. In most aggressive mitigation models, production of electricity using bioenergy with CCS in the second half of the century functions to supplement afforestation in providing sufficient capacity for withdrawal and storage of atmospheric CO_2. Models that limit future surface warming to less than 2°C in the absence of CCS require massive programs of afforestation that may not be possible given land requirements for food production in the future. In the absence of CCS, the costs and practicality of compensatory measures are uncertain but could double or triple total mitigation costs between 2015–2100.[469]

In the immediate future, delays in arriving at peak emissions and transitioning to an orderly pathway of annual progressive cuts in emissions will have profound effects on the viability and costs to limit future surface warming to less than 2°C. A short-term continuation of business as usual practices extending to 2030 will quickly lead to an overexpenditure of the 1 trillion tonnes of CO_2 equivalents remaining in the 2011–2100 budget of total emissions. Under this scenario, future attempts to limit atmospheric GHG concentrations to about 450 ppm of CO_2 will require an abrupt and costly changeover of practices and an excessive longer-term reliance on BECCS. Total mitigation costs to effectively limit future surface warming will increase in direct correlation with the magnitude of the year 2030 emissions gap. Models of limited immediate action on climate change are consistent with a delay in peak emissions past 2030 and, by mid-century, a 30–80% increase in total mitigation costs.[267]

6.8.4 International Climate Funds and Developing Countries. Serious levels of international climate financing did not begin to flow until after the 2009 commitment by developed countries to mobilize $100 billion US annually by the year 2020. Between 2010 and 2012, $30 billion was delivered and, as of 2014, total flows from advanced economies had reached $61.8 billion per year with 70% of this total coming from public funds.[470,471]

Distribution and use of international funds to mitigate and adapt to climate change is at the early stages. Thus far, funds have been allocated based on the priorities of capability to implement and emissions abatement potential. Half of the dispersed climate funds have gone to support climate change mitigation projects primarily in fast growing economies.[359] These projects are focused on implementation of clean energy technology and improvement in energy efficiencies. India has been the leading recipient of climate

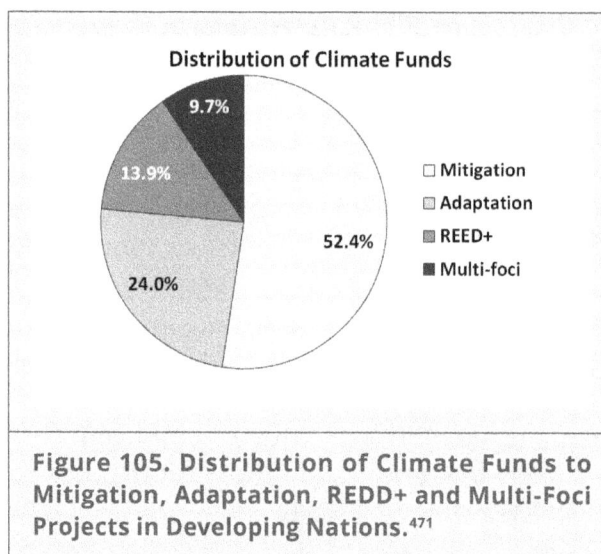

Distribution of Climate Funds

9.7%
13.9%
52.4%
24.0%

- Mitigation
- Adaptation
- REED+
- Multi-foci

Figure 105. Distribution of Climate Funds to Mitigation, Adaptation, REDD+ and Multi-Foci Projects in Developing Nations.[471]

change mitigation funds, followed by Brazil, Morocco, Mexico and South Africa.[359] Most of this funding has been in the form of long-term concessional loans.[470] For each of these countries, the level of ambition to limit year 2030 emissions either approximates or exceeds the population-based concept of a fair share contribution to global efforts to combat climate change.

Morroco has successfully accessed considerable funding from the Clean Technology Fund and has become a world leader in renewable energy. Morocco has no fossil fuel resources, and the nation of almost 40 million is entirely dependent on imports. In 2016, Morocco launched the first phase of what will be the world's largest concentrated solar power plant (CSP).[472] The Noor-Ouarzazate CSP plant will have the capacity to provide zero-emissions power to 1.1 million people when fully operational in 2018.[472]

Climate Action Tracker rates the combination of Morroco's NDC and climate change policies and measures as "sufficient" and aligned with a fair share contribution to limit future surface warming to less than 2°C.[215] By the year 2030, Morocco's per capita emissons are projected to be well below the year 2030 global average of 4.8 tCO$_2$eq/person aligned with restricting future surface warming to less than 2°C.[215]

Funding for adaptation to climate change is widely dispersed among highly vulnerable, generally low income countries. Bangladesh, Niger, Cambodia, Mozambique and Nepal are the leading recepients of funding for adaptation projects.[359] Smaller projects are generally funded by grants from various climate funds.[359]

To date, REDD climate funding has been allocated on the basis of a combination of readiness and mitigation potential. Brazil has received 30% of the total funding.[359] Remaining funds to reduce emissions from deforestation and forest degradation and have been widely dispersed to cover projects in Indonesia, the Democratic Republic of Congo, Mexico, Guyana, Ghana, Colombia and a multiple of other developing countries.[359]

By 2030, climate funds will be integral to global efforts to uncouple GHG emissions from economic growth in developing and emerging economies. These funds will be vital to curtail and eliminate emissions from deforestation and, eventually, to increase carbon-sequestering forest biomass in developing countries. Finally, climate funds will be needed to increase capacity for climate change adaptation within least developed countries. In an idealized future global scenario, the combination of mitigation and adaptation will be sufficient to avoid significant climate change damage within the most vulnerable countries and regions on Earth.

6.8.5 Risk-Benefit Analysis of Climate Change Mitigation and Adaptation?
Investment decisions by companies and governments are often based on cost-benefit analysis and threshold rates of return. This type of analysis using estimates of the Social Cost of Carbon and Marginal Abatement Costs is complex and marred by uncertainty, regional differences, long time spans and a multitude of ethical dilemmas. Applying typical cost-benefit analysis as a decision-making tool for the implementation of climate change action plans is a suspect approach at best.

Sir Nicholas Stern served as Chief Economist and Senior Vice-President of the World Bank before his appointment to the Chair at the Grantham Institute and is considered among the world experts on the economics of climate change. The conclusions of Stern's nearly 700 page review[469] on the cost of future climate change damage relative to the projected costs of effective mitigations and adaptations are mostly qualitative and can be summarized as follows:

1. Climate change will affect the basic elements of life for people around the world —access to water, food production, and the environment.
2. All countries will be affected. The most vulnerable—the poorest countries and populations—will suffer earliest and most, even though they have contributed

least to the cause of climate change. The costs of extreme weather, including floods, droughts and storms, are already rising, including for rich countries.

3. The overall costs and risks of climate change will be equivalent to losing at least 5% of global GDP each year, now and forever. If a wider range of risks and impacts is taken in account, the estimates of damages could rise to 20% of GDP or more.

4. The investment that takes place in the next 10–20 years will have a profound effect on the climate in the second half of this century and in the next. Our actions now and over the coming decades could create risks of major disruptions to economic and social activity, on a scale similar to those associated with the great wars and the economic depression of the first half of the 20th century. And it will be difficult or impossible to reverse these changes.

5. The costs of stabilizing the climate are significant but manageable; delay would be dangerous and much more costly.

6. The world does not need to choose between averting climate change and promoting growth and development. Changes in energy technologies and in the structure of economies have created opportunities to decouple growth from greenhouse gas emissions. Indeed, ignoring climate change will eventually damage economic growth.

7. With strong, deliberate policy choices, it is possible to reduce emissions in both developed and developing economies on the scale necessary for stabilisation in the required range while continuing to grow.

Stern extends his conclusions with recommendations to expand and link emissions trading schemes around the world, to boost investments in innovation, to take strong action to reduce deforestation, and to fully integrate climate change into development policy with rich countries honouring their pledges to support developing countries.

Basically, Stern concludes that a lack of effective immediate action on climate change has no moral or economic justification. The damage that will be imposed on future generations will be extreme such that a consideration of returns on investment becomes absurd. The current generation and political leadership are compelled to implement effective and immediate climate action plans. Discussions should and must move well past any consideration of the value of action and focus instead on how to achieve emissions reduction targets and progressively decarbonize economies as efficiently and cost-effectively as possible.

6.9 The Politics of Carbon (The US Example)

On June 26, 2009, the US House of Representatives narrowly approved the American Clean Energy and Securities Act of 2009 (ACES).[473] The bill contained provisions for a national emissions trading scheme and other measures designed to achieve emissions reductions targets of 42% by 2030 and 83% by 2050 relative to a 2005 benchmark.

In theory, the US-ETS would have been the market-based carbon pricing mechanism required to achieve the ambitious climate change agenda put forward by the Obama Administration. The act was met with stiff unified Republican opposition in the Senate and was never brought to the floor for discussion or vote.[474]

Over the past decade, the United States has become the world's largest producer of oil and natural gas and has continued to produce significant quantities of coal for power production.[475] With advances in exploration, and in fracking and horizontal drilling technologies, the country has become increasingly energy self-sufficient such that the fossil fuel industry has gained economic prominence and political clout. This fact has created a substantial barrier to the implementation of policies and measures to curb emissions.

The politics of carbon become complicated at the ground level in constituencies where the fossil fuel industry is a significant contributor to wealth creation and employment. These jurisdictions are not encumbered by energy insecurity, and they benefit from the production, use, and export of fossil fuels. In isolation, climate change mitigation measures will lead to a downturn in emissions-intensive subsectors of the economy. Barriers to the implementation of climate change action plans are magnified in countries and regions with substantial oil, gas, and coal activities. Political indifference or outright denial of climate change is often seen as serving the best interest of the electorate through support of existing industries.

Over the eight years of the Obama Administration, Congress basically stonewalled any significant climate change legislation. The only viable path to bring about climate change policies was through the EPA and the Clean Air Act. The US-Canada agreement to cut methane emissions from the oil and gas sector and the CPP are based on EPA development and imposition of GHG emissions regulations to protect the health of the population. These measures are open to legal challenges and are far less efficient and effective means to drive down emissions than market-based carbon pricing mechanisms. As previously described, lawsuits filed by opponents of the CPP led to the Supreme Court decision to stay implementation pending lower court judicial reviews.

Short-term economic arguments against climate change action ignore the fact of extreme hardship and costs that will be imposed upon future generations, following a business as usual continuation of current practices. Further, an orderly transition of economies using market-based carbon pricing mechanisms will lead to a reallocation of jobs to growing low-carbon industries with minimal effects on economic performance.

At the national level in the US and in other countries with substantial fossil fuel-based energy sectors, the political challenge to implement effective emissions reduction measures is immense. A well-informed electorate in combination with competent and visionary political leadership is required at all levels of government if these barriers to implementation are to be overcome.

6.9.1 2016 United States Presidential Election. By October 5, 2016, at least 55 parties to the UNFCCC, accounting for at least 55% of total global greenhouse gas emissions, had ratified the Paris Agreement.[265] Thirty days after reaching this milestone, the agreement automatically entered into force. Four days later, on November 8, 2016, Donald J. Trump became president-elect of United States of America.

Prior to the election, Mr. Trump had called climate change a hoax invented by the Chinese to make US manufacturing non-competitive[476], and had stated that, if elected, he would cancel the Paris Agreement and withdraw any funding for UN programs related to global warming.[477] Since the election, Mr. Trump has attempted to backtrack by stating that his tweet claiming that climate change was a hoax was in jest. However, the appointment of Scott Pruitt to head the EPA is consistent with the intensions of the president to cancel US national and international climate change programs and commitments. Mr. Pruitt is on record as questioning the contribution of anthropogenic carbon dioxide emissions to global warming, and his stated opinion conflicts with the clear science-based conclusions reached by the organization that he was appointed to lead.[478] All indications are that a Trump-era EPA will no longer function to develop regulations governing GHG emissions. In announcing a withdrawal from the Paris Climate Agreement, Donald Trump confirmed that, under his leadership, the current federal government has no intention of implementing meaningful climate action policies or honoring international commitments.

Without doubt, Donald Trump's election victory is a setback to US and global efforts to combat climate change. However, the doom and gloom scenarios of the impact of a Trump presidency on global progress to curtail GHG emissions may be overstated.

During the Obama Administration, impediments were such that Congress did not pass any meaningful climate change legislation. From 2008 to 2016, federal capabilities to act on the climate change file were limited and dependent upon EPA implementation of regulations to limit GHG emissions.

Despite an inability to deliver at the highest levels of government, there has been progress within the country to curtail emissions. California has the largest economy in the union and has unilaterally implemented a broad spectrum ETS as a cornerstone policy of one of the most ambitious climate change action plans in the world. Successive state governments in California have acted responsibly in acknowledging the realities of climate change and have implemented policies and measures to cut emissions and progressively decarbonize the state economy as required to meet regional obligations to limit future surface warming to less than 2°C. Going forward to mid-century, California will benefit from taking a leadership role in developing and implementing low emissions practices and will be positioned for success as an early-stage decarbonized economy.

California and New York have established targets to cut emissions by 40% in 2030 and 80% in 2050 relative to a 1990 baseline. Recently, New York State Governor Andrew Cuomo announced the Clean Energy Standard mandating that by 2030, 50% of the total electricity supply will be generated by renewables.[479] State-run nuclear and hydro power plants will be maintained as New York transitions to decarbonized power production as a core component of an overall emissions reduction plan. Reportedly, Governor Cuomo has ordered state agencies to engage with partners in the Regional Greenhouse Gas Initiative and the California-Quebec-Ontario ETS with the intent of exploring the potential of an expanded partnership that could evolve into a large, linked North American carbon market.[480]

The formation of the US Climate Alliance, led by California, and the states of New York and Washington, is indicative of a growing movement in the US to implement effective climate change action programs at the state and municipal levels, largely independent of federal initiatives. The intention of the US Climate Alliance is to form a consortium with the objective of cutting GHG emissions as specified in the NDC submitted by the US under the Paris Agreement. This process is at the early stages, and outside California, considerable work is required to implement serious policies and measures to curtail emissions.

Texas and other states with substantial fossil fuel industries and similar political inclinations are unlikely to join the US Climate Alliance or to otherwise show support for the principles of the Paris Agreement.

The current political leadership in the US has essentially fragmented the country into states and municipalities that recognize the importance of meaningful action to prevent future global warming and those that do not. Under this situation, the overall national rate of GHG emissions will be dependent upon a bottom-up process of state by state implementation of climate change action plans and not on top-down federally enacted policies and regulations.

Carbon pricing mechanisms and other government initiatives implemented by states within the US Climate Alliance will drive innovation and change practices toward lower emissions options. With time, demand for fossil fuels will diminish, leading to a decline in US production and reduced emissions from the fossil fuel sectors of the national economy. Further as the cost of renewables and products such as zero-emissions battery electric vehicles continues to decline, low emissions options will become increasingly cost-competitive.

Texas leads all other states in the production and use of oil and natural gas, and 30% of the total oil refining capacity of the country is located in Texas.[481] Fossil fuels dominate overall energy consumption in Texas, and the fossil fuel sector is a major contributor to the state economy. However, Texas is rich in wind and solar energy potential, and as costs become increasingly competitive, renewables have begun to penetrate the energy mix. As of March 2017, Texas led the nation in wind power generation capacity and renewables, primarily wind energy, accounted for 22% of the production of electricity.[481]

Currently, the US provides approximately $5.8 billion per year to climate funds.[259] This level of funding equals 9.3% of total flows into the funds from developed countries.[359]

The UK and Norway both contribute more dollars to climate funds than does the US.[363] Indications are that the US federal government intends to withdraw from UNFCCC financial obligations. The resultant shortfall to climate funds would be significant but not devastating.

The actions and inactions of the current US president and Congress should not override or impede state, municipal, or international efforts to curb emissions and combat climate change. Constitutionally, it may be difficult for the US Climate Alliance to ratify the Paris Agreement. However, these states could abide by the guidelines and have their contributions recognized outside the formal agreement. International climate funding shortfalls caused by the exodus of the US federal government could be compensated by other developed countries that have acknowledged the realities and responsibilities of climate change. Potentially, the US Climate Alliance could contribute to climate funds and thus partially cover the obligations that the United States originally assumed when President George H.W. Bush signed onto the UNFCCC in 1992.

The politics of carbon emissions and barriers to change vary between countries and regions because of political systems, levels of prosperity, natural resources, and public understanding of the issues and engagement in the decision-making process. It is beyond the scope of this narrative to consider the unique landscape of conditions that define the level of ambition to combat climate change for each country. The US example is exactly that—an example of the politics involved and barriers to implementing climate actions plans within one country.

6.10 Activism and the Price of Carbon

The climate movement can be described as a loose collection of non-governmental organizations engaged in activism related to climate change issues. The movement includes a multiple of organizations and groups, some of which have environmental and social agendas that extend beyond climate change activism. Greenpeace, 350.org, and the Sierra Club are among the better-known NGOs with a significant presence in the climate movement.

Typically, climate change activists tend to focus their efforts on disrupting the building of new energy supply infrastructure projects such as oil and gas pipelines and power plants deemed to be damaging to the environment.

The proposed Keystone XL pipeline would bring Alberta bitumen to refineries in Texas. Keystone XL has been highly controversial since inception and environmental groups have been active and vocal in expressing opposition to the project. In 2015, President Obama rejected TransCanada's application to construct Keystone XL stating that the pipeline would not serve the national interest of the US.[482] However, President Trump has reversed this decision and reopened the potential for building the pipeline.

The fixation of climate activists on preventing the construction of oil and gas pipelines will have little to no impact on decreasing future emissions of GHGs. Emissions during pipeline transport of oil and gas are of no consequence and arguments can be made that pipelines are relatively efficient and safe in comparison to rail. As long as demand continues, fossil fuels will be delivered to refineries. A new pipeline provides an additional option for transport but has little to no influence on global demand or on the total volume of oil and gas produced. In the case of Keystone XL, activists rightly point out that production of Alberta bitumen comes with a higher intensity of emissions than other sources of crude oil. However, complete well-to-wheel life cycle GHG emissions are only 23% higher for oil sands bitumen in comparison to conventional crude.[483] Most of the CO_2 emissions come from end-use combustion, which is identical regardless of origin of the crude oil.[483]

Environmental activists are often opposed to nuclear power and, activism has contributed to moratoriums on new builds and to the shutdown of existing nuclear plants. As discussed, nuclear power is among the safest and cleanest of energy sources used for the production of electricity. Many regions of the world are lacking in hydro or other low emissions dispatchable energy options. A ban on nuclear power could lead to a greater reliance on fossil fuels and thus an increase in the emissions of GHGs and non-GHG pollutants.

Construction of large scale hydro projects can disrupt water flow and destroy natural habitats. In some locations, unavoidable damage to ecosystems may be such that the project is not warranted. However, in many circumstances, hydro projects can be designed with minimal ongoing negative impacts to local ecosystems. Activists must consider the facts of objective science-based environmental assessments, along with the benefits of a continuous source of very low emissions electricity before reaching conclusions as the net value of a given hydro project.

Extremists in the climate movement can undermine public understanding and confidence in the science of anthropogenic climate change and mitigation pathways to limit future surface warming. The activist organization 350.org publicly states that atmospheric CO_2 concentrations must be reduced from current levels of 400 ppm down to 350 ppm by the end of the century or "we risk triggering tipping points and irreversible impacts that could send climate change spinning truly beyond our control."[484] The suggestion by 350.org is that mankind has already seriously overspent the total budget of anthropogenic GHGs that can be safely released to the atmosphere. Somehow, to minimize further accumulation, world economies would need to complete the near impossible and economically devastating task of a complete transition to net zero-emissions within 10-15 years. Subsequently, to drawdown atmospheric CO_2 to 350 ppm, the planet would need to be run on a negative emissions balance for the duration of the century.

The climate movement can play a significant role in holding governments and industries to account and in promoting public awareness and understanding of the issues.

However, activists and NGOs must themselves be objective and informed and must present a clear message that is consistent with the science. In addition, activism should be directed toward achieving meaningful outcomes. A refocus of efforts to the demand side of the fossil fuel equation would be a far more effective use of human resources than protesting pipeline projects. Public discussion and pressure on governments to implement carbon pricing mechanisms could become a focal point of the climate movement. Among a plethora of meaningful demand side actions, governments could be pressured to change building codes to high efficiency standards, to implement measures to improve energy efficiencies, and to create incentives for the switchover of road vehicles to electric propulsions systems.

Competent climate change political leadership is characterized by visionary policies and measures to curtail emissions and to begin the process of an orderly decarbonization of economic sectors. Activists can play an important role in the politics of climate change by promoting climate change competence in elected officials and recognizing incompetence as evidenced by a failure to act on one of the key issues of our times.

6.11 Summary of Current Policies and Emissions Projections to 2030

6.11.1 China. China's NDC is calculated on the basis of intensity of emissions per unit of GDP and, following best estimates of economic growth, China would emit 14–15 Gt of CO_2eq in 2030 if emissions were to match the NDC target. Current policies are consistent with a model whereby China would reach peak emissions in 2025 and, by the year 2030, total emissions would be 5.5% below the NDC target. However, even when considering policy-based projections, the magnitude of China's future emissions gap remains immense and well in excess of any concept of a fair share ceiling to emissions. Without a concerted effort to enhance the level of ambition to curtail emissions, it remains largely inconceivable that compensatory negative emissions gaps from other nations could possibly balance China's excess emissions by the year 2030.

6.11.2 United States of America. In 2012, the intensity of GHG emissions per capita in the US grossly exceeded that of the aggregate of member states of the European Union. In comparison to the EU, the US has lagged in efforts to enact meaningful climate change polices at the national level. The current high intensity of emissions from the US precludes an orderly process of emissions reduction that would arrive at a fair share contribution over the short-term mitigation period to 2030. However, if the US were to achieve the level of ambition contained in the current NDC, the annual rate of emissions curtailment would, over a longer period, allow the US to catch up and approach a fair share contribution. The larger issue is not so much the level of ambition contained in the US NDC but rather the policies and measures in place to achieve this level of ambition. Climate Action Tracker estimates that, based on current policies and the lack of a national climate change action plan, emissions from the US will be 40% over the NDC target ceiling. If this scenario were to transpire, the resultant 3.8 billion tonnes of excess emissions would be a major contributor to the year 2030 global emissions gap.

6.11.3 European Union. Projections for year 2030 emissions from the European Union are an exception among advanced economies. This region has established the policy framework required to progressively ratchet down emissions in accordance with the principle of a fair contribution to limit future surface warming. Recent adjustments to the EU ETS and to binding targets for GHGs not covered by the ETS are designed to achieve at least a 40% cut in total emissions by 2030. Going forward to mid-century, the EU should be well positioned to effectively decarbonize national economies as required to limit emissions to less than 20% of 1990 rates.

6.11.4 India. By 2030, India will be the world's most populous nation, and the total rate of GHG emissions from India will have surpassed that of the United States. India has enacted an ambitious set of National Missions to combat climate change and has implemented a market-based system of tradable energy saving certificates to promote efficiency in energy use. Overall, government policy is designed to achieve sustainable economic growth while responsibly controlling GHG emissions. Projected emissions based on India's NDC are well below the fair share ceiling and would result in a significant "negative emissions gap." After assessing India's climate change mitigations policies, Climate Action Tracker concluded that India will outperform NDC targets and actual year 2030 emissions will be between 5.2–5.5 of Gt CO_2eq. If this were to transpire, India's "negative emissions gap" would be sufficient to offset the projected excess in emissions from Russia.

6.11.5 Other Advanced Economies. The remaining small to mid-sized advanced economies outside of the US and the EU can be considered together. This group of "other advanced economies" consists of Japan, Canada, Australia, South Korea and 5 other smaller countries. Based on current national policies and measures, actual year 2030 emissions for Japan, Australia and South Korea will exceed the target ceiling specified in the respective NDCs. With the notable exceptions of Switzerland and Norway, the NDCs and policies of these countries are lacking in sufficient ambition to approach a fair share contribution to global emissions reduction by the year 2030. The estimated year 2030 total emissions from other advanced economies will exceed the aggregate fair share ceiling of these countries by about 2.1 Gt.

6.11.6 Petro States. Russia, Saudi Arabia, Iran, Venezuela, United Arab Emirates, Iraq, Kuwait, Libya, Oman, Qatar, and Bahrain can be grouped together as narrow economy oil and gas exporting nations. These countries have some of the highest rates of per capita emissions in the world and disproportionately contribute to the current imbalance in the global carbon cycle. There is no evidence of ambition to implement meaningful measures to control future emissions among these countries. If the global level of ambition to curtail emissions does not change, the demand for fossil fuels will be maintained, and the extreme rates of emissions from narrow economy petro states will continue. By 2030, under a current policies scenario, the rate of emissions from Russia, Saudi Arabia and the other countries in this group will be over 5 Gt in excess of the aggregate fair share ceiling to emissions. The size of this emissions gap will be second to that of China and will be a major contributor to the global excess in emissions by 2030.

6.11.7 Climate Vulnerable Least Developed Countries. The Climate Vulnerable Forum (CVF) provides a unified voice for 48 of the most climate change at-risk countries.[485] The combined population of these countries is over 1 billion and by the year 2030 will have expanded to 1.5 billion. Member countries consist of a mix of heavily populated low GDP per capita nations such as Bangladesh, Ethiopia, the Democratic Republic of Congo, the Philippines, and Vietnam and a long list of low to middle income, mid-sized and small countries. Most of these countries have limited capabilities for adaptation to climate change and are located in some of the most at-risk regions on the planet.

Indonesia, Nigeria, all 48 members of the Climate Vulnerable Forum and another 18 countries can be grouped together and defined as Climate Vulnerable Least Developed Countries on the basis of low GDP per capita income and vulnerability to climate change. These countries are lacking in the emissions intensive infrastructure that is characteristic of advanced and emerging countries. Current emissions from the power supply, transportation, industry and buildings sectors of the economies of these countries are a minor contributor to the global total. In 2012, per capita emissions from climate vulnerable least developed countries were only 53% of the global average. With the exception of Indonesia, the current per capita intensity of emissions for all low-income countries are well below the year 2030 global average for the cost-effective pathway to limit future surface warming to less than 2°C.

Indonesia has a long history of rampant deforestation and degradation of peatlands. Indonesia's NDC and national climate change policies are lacking in ambition and excessive rates of emissions are projected to continue going forward to 2030. Unless actions are undertaken, deforestation and land use practices will be the primary source of future emissions from Indonesia.

Excessive deforestation and forest degradation is common to many least developed countries and often is the major source of GHG emissions. Deforestation and degradation accounted for 45% of the total emissions from these countries. By 2030, this group of countries will be home to approximately 29% of the global population. Based on current policies, emissions from these countries would account for only 12% of the global total and would be 5.5 Gt below the aggregate fair share ceiling. The size of this negative emissions gap would be sufficient to offset 80% of the excess emissions from China.

As is the case for India, the challenge, and indeed the opportunity, for other least developed countries is to grow their economies while responsibly controlling emissions. Many of these lower income nations submitted NDCs with unconditional and conditional targets. Pooled emissions projections that were based on conditional NDCs were 9% less than projections from unconditional NDCs. Access to international climate funds will be critical for climate vulnerable least developed countries to implement REDD programs and to further uncouple economic growth from GHG emissions.

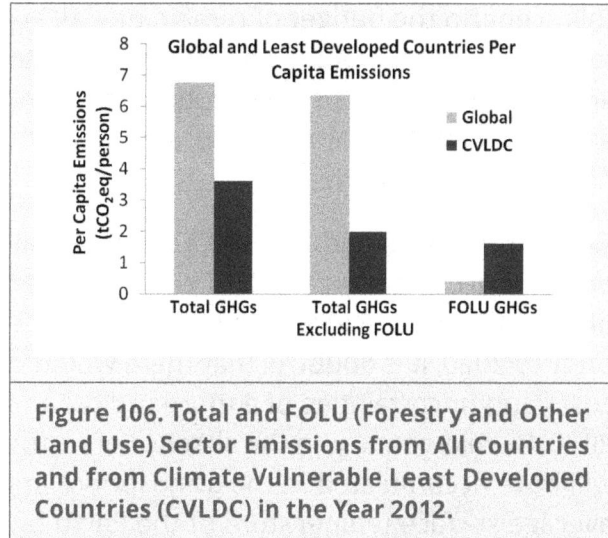

Figure 106. Total and FOLU (Forestry and Other Land Use) Sector Emissions from All Countries and from Climate Vulnerable Least Developed Countries (CVLDC) in the Year 2012.

6.11.8 Other Emerging Economies. The remaining countries are a diverse collection of small to medium sized nations that do not fit into the categories of advanced economies, narrow economy petro states or least developed countries. There are no commonalities among these countries other than per capita GDP incomes somewhere between those of least developed countries and advanced economies. Based on an assessment of current policies, Brazil, Mexico, Pakistan, Turkey and South Africa will have the highest rates of emissions within this group by the year 2030.

Since 2004, Brazil has introduced aggressive and successful measures to cut the rate of deforestation of the Amazon by over 80% relative to the peak deforestation rates of 2004. REDD+ programs to further reduce and ultimately eliminate emissions from forestry and other land use are central to Brazil's NDC and to national and subnational climate change mitigation polices. Brazil's projected emissions for the year 2030 are aligned with the fair share ceiling required to meet regional obligations to limit future surface warming to less than 2°C.

By 2030, an estimate 1.6 billion people will live in countries in the group of "other emerging economies." Based on available information from NDCs and current climate change policies, by 2030, emissions from this group of countries will be about 1.2 Gt in excess of the total aggregate fair share ceiling.

6.11.9 2030 Global Emissions Gap. In the absence of changes to current climate change policies, China, the US and other advanced and emerging economies will continue to emit GHGs at rates well in excess of any concept of a fair share emissions ceiling. By 2030, a negative emissions gap will exist between the projected rates of emissions and the fair share ceiling to emissions from India and the group of climate vulnerable, least developed countries. However, emissions rates from low intensity countries will not offset the excess in emissions from high intensity countries. Under this scenario the budget of remaining GHG emissions that could be safely released to the atmosphere would be seriously depleted by 2030. The only possibility to avoid an over expenditure of what would remain in the carbon budget would be an abrupt and costly implementation of drastic abatements across all economic sectors in all high intensity countries.

Going forward to mid-century, the practically of achieving a disjointed pathway to deep decarbonization of world economies is questionable. If the political leadership of the world was unwilling to initiate cost-effective and orderly mitigation pathways prior to 2020, it is doubtful that there would be an appetite to incur the much higher costs and uncertainties of a more radical and abrupt process to cut emissions post 2030. By mid-century, in the absence of further ambitions to curtail emissions, GHGs will have accumulated in the atmosphere at concentrations that would elevate the average surface temperature of the Earth by 3–4°C relative to pre-industrial times. This degree of future surface warming will result in serious climate change damage over the course of the century.

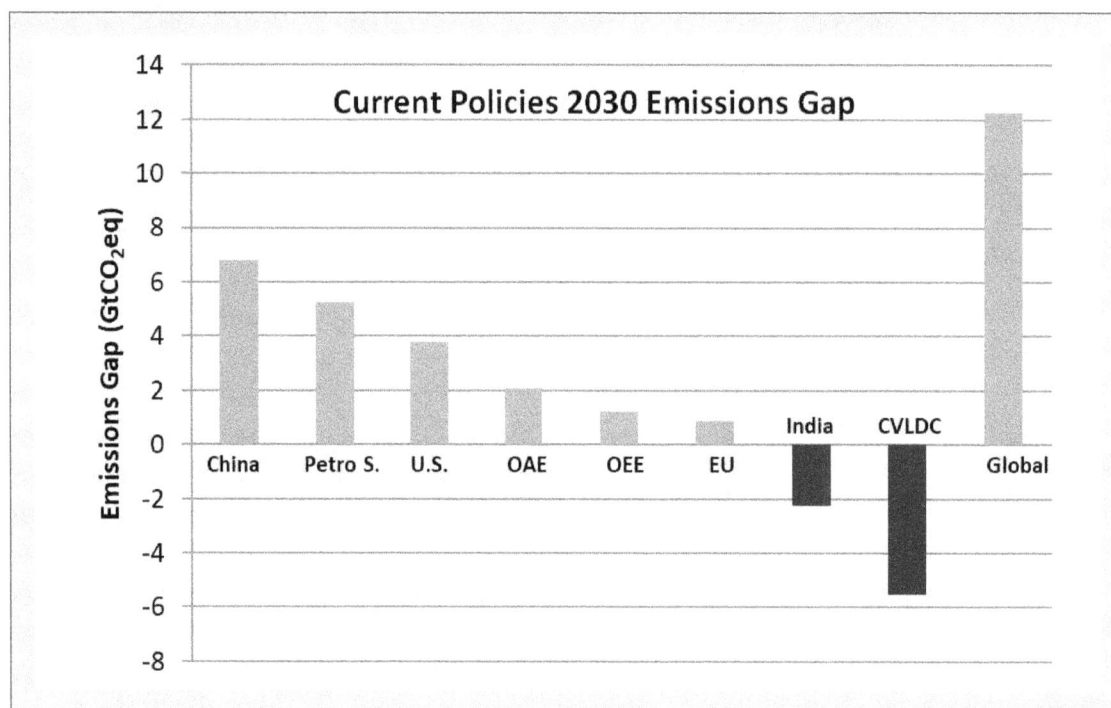

Figure 107. Estimates for Year 2030 Emissions Gaps Based on Current Policies. Year 2030 per capita global emissions of 4.8 tCO_2eq per person aligns with the core objective of the Paris Agreement to limit end-of-century surface warming to well below 2°C. Fair share ceilings for each country are calculated based on estimates of future populations. Emissions gaps are the difference between fair share ceilings and projected emissions with the implementation of current climate change policies. **Petro S.** – Narrow economy petro states. **OAE** – Other Advanced Economies. **OEE** – Other Emerging Economies. **CVLDC** – Climate Vulnerable Least Developed Countries.

Chapter 7: HOW WE GET THERE – MITIGATION PATHWAYS TO A SUSTAINABLE FUTURE

The current set of Nationally Determined Contributions submitted by member states of the Conference of Parties of the United Nations Framework Convention on Climate Change lack sufficient ambition to progressively reduce anthropogenic GHG emissions as required to transition world economies to net zero-emissions status during the second half of the century. Based on achieving but not exceeding the pledges contained in these NDCs, end-of-century average surface temperatures will disrupt climate systems and weather patterns leading to loss of life and livelihood, infrastructure damage, crop failures, and damage to natural ecosystems.

The cost of extensive climate change damage to future generations will be immense. Applying a fraction of this cost to limit emissions and thereby avoid the consequences of excessive global warming is not only ethical but an exercise in common sense economics. Carbon is costly and should be priced as such. Free market economies will adapt to effective carbon pricing by transitioning to more profitable, cost saving, lower emissions options. Carbon pricing mechanisms, either as carbon taxes or free market exchange of emissions permits, will be core economic drivers within national programs to reduce emissions. Carbon pricing must be supplemented with additional incentives and legislation to improve efficiencies and reduce emissions within the energy supply, industry, AFOLU, transportation and buildings sectors of national economies.

As a first order of business, the level of ambition to reduce emissions put forward by most of the nations on Earth must be increased such that projected future atmospheric GHG concentrations are aligned with the core objective of the Paris Agreement. Most NDCs are simple statements of intent that facilitate an accounting mechanism to quantify projected emissions. These statements are meaningless in the absence of effective national action plans to combat climate change.

National climate change mitigation programs and approaches to curtailing emissions will vary between countries depending upon the state of economic development and opportunities within economic sectors. Nations with considerable potential for reforestation and afforestation may choose to focus on REDD+ programs to achieve atmospheric carbon dioxide reduction through an expansion of forest biomass. Advanced, heavily industrialized nations will need to focus on improved efficiencies of energy and materials use and decarbonization of existing electricity supply, industry, transportation, and buildings sectors. Established economies will need to emphasize retrofits while developing countries must decouple economic growth from emissions.

7.1 Short-Term Climate Change Mitigation (2015–2030)

Over the short-term, the highest probability of success and most cost-effective route to limit future surface warming to less than 2°C requires that global emissions peak prior to 2020 and then decline by about 1.2 Gt/year. This pathway arrives at a year 2030, rate of GHG emissions in the range of 39–42 Gt of CO_2 equivalents and sets the table for an orderly deep decarbonization of world economies going forward to mid-century and beyond.

The magnitude of the year 2030 emissions gap between the global ceiling on emissions consistent with a less than 2°C scenario and projections for emissions based on current NDCs and national policies is daunting, and can lead one to question the commitment of world governments to effectively combat climate change. However, there are effective and aggressive climate change action plans in force in many countries and regions, and there are indications that some key nations are in the process of developing more ambitious targets and policies to curtail emissions. A workable roadmap arriving at an adequate level of global ambition may yet evolve.

7.1.1 All Climate Change Roads Go Through China. Historically, China has shown little interest in combatting climate change and over the past 30 years has followed the same emissions-intensive fossil-fuel-based route of economic development taken by advanced nations during the golden age of capitalism after the Second World War. The industrialization of the world's most populous country has driven global emissions on a continuously upward trajectory since 1990 and has been a major contributor to the accumulation of atmospheric CO_2 concentrations in excess of 400 ppm.

China's NDC is based on achieving a targeted decrease in the emissions intensity per unit of GDP. If actual emissions were to approximate this future target, then best estimates for economic growth are consistent with a 34% increase in emissions from China by 2030 relative to 2012. At this rate of emissions, it becomes highly improbable that compensatory deep cuts from all other countries combined could overcome China's emissions gap. A roadmap to bridge the global emissions gap to a sustainable future begins with China.

Whatever China does or does not accomplish over the next 13 years to combat climate change will be of far greater significance in defining future surface temperatures than the outcome of the recent US presidential election or the actions and inactions of governments in other countries around the world. When compared to China's current NDC target, there are indications of intent to further uncouple economic growth from energy use and emissions intensity. China has begun to develop and implement a significant climate change action plan and has stated an intention to contribute to international efforts to control emissions.

China's national ETS remains on schedule for launch in 2017. The system will cover 33% of total emissions with mandatory participation by power producers, heavy industries, and aviation. As of this writing there are no indications of the total allowances to be issued; however, it would seem unlikely that China would institute an ineffective national system after committing substantial resources to build capacity through the operation of large pilot scale-trading systems.

China's landmark decision to launch a national ETS provides an effective domestic tool to control emissions and sends a strong message to other countries that the world's largest emitter and a major player in international trade has begun to acknowledge responsibility and to take effective action to curtail emissions. As China's ETS matures and allowances are sold by auction, the potential for linkage to other cap and trade systems will emerge.

The results of the 2016 US presidential election do not appear to have negatively affected China's climate change agenda. On the world stage of the 2016 COP meeting in Marrakech, China stated that any change in US policy on climate change "won't affect China's commitment to support climate negotiations and also the implementation of the Paris Agreement."[486] A US withdrawal from international trading in low carbon goods and technologies may well provide an opportunity for China to assume a larger role as an innovator, manufacturer and exporter in an evolving low carbon global economy. At present, China has nearly 50% of installed global wind power[487] and is the world leader in photovoltaic solar installations.[488] China's latest five-year plan, while short on details, is consistent with a substantive change in the direction of future economic growth to lower emissions alternatives.

The juggernaut that is China's energy and emissions intensive industrialized economy cannot be reconfigured overnight. The International Energy Agency estimates that by the year 2030, China can readily reduce annual emissions from the production and use of energy by 1.3 Gt relative to the current NDC target.[489] These cuts can be achieved primarily through energy efficiencies. China's ETS can be an effective driver of emissions reduction through improved efficiencies in industry while additional policies will be required to increase the efficiency of energy use in buildings and other sectors. China's state-run power supply network and socialist market economy may facilitate implementation of effective mitigation policies.

Under the articles of the Paris Agreement, revisions to NDCs must be submitted by 2020. With rollout of the national ETS in 2017 and continued strengthening of emissions reduction policies and measures, China should be positioned to markedly increase the level of ambition contained in the current NDC. Hypothetically, China's revised NDC could aim for arrival at peak emissions well in advance of 2025 followed by about a 7% cut from peak emissions by 2030. China's rate of emissions under this more aggressive scenario would still be 62% above a fair share ceiling to emissions as a regional contribution to global efforts to maintain future surface warming to less than 2°C.

A significant increase in China's ambition to curtail emissions over the short-term is highly speculative but would provide a further indication of China's commitment and intent to take a leadership role in a low carbon global economy. In addition, China could, cost-effectively, contribute to international efforts to control emissions through participation in climate funds including REDD+ programs. China could be credited with further emissions reduction under the provision for the transfer of mitigation options

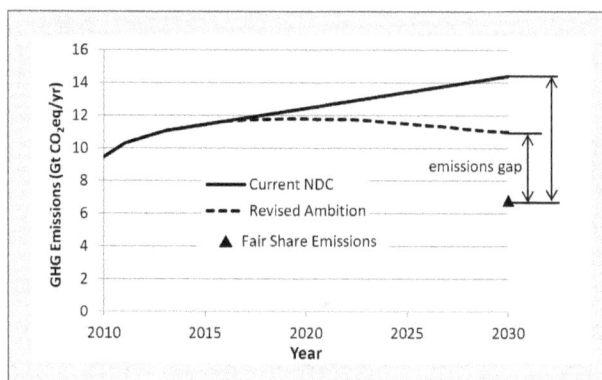

Figure 108. China's Historical Emissions and Possible Scenarios for Future Emissions over the Short-term Mitigation Period to 2030. Future projections do not include possible credits for international transfer of mitigation options.

contained in the Paris Agreement. Participation in international climate change programs would be consistent with the ambitions of China as world economies progressively decarbonize.

A practical roadmap to limit future surface warming to less than 2°C is unlikely to follow a path whereby emissions from all countries converge to a common fair share per capita rate by 2030. If China were to cut emissions by 20% relative to current NDC projections, it becomes feasible for compensatory low emissions rates from other countries to balance the continued year 2030 excess in emissions from China.

7.1.2 The Strange Case of Predicting Future GHG Emissions from the United States.
As of this writing, there are too many variables in play to allow for any degree of precision in predicting US GHG emissions over the short-term mitigation period extending to 2030. US withdrawal from the Paris Agreement and a roll back of climate change initiatives at the federal level are clearly setbacks to national and international efforts to combat climate change. However, prior to the 2016 presidential election, the US climate change agenda was largely driven by policies developed and enacted at the state and municipal levels of government.

It is impossible to quantify how much of a detriment a climate change-inactive president and Congress will be to continued policy development at lower levels of government. On one hand, the lack of incentives and synergies between levels of government may be an impediment to action. Alternatively, inactivity at the national level may become a motivating factor for compensatory responsible governance at the state and municipal levels.

The formation of the US Climate Alliance provides some indication of the potential for a coordinated effort among states to implement a broader climate action plan. Conceivably, other states could join the California ETS or the RGGI. An extensive network of American states and Canadian provinces could function as a carbon club with emissions covered by linked cap and trade systems. Carbon pricing mechanisms would cost-effectively drive down emissions for participants. However, the opportunity for states to join the California ETS has existed for some time without uptake. Inaction at the federal level could either motivate or further dissuade states from joining or otherwise implementing carbon pricing mechanisms. Undoubtedly, in the absence of national programs, some states will remain largely inactive or will introduce ineffectual token gestures under the guise of climate change action plans.

Even when considering a best-case scenario of ambitious state level climate change initiatives, the absence of leadership at the federal level will limit what the US can accomplish over the short-term mitigation period. Other countries will explore options to further cut emissions and advance ambitions prior to the 2020 deadline for submission of revised NDCs. However, in the case of the US, the federal government is no longer engaged with international commitments under the UNFCCC or the Paris Agreement. Member states of the US Climate Alliance and other states that have indicated continued support for the Paris Agreement, would need to compensate for the absence of federal initiatives such as the Clean Power Plan that, potentially, could have been implemented at a national level.

A change in government following the 2018 mid-term elections and the 2020 presidential election could see a reboot of climate change activity at the federal level. Legally, the US cannot withdraw with the Paris Agreement until 2021, and a new administration could simply state an intention to re-engage within the agreement and to honour US obligations.

Given the current circumstances of governance in America, it is difficult to envision an outcome whereby the rate of national emissions would be less than the target ceiling contained in the NDC as submitted under Paris Agreement. A 26–28% cut in emissions by 2025 relative to a 2005 baseline can be considered as the upper limit of the short-term US emissions reduction potential.

Any attempt to predict where the United States is heading on the climate change agenda is wildly hypothetical. However, we can consider three possible scenarios as outcome boundaries. The actual rate of US GHG emissions in the year 2030 should fall somewhere within these boundaries and will depend on how the complex mix of climate change politics plays out over the next 13 years.

High-Emissions Scenario. The US formally withdraws from the UNFCCC and the Paris Agreement. Subsequent federal governments are no longer engaged in combatting climate change and actively promote the continued use of fossil fuels. This combination of factors leads to a stagnation in climate change policy development at the state and municipal levels that extends well into the next decade. Year 2030 rates of emissions exceed Climate Action Tracker projections under the "current US policies without the CPP" scenario.

- By 2030, emissions have increased to 2005 levels.
- Rates of emissions are 3.5-fold greater than the fair share ceiling to emissions from the US.
- By 2030, a 4.3 Gt/year gap exists between actual emissions and the US fair share ceiling. This emissions gap is imposed on other countries. These countries must then achieve deep (likely impractical) compensatory emissions reductions if future global warming is to be limited to less than 2°C.

Current Policies Scenario. The president and Congress remain inactive on the climate change agenda; however, state initiatives advance to a level sufficient to compensate for the cancellation of the CPP. The emissions outcome is comparable to Climate Action Tracker estimates that are based on current policies.

- Year 2030 emissions are 12% less than 2005 levels.
- Emissions are 3.2-fold greater than the US fair share ceiling.
- A 3.8 Gt/year emissions gap is imposed on other countries.

Low-Emissions Scenario. Following a change in government at the federal level, the US announces that it will remain within the UNFCCC and the Paris Agreement. The CPP comes into force, along with the US-Canada agreement to limit methane emissions from the oil and gas sector. Other states within the US Climate Alliance bring into force emissions trading systems modelled after the California ETS. These systems

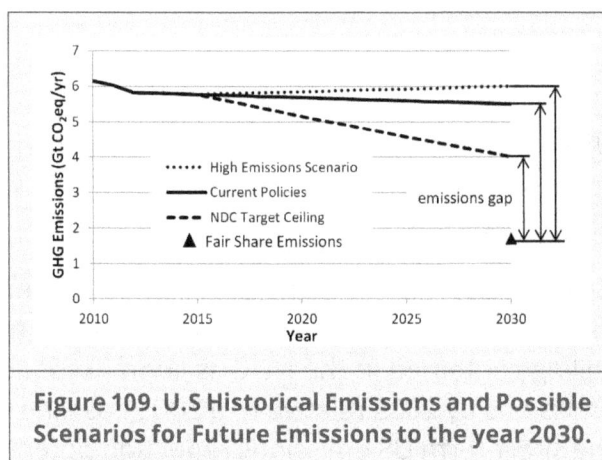

Figure 109. U.S Historical Emissions and Possible Scenarios for Future Emissions to the year 2030.

become linked and operate with a common year 2030 objective to cut emissions by 40% relative to 1990 baseline levels. This target aligns with regional obligations to limit future surface warming to less than 2°C. National emissions and domestic demand for fossil fuels decline. The low rate of emissions from the collective of "green states" balances continued excess emissions from other states such the net effect is to approximate the year 2030 extrapolation of the current US NDC target.

- Overall, US emissions decline by 35% relative to 2005 levels.
- Emissions are 2.4-fold greater than the US fair share ceiling.
- A 2.3 Gt/year emissions gap is imposed on other countries.

There are multiple variations to these simple scenarios for national climate change policy development going forward to 2030. The high-emissions scenario would likely undermine the global effort to combat climate change. The extent of compensatory cuts in GHG emissions from other countries as required to avoid a serious overexpenditure of the remaining carbon budget may well be impractical to achieve. The net result would be a catastrophic increase in surface temperatures by the end of the century. Alternatively, the current actions and inactions of the president and Congress could be a bump in the road toward progressive decarbonization of economic sectors in the United States that is largely driven by actions at the state and municipal levels.

7.1.3 Emissions from the European Union over the Short-term Mitigation Period. Since conclusion of the Paris Agreement, the EU-ETS, and the complementary system of binding national allowances have been fined tuned to achieve the year 2030 NDC target of at least a 40% cut in emissions relative to a 1990 baseline.

The world leadership role assumed by EU in combating climate change could be further advanced by increasing the level of ambition in the 2020 revision of the NDC. A new target to cut year 2030 emissions by 50% would put the EU on a pathway toward an emissions rate that would approximate the fair share ceiling for the region. Going forward to mid-century, member states of the EU would be well advanced in the transition to decarbonized economies. This scenario would provide the EU with a technology and trading advantage when compared to countries and regions that are on a slower trajectory toward lowering emissions.

7.1.4 Projected Year 2030 Emissions from Other Advanced Economies. As of this writing there is no evidence that Japan intends to step back into a leadership role among advanced economies in the global effort to combat climate change. Without a substantial change in direction, the best possible outcome would be a strengthening of current policies to achieve the NDC target. This outcome would result in a year 2030 emissions gap of about 0.4 Gt above Japan's fair share ceiling.

Since signing the Paris Agreement, Canada has made good progress in developing a national climate change action plan. Carbon pricing mechanisms are either in place or will come into force in each province, along with a complete phasing out of coal-fired power plants by 2030. Canada is positioned to increase the level of ambition to cut emissions from 30 to 40% relative to the 2005 baseline. If year 2030 emissions from Canada were to match this hypothetical revised NDC, Canada's emissions would drop to 0.3 Gt above the fair share ceiling.

Australia's NDC is among the least ambitious submitted by an advanced diverse economy under the Paris Agreement. There is little evidence of consistency in policy development within Australia to combat climate change. CAT estimates that, based on current policies, actual year 2030 emissions would exceed Australia's NDC target by about 40%.[215] However, Australia has implemented a cap and trade system and potentially could use this carbon pricing mechanism to achieve the low level of ambition specified in the current NDC. Under this scenario, Australia's emissions gap would be reduced from current policy projections down to 0.3 Gt/year.

Climate Action Tracker estimates that South Korea will emit about 0.7 Gt of CO_2 equivalents in 2030.[215] On a per capita basis, emissions would be almost 3-fold greater than the fair share ceiling and would be well above the limits set forth in South Korea's current NDC. South Korea's energy policy included plans to bring 20 new coal-fired power stations online by 2020.[215] Recently, there are some indications of a move away from coal in the South Korea's future energy mix.[490] In 2015, South Korea

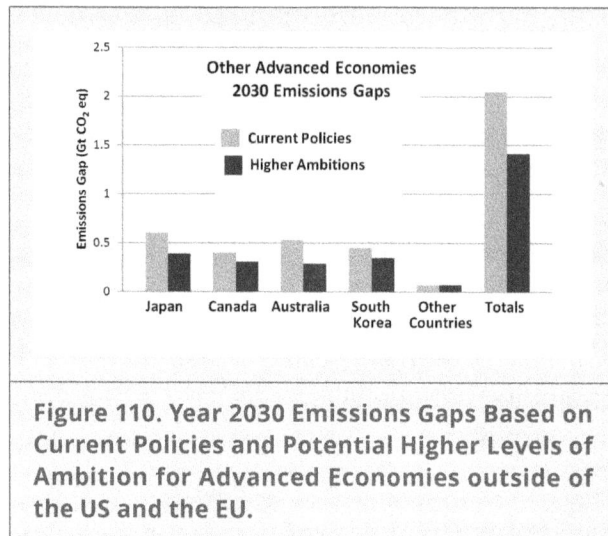

Figure 110. Year 2030 Emissions Gaps Based on Current Policies and Potential Higher Levels of Ambition for Advanced Economies outside of the US and the EU.

introduced a broad spectrum ETS covering 68% of national emissions.[339] South Korea has a carbon pricing system in place that could be used to effectively ratchet down emissions on an annual basis. Potentially, South Korea's climate action policies could be tightened to at least achieve the current NDC target. If this scenario were to transpire, emissions from South Korea would exceed the fair share ceiling by about 0.35 Gt in 2030.

As an aggregate, advanced economies outside of the US and member states of the EU are lacking in sufficient ambition to mitigate climate change. Based on current polices, year 2030 emissions from these countries would be 2 Gt in excess of the aggregate fair share ceiling. Potentially, policies could be advanced to cut this emissions gap to about 1.4 Gt by 2030.

7.1.5 The Important Leadership Role of India in Limiting Future Surface Warming to Well Below 2°C. Based on current policies, Climate Action Tracker estimates year 2030 GHG emissions from India in the range of 5.1-5.4 Gt of CO_2 equivalents. India should easily exceed the level of ambition to control future emissions contained in the current NDC. This rate of emissions is would be over 2 Gt less than the population-based calculation of India's fair share ceiling.[215]

Current policy projections for 2030 emissions developed by CAT assume a substantial increase in power production capacity in India with new coal-fired power plants coming on line along with nuclear and renewables to meet the growing demand for electricity.[215] In agreement with this assessment, the International Energy Agency forecasts that between 2016 and 2040, India will have added nearly 900 GW of power production, with 50% of this new capacity coming from renewables.[172] The IEA predicts that coal will still be an important energy source for India, accounting for about 34% of the total capacity of new installations.[172]

India's ambitious climate action plan is based on eight national missions to improve energy efficiencies and to establish low emissions practices across multiple economic sectors. Basically, India's intentions are to maintain a high rate of economic growth with a minimal carbon footprint. However, India has been criticized for plans to bring new coal-fired power plants online to meet the growing demands for electricity.

Since CAT and the IEA completed their assessments, a series of recent government announcements indicate an intention to pull back on the inclusion of coal in India's future energy mix. In June 2016, India announced cancellation of plans to build four Ultra Mega coal-fired power plants.[491] These plants would have consumed an estimated 46 million tonnes of coal with release of about 0.13 Gt of CO_2 on an annual basis.[491] Plans to add 78 GW of thermal power production by 2022 have been scaled back by 64%.[491] India also announced a shutdown of older heavy polluting thermal power stations accounting for 20% of current coal-fired capacity.[491]

In December 2016, India released a draft of its National Electricity Plan.[492] The study concludes that no additional coal fired power stations beyond those under construction will be needed over the period extending to 2027. Nuclear, hydro and other renewables will account for 47% of total power production capacity by the end of 2021-22 and this will increase to 57% by 2026-27.[492]

On April 27, 2017, India's Minister of State announced that all vehicles sold in India by 2030 will be electric.[493] This policy establishes a path to the rapid electrification of India's fleet of road vehicles. At present, India is the world's third largest oil importer, and complete elimination of sales of new internal combustion engine vehicles would result in an annual savings of $60 billion in energy costs and would significantly impact the global demand for oil.

Since the submission of its NDC to the UN in October 2015, India has further advanced a set of policies designed to achieve low carbon economic growth. The pace of policy development has led to a growing disconnect between the year 2030 ceiling to the intensity of emissions per unit of GDP as specified in India's NDC and estimates of future emissions based on the implementation of new policies. CAT acknowledges this disconnect by stating that if India were to revise its NDC target ceiling to reflect the potential of new policies to limit future emissions, India's efforts to combat change could be upgraded to a "sufficient" rating. India would be the first major economy to commit to a climate action plan with a sufficient level of ambition to achieve the core objective of the Paris Agreement to limit future surface warming to less than 2°C.

By 2020, India should be positioned to submit a revised NDC with a much higher level of ambition. This NDC could include a pullback on new builds of coal-fired power plants, along with further saving from improved energy efficiencies and other initiatives such as the rapid electrification of road vehicles. Potentially, actual year 2030 emissions from India could be a full 3 Gt/year below India's fair share ceiling.

India has an obligation, and indeed an opportunity, to build a low carbon economy. Developing nations including India will need access to international climate funding mechanisms to achieve the highest levels of ambition. India is on course to be a global leader among low GDP per capita countries in planning and implementing low emissions pathways of economic growth. A successful outcome to this process is an essential component of a possible global pathway to achieve the

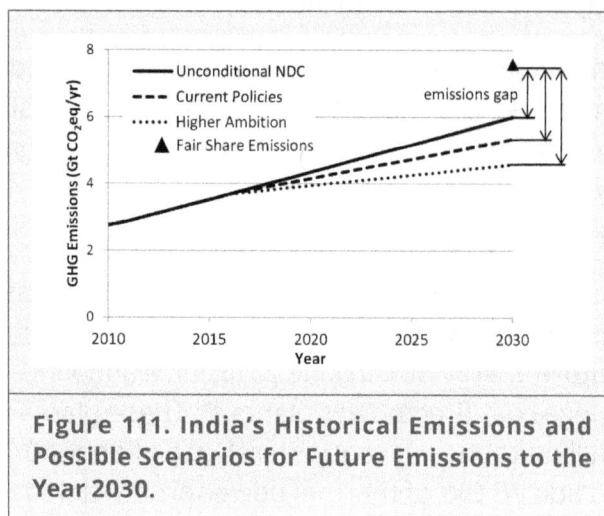

Figure 111. India's Historical Emissions and Possible Scenarios for Future Emissions to the Year 2030.

core objective of the Paris Agreement. By 2030, India will be the world's most populous country and a substantial negative emissions gap will be an absolute requirement to compensate for excessive emissions coming primarily from China, the US and Russia.

7.1.6 Energy Demand and Future Emissions from Narrow Economy Petro States.
There are no indications of serious efforts to step up the level of ambition of Russia, Saudi Arabia, Iran or any other country with an economy that is highly focused on fossil fuel exports. As a group, these countries appear to be intent on maintaining their respective positions as major exporters of conventional energy products. Patterns of emissions-intensive production and high levels of domestic use of fossil fuels for this group of countries are likely to continue over the short-term mitigation period extending to 2030.

Based on current policies and NDC targets, Climate Action Tracker has published estimates for year 2030 emissions from Russia, Saudi Arabia and the United Arab Emirates. Similar predictions for Iran, Venezuela, Iraq, Kuwait, Libya, Oman, Qatar and Bahrain are not provided, and the NDCs of these countries are lacking in detail. However, a rough estimation of future emissions can be calculated by assuming that the current intensities per unit of GDP will extend to 2030. In total, with no changes to current global energy strategies, the aggregate of narrow economy petro states will emit over 7 Gt of CO_2eq in 2030. The combined emissions gap above the fair share ceiling of these countries is estimated at 5.3 Gt and second only to China.

The International Energy Agency's new policy scenario is based on the global aggregate of national policies and NDCs in place as of early 2016. Under this scenario, global demand for oil will slow but continue to steadily increase up to 7% above current demand by the year 2030.

Global implementation of effective climate change action plans will have profound effects on the demand for traditional energy products from narrow economy petro states. Under the IEA 450 Scenario, demand for oil will follow a similar trajectory as global emissions and will peak prior to 2020 followed by a steady decline over the next decade. A global revision of national energy strategies, as required to achieve the core objective of the Paris Agreement, will result in a 25% cut in the demand for oil by the year 2030 when compared to the IEA new policies scenario.

Long-term oil and gas sector business plans are based on models of future demand and supply. New production is brought online in a cost-effective manner as output from active sources becomes depleted. Oil and gas sector companies can develop business plans to avoid costly investments in what would become stranded assets as the future demand for oil declines in a post Paris Agreement world. The IEA estimates a global cost of $11 trillion dollars to develop the 580 billion barrels of new resources that would be required under the new policy scenario.[172] This investment

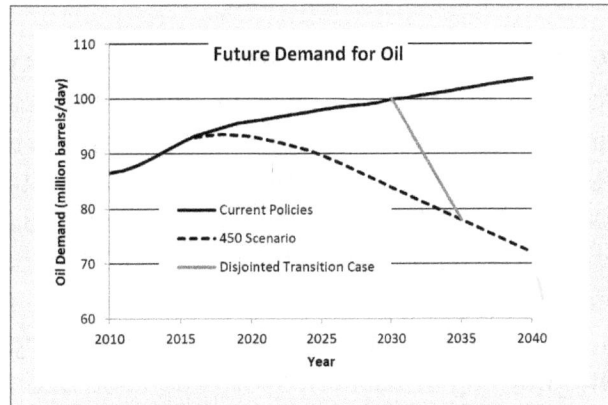

Figure 112. IEA Estimates of Future Demand for Oil (million barrels/day) under the Current Policies, the 450 Scenario, and a Disjointed Transition to Low Emissions by 2035. Adapted from International Energy Agency. World Energy Outlook 2016.[172]

drops to $6.8 trillion under the 450 Scenario. The IEA has also modelled a disjointed transition case whereby peak demand is delayed until 2030. In this model, drastic measures are abruptly introduced to sharply cut emissions over a very short period of 5 years. If the world was to follow this emissions reduction pathway, oil and gas compan-

ies would be unable to effectively plan for changes in demand and would incur massive financial losses associated with the construction of stranded assets and wild fluctuations in supply and demand. The disjointed transition case for the oil sector is illustrative of the broader economic consequences of climate change mitigation pathways that would delay peak emissions past the year 2020.

The extreme rates of emissions from Russia, Saudi Arabia, and other

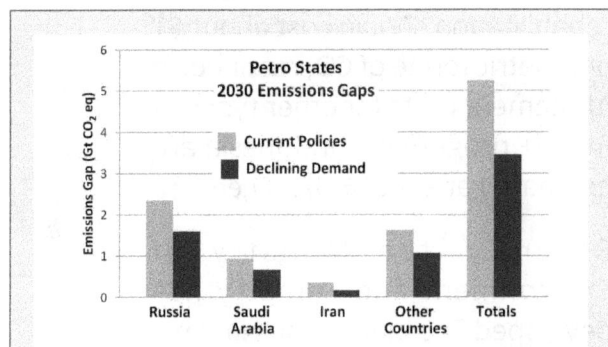

Figure 113. Potential year 2030 Emissions Gap for Petro States Based on Current Global Energy Policies and Declining Demand for Product with Global Implementation of Higher Ambition Climate Change Action Plans.

economies that are highly focused on oil and gas exports will not be maintained if the demand for fossil fuels declines as predicted under the IEA 450 Scenario. As carbon pricing and other policies to combat climate change become increasingly effective, demand for oil, gas, and coal will diminish and emissions from fossil fuel exporting countries will decline. By the year 2030, emissions from these countries could drop by over 20% relative to the IEA new policy scenario. With declining demand for product, the total emissions gap from narrow economy petro states could be cut by about 1.8 billion tonnes of CO_2 per year.

7.1.7 The Emissions Abatement Opportunity in Indonesia. In 2012, emissions from forestry and other land use in Indonesia equalled 1.22 Gt or 2.6% of the total global anthropogenic release of greenhouse gases to the atmosphere.[124] Drainage of peatlands for timber extraction and other land use accounts for about 38% of these emissions.[494] Dry peatlands become aerated leading to oxidation, decomposition, and release of CO_2 that had accumulated in the biomass over thousands of years. Dry peatlands are also prone to fires. Deforestation, forest degradation, and forest fires are the second major source of emissions from Indonesia. Poor land use management such as the draining and burning of peatlands for expansion of palm plantations is the primary cause of the massive release of GHGs from the forestry and other land use sector in Indonesia.

Indonesia's NDC is based on an unconditional target to cut emissions by 26% relative to a business as usual scenario.[254] A second more ambitious target of a 41% cut in emissions is conditional to accessing international climate funds.[254] However, these targets do not approach Indonesia's considerable emissions abatement potential. A detailed study on abatement opportunities concluded that by 2030, Indonesia could potentially cut emissions by 67% relative to a 2005 baseline.[494] Over 75% of this potential can be realized through the implementation of a broad spectrum of measures including REDD+ programs, fire prevention, peatland preservation and rehabilitation, and sustainable forest management practices. Globally, these abatements are among the most cost-effective opportunities available to cut emissions. Programs of peatland preservation and rehabilitation should cost about $1 per metric tonne of CO_2 retained.[494] Abatement costs for other types of REDD+ programs in Indonesia are estimated at about $11/t$CO_2$eq.[494]

Recently, the Ministry of Environment and Forestry has developed The Indonesian National Carbon Accounting System (INCAS) as a mechanism to accurately monitor, verify and report on changes in emissions from all possible land sources.[495] Indonesia is

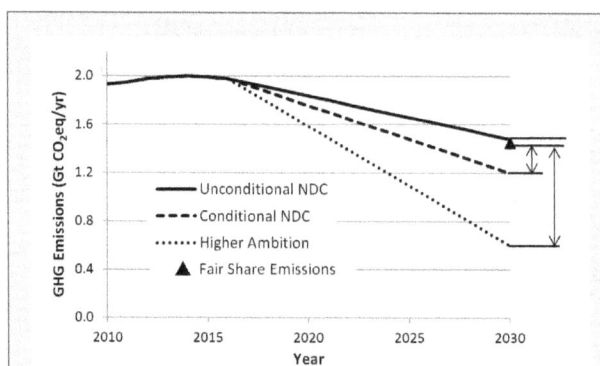

Figure 114. Indonesia Historical Emissions and Possible Scenarios for Future Emissions over the Short-term Mitigation Period to 2030.

suitably advancing toward REDD readiness. The low abatement costs in Indonesia should be economically attractive for developed nations to underwrite these programs in exchange for carbon credits. The $1 billion Norway-Indonesia REDD agreement is indicative of the potential for a substantial flow of funds to access cost-effective mitigations opportunities within Indonesia.

It is difficult to envision a global pathway to effectively limit future surface warming that does not include a substantial curtailment in emissions from peatland and forest degradation in Indonesia over the short-term mitigation period. International carbon funding has the potential to assist Indonesia in achieving a high level of ambition. By 2030, a negative emissions gap of up to 0.85 Gt/year below Indonesia's fair share emissions ceiling would partially compensate for excess emissions from other countries.

7.1.8 The High Abatement Potential of Climate Vulnerable Least Developed Countries. Nigeria's potential to abate future release of GHGs is comparable to that of Indonesia. Nigeria's NDC is among the most ambitious and well-constructed statement of intent to control future emissions thus far submitted by a developing country to the UNFCCC as part of the Paris Agreement. The document lays out a pathway for low carbon economic growth plus forest preservation and rehabilitation and concludes with an ambitious unconditional target and a more ambitious target to limit future emissions that is contingent upon access to carbon funds and international cooperation. If actual year 2030 emissions were to approximate Nigeria's conditional target, per capita emissions would be about 1.9 tCO_2eq/person. This rate of emissions would be 40% below the global average that aligns with the short-term mitigation pathway under a less than 2°C surface warming scenario. On a per capita basis, Nigeria's conditional NDC target is almost identical to the emissions abatement potential of Indonesia.

There are few high abatement potential countries outside of Indonesia and Nigeria for which a detailed sector by sector analysis of low carbon policy options is available. Ethiopia, Morocco, Bhutan, Costa Rica and The Gambia are exceptions. These are the only countries in the world with NDCs and climate changes polices that are rated as "sufficient" by Climate Action Tracker. These countries are members of the Climate Vulnerable Forum (CVF).

The CVF made perhaps the most impactful declaration at the 2016 COP22 meeting in Marrakech.[496] As the first meeting of the Conference of Parties since the Paris Agreement, the agenda and objectives were primarily technical and focused on improving the mechanisms of the agreement after ratification by member states. However, the CVF released a significant policy statement that specifically and strongly addressed the issue of global ambition to decarbonize economies and the role of least developed and small vulnerable countries in the process.[485] The CVF Vision Statement set forth a position that member states would strive to meet 100% of domestic renewable energy production as rapidly as possible with the objective of maximizing their contribution to limit future surface warming to less than 1.5°C. As part of a series of member actions, NDCs are to be updated as early as possible such that the CVF can take a leadership role in triggering a higher level of ambition from all countries. Member countries will strive toward a significant increase in climate investments and will make use of appropriate carbon pricing mechanisms.

The CVF Vision is more than a token gesture from a group of low emitting countries that are minor contributors to climate change. These countries have declared that they will not embark on carbon-based pathways of economic development. CVF countries have concluded that they can access an early stage opportunity to build decarbonized economies and in so doing avoid contributing to global emissions.

Realizing the abatement potential of CVF countries is a vital component to a potentially viable pathway to bridge the year 2030 emissions gap. It has become essential that CVF and other high abatement potential countries minimize future emissions to compensate for what will certainly be an excess in emissions from more advanced countries over the short-term mitigation period extending to 2030. A leadership role is required from least developed countries to maximize abatement potentials and thus minimize emissions in the post-Paris Agreement world.

An assessment of the abatement potential for most countries other than Indonesia and Nigeria requires some assumptions and extrapolations. These countries all share a common characteristic of an existing low intensity of emissions and thus do not have extensive fossil fuel and carbon-emitting infrastructure in place at the present time. In addition, many of these countries have high rates of emissions from deforestation and land degradation. Given the unified position put forward by the CVF to build decarbonized economies and to combat climate change, it is reasonable to extrapolate the abatement potential of Nigeria, as an example of a heavily populated Sub-Saharan African country, and Indonesia, as representative of South East Asia, to other high abatement potential countries. This extrapolation can be modelled by simply assigning Nigeria's conditional NDC year 2030 target of a per capita emissions rate of 1.9 tonnes of CO_2 equivalent per person to all high abatement potential countries. The net result of this exercise is to expand the total negative emissions gap of these countries to 7.35 Gt per year below the combined population-based fair share ceiling.

By 2030, nearly 30% of the global population will live in high abatement potential countries. A massive avoidance of emissions from these nations can be achieved through a combination of low carbon pathways of economic growth and implementation of aggressive programs to halt deforestation and degradation of forests and peatland. Many high abatement potential countries will

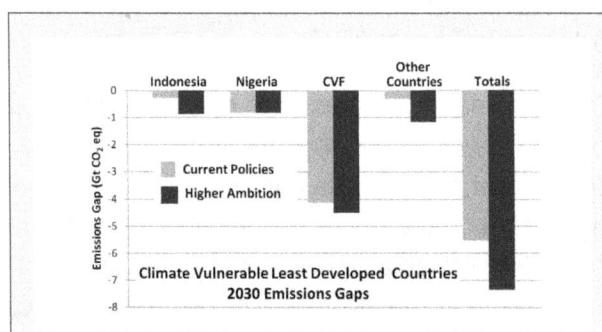

Figure 115. Potential Year 2030 Negative Emissions Gaps from Climate Vulnerable Least Developed Countries. CVF – Aggregate of countries in the Climate Vulnerable Forum.

benefit from a transition to sustainable intensive agricultural practices and a modernization of energy use for home cooking and heating purposes. Improved efficiencies in rural economies and practices within some of the least developed countries will be synergistic to forest and peatland preservation and rehabilitation. If this ambitious agenda can be put into action, the low rates of emissions from high abatement potential countries can serve to balance excess emissions from emerging and advanced economies over the short-term mitigation period extending to 2030.

7.1.9 Abatement Potential of Other Emerging Economies. The remaining countries do not fit into the broad categories of advanced economies, petro states, or climate vulnerable, least developed countries. The economic circumstance varies considerably among this diverse group of nations; however, on a country by country basis, it is possible to roughly estimate the potential to further ambitions to curtail greenhouse emissions over the short-term mitigation period.

Brazil has made considerable progress in halting deforestation of the Amazon and has submitted an NDC with an emissions target that aligns with a population-based concept of a fair share ceiling. Overall, Brazil has demonstrated a commitment to the development and implementation of effective climate change policies. This country is well positioned to further advance these policies such that year 2030 emissions could be about 10% below the fair share ceiling.

Mexico's NDC arrives at a year 2030 target for emissions that approximates this country's fair share ceiling. The recent initiative to implement a national ETS provides further evidence of Mexico's commitment to combat climate change. Potentially, Mexico can further increase the level of ambition in domestic policies such that year 2030 emissions are about 10% less than the population-based calculation of a fair share ceiling.

By 2030, Pakistan will be the most populous nation in the loosely defined collection of "other emerging economies." In 2012, Pakistan's per capita GHG emissions rate was only 1.9 tCO_2eq per person, and the country's GDP per capita was 14% of the world average. These statistics would suggest that Pakistan should be classified as a high abatement potential country. However, there is little opportunity to control emissions from Pakistan's land use and forestry sector, and available information on government policies is not consistent with a strategy of low carbon economic growth. In the post-Paris Agreement world, it is difficult to envision any country following a pure emissions-intensive fossil fuel dependent pathway of economic development. Over the next 13 years, the advantages of low carbon options to build economic sectors may direct Pakistan toward a more ambitious pathway to limit future emissions. Pakistan could double current GHG emissions by 2030 and still remain well below the fair share ceiling. On a per capita basis, this rate of emissions would exceed that assigned to climate vulnerable least developed countries but would be similar to the year 2030 rate of emissions assigned to neighbouring India.

In 2012, Turkey's emissions per capita were 23% below the world average, and the GDP per capita was about 87% of world average. Turkey submitted an NDC with a low ambition to cut 2030 emissions by 21% relative to a business as usual scenario. When Turkey's projected rate of economic growth is considered, achieving the NDC target would result in an actual rate of emissions that would be 2.3-fold in excess of the country's fair share ceiling. Turkey defends the weak NDC as a requirement to facilitate development of the country's economy. This argument has little merit in the post-Paris Agreement world and is not supported by evidence from the EU and other countries that have cut emissions without affecting economic performance. Clearly, there is an opportunity and an obligation for Turkey to increase the level of ambition to contribute to global efforts to combat climate change. Turkey's GDP per capita is 15% greater than that of China and it is reasonable to assign similar year 2030 per capita rates of emissions to China and Turkey. Under this scenario, year 2030 emissions from Turkey would be 25% below the current NDC target but would still be 70% above the fair share ceiling of this country.

In assessing available information on the policies of South Africa, there is little evidence of any serious intent to combat climate change. Current per capita emissions from South Africa are 30% above the world average while the GDP per capita is 45% below the world level. South Africa's NDC has no defined targets and CAT estimates that, based on a continuation of current polices, emissions will double from current levels. If this were to transpire, emissions from South Africa would be 3-fold in excess of the fair share ceiling. As is the case for Turkey, South Africa has an obligation and responsibility to contribute to global efforts to limit future surface warming. Assuming that these obligations are at least partially addressed upon submission of a revised NDC prior to 2020, it is reasonable to estimate South Africa's abatement potential using the same per capita rate of emissions assigned to China and Turkey.

Going down the list of smaller remaining countries, where specific policy information to estimate future emissions is lacking, the process of applying per capita emissions rates from comparable economies can be used to arrive at a rough estimate of abatement potentials.

Overall, if the level ambition of other emerging countries were to approximate abatement potentials, year 2030 emissions could be cut by about 25% relative to estimates based on current NDCs and climate change policies. This aggregate rate of emissions would be about 1 Gt below the collective fair share ceiling of emissions aligned with the highest probability to limit future surface warming to less than 2°C.

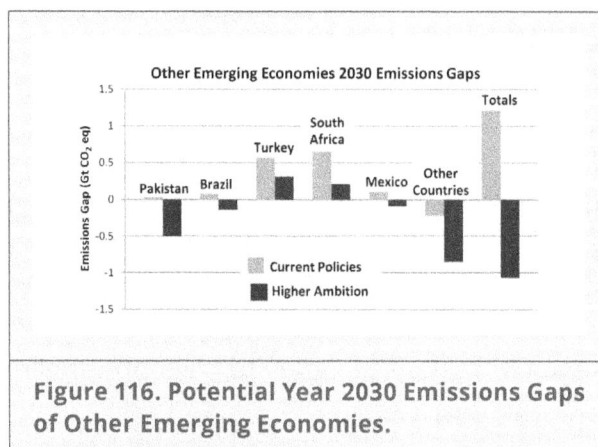

Figure 116. Potential Year 2030 Emissions Gaps of Other Emerging Economies.

7.1.10 Summary: Short-term Climate Change Mitigation Pathways. A successful outcome over the short-term mitigation period requires that almost all nations increase their respective levels of ambition to combat climate change. Going forward to 2030, a process whereby all countries would converge to a common per capita rate of emissions of about 4.8 tonnes of CO_2 equivalents per person is no longer feasible. China, the US, Russia, and the majority of other advanced economies other than the member states of the EU have failed to seriously reduce emissions since the original United Nations Framework Convention on Climate Change was adopted in 1992. Even with aggressive implementation of ambitious climate change action policies, the time frame is such that these countries will continue to emit GHGs at per capita rates in excess of the year 2030 target global average.

Overcoming the current policy-based estimates of a 12 Gt global emissions gap requires that, by the year 2030, the aggregate negative emissions gaps for developing countries approximates the excess in emissions beyond the fair share ceiling for advanced and emerging economies.

The concept of a fair share ceiling of emissions over the short-term mitigation period to 2030 is somewhat misleading. A ceiling implies an allowance for very low emitting countries to increase per capita emissions up to a target global average by 2030. India and high abatement potential countries have an opportunity, and indeed an obligation, to follow low carbon pathways of economic growth. By 2030, about half of the global population will reside in high abatement potential countries. The current massive imbalance in global emissions will still exist. However, excess emissions from China, the US, Russia and other advanced and emerging economies can potentially be offset by very low rates of emissions from India and other climate vulnerable least developed countries.

There is a limit to the abatement potential of low GDP per capita countries. Access to substantial international climate funding is an absolute requirement for these countries to fully realize the opportunity to grow their economies following low carbon pathways. China, the US and other advanced and emerging countries with projected excess rates of emissions must strengthen climate change policies such that year 2030 total emissions are within a range that can be realistically offset by India and other developing countries.

A viable pathway to a successful outcome over the short-term mitigation period extending to 2030 does exist. The responsibility is global and every country must contribute through the implementation of abatement measures according to the specifics of circumstance. Effective political leadership will be needed, and there is no longer any luxury of time for further delays in the implementation of climate action plans. Reaching peak emissions prior to 2020 followed by progressive annual curtailments arriving at a 20% cut from peak emissions by the year 2030 is required to establish the foundation for subsequent deep decarbonization of world economies going forward over the mid-term mitigation period from 2030 to 2050.

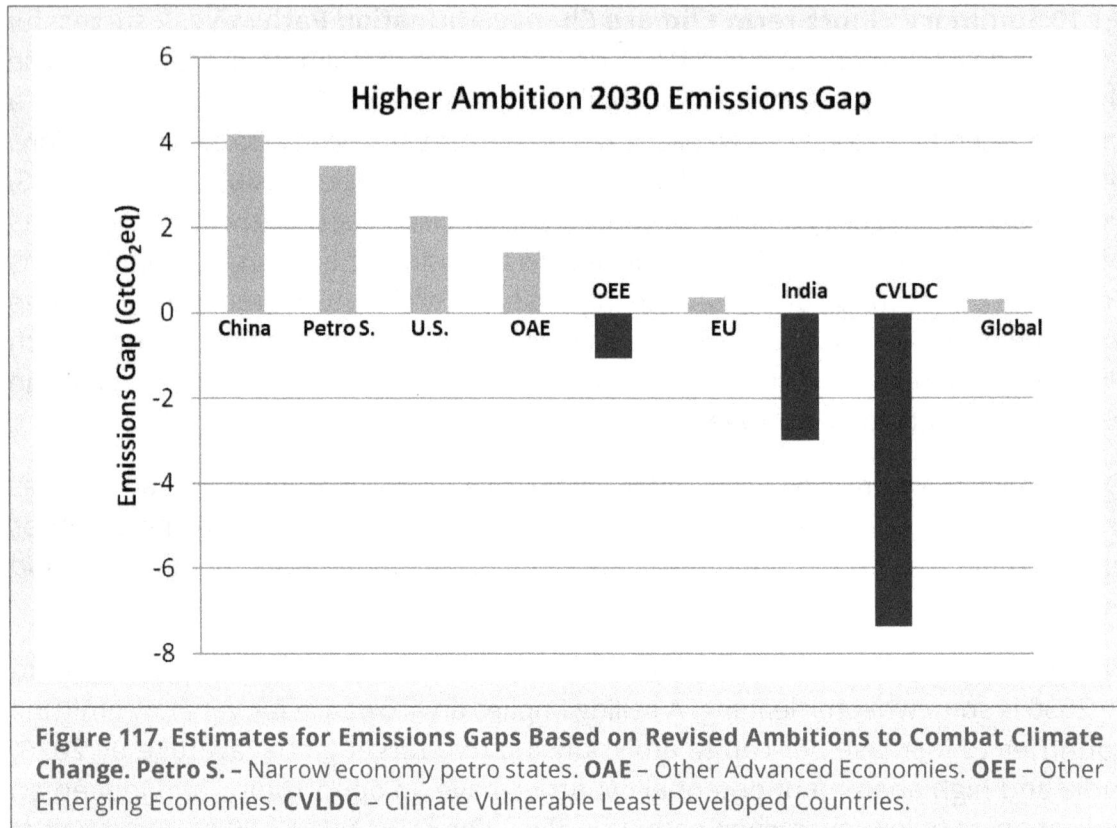

Figure 117. Estimates for Emissions Gaps Based on Revised Ambitions to Combat Climate Change. Petro S. – Narrow economy petro states. **OAE** – Other Advanced Economies. **OEE** – Other Emerging Economies. **CVLDC** – Climate Vulnerable Least Developed Countries.

7.2 Mid Term Global Pathways to Climate Change Mitigation (2030–2050)

The year 2030 can be considered as an inflection point for global and national emissions reduction pathways designed to achieve the core objective of the Paris Agreement. NDCs are almost all based on year 2030 targets for emissions and specific policies and measures to cut emissions are focused on this short-term mitigation period. Many countries and sub-national regions have provided longer term targets, but specific policies and measures to achieve these targets have yet to be articulated. The details of climate change action plans over the mid-term mitigation period from 2030 to 2050 will be dependent on the combination what was achieved by 2030, what new innovations have advanced to commercial potential, and on the ambitions of future governments. As the world progresses through the short-term mitigation period, climate change action plans will be revised and extended toward mid-century.

Assuming that emissions peak at approximately 52 Gt per year prior to 2020, then going forward to mid-century, this rate of net GHG release to the atmosphere must be cut by about 56% from peak under a less than 2°C pathway and by about 85% to provide a 50% chance that future surface warming will be limited to 1.5°C above pre-industrial temperatures. The profile of emissions will shift away from the predominance of CO_2 over the course of a successful mid-term mitigation period. By 2050, emissions of methane and nitrous oxide associated with the increase in agricultural productivity as required to feed an additional 3 billion people will offset a significant portion of the emissions abatements for these non-CO_2 GHGs. On a percentage basis, CO_2 emissions will decline at a much steeper rate than emissions of CH_4 and N_2O. In this model, under a 1.5°C scenario of future surface warming, methane and nitrous oxide from agriculture would contribute 43% of residual emissions by mid-century.

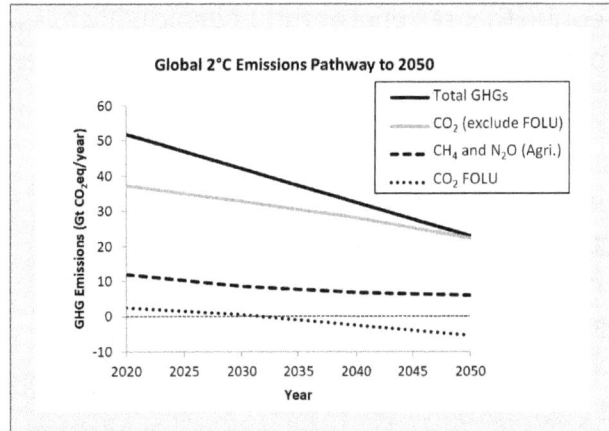

Figure 118. Potential Emissions Trajectories under a Less than 2°C Surface Warming Scenario Extending to Mid Century. Total GHGs – Net CO_2 plus emissions of methane (CH_4) and nitrous oxide (N_2O). **CO_2 (exclude FOLU)** – Net CO_2 emissions from all sectors less emissions/withdrawal from forestry and other land use. **CH_4 and N_2O (Agri.)** – emissions of methane and nitrous oxide primarily from the agriculture sector. **CO_2 FOLU** net carbon dioxide emissions/withdrawal from the global forestry and other land use sector.

Between 2020 and 2050, the global forestry sector should transition from a source of emissions to net carbon dioxide reduction. Forest preservation, along with programs of reforestation and afforestation, have the potential to increase carbon-sequestering forest biomass such that the balance between CO_2 release and sequestration from forests equates to a net withdrawal of up to 5 Gt per year from the atmosphere. In an example of a successful less than 2°C climate change mitigation model, CDR from forest biomass would offset GHG emissions (as CO_2 equivalents) from agriculture and residual CO_2 emissions from other economic sectors would approximate 22 Gt/year by mid-century. More aggressive mitigations to limit surface warming to 1.5°C would

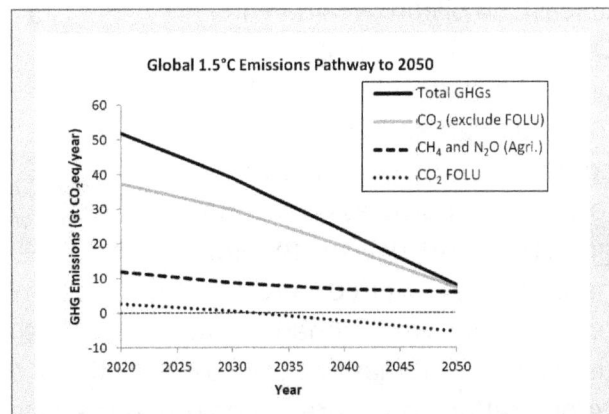

Figure 119. Potential Emissions Trajectories under a 1.5°C Surface Warming Scenario Extending to Mid Century.

require that residual anthropogenic CO_2 emissions from sectors outside of agriculture, forestry, and other land use drop to less than 10 Gt/year by the year 2050.

Going forward to mid-century, emissions from all countries and groups of countries must converge to a common per capita rate. However, the pathway to arrive at convergence will differ with circumstances of economic development. Assuming a successful outcome to the short-term mitigation period, economic growth in India and other developing countries will have proceeded along low carbon pathways. Going forward to mid-century, the specifics of emissions abatement

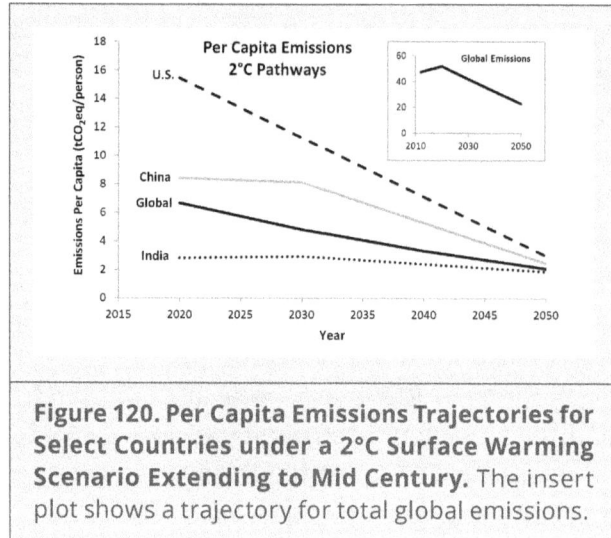

Figure 120. Per Capita Emissions Trajectories for Select Countries under a 2°C Surface Warming Scenario Extending to Mid Century. The insert plot shows a trajectory for total global emissions.

measures will continue to differ between low GDP per capita developing countries and emissions intensive advanced and emerging economies.

Low income, developing countries will account for the majority of the projected increase in population between 2020 and 2050. The total population of Sub-Saharan African countries will more than double to at least 2.4 billion by mid-century.[497] Over the mid-term mitigation period, what are currently defined as low income, high abatement potential countries must continue to follow low carbon pathways of economic growth. Land use in developing

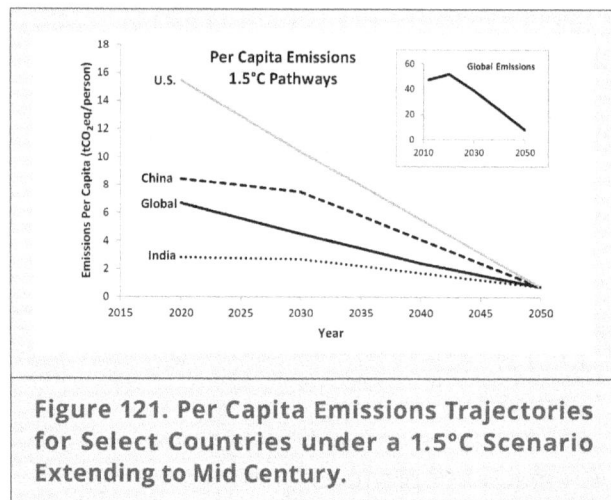

Figure 121. Per Capita Emissions Trajectories for Select Countries under a 1.5°C Scenario Extending to Mid Century.

nations must be effectively managed to optimize the production of food and possibly energy crops while enhancing forested areas as required to build up carbon-sequestering global biomass. Access to international climate funds will be required to support reforestation and afforestation programs. Overall, under a less than 2°C scenario, per capita emissions from India and other high abatement potential countries are unlikely to change during the mid-term mitigation period. More aggressive mitigations aligned with a 1.5°C outcome will require that the poorest countries further decarbonize their economies or advance forestry programs such that the per capita intensity of emissions actually declines as the population increases.

Prior to ratification of the Paris Agreement, most mid to high GDP per capita countries outside of the EU lagged in efforts to curtail emissions and combat climate change. Emissions from these countries will be well above any concept of a fair-share ceiling by the year 2030, and this imbalance must be largely corrected by mid-century.

Under a less than 2°C scenario, total GHG emissions from the US should be reduced to about 1.2 Gt/year by mid-century. This equates to an 81% cut from the year 2005 baseline used by the US in determining the NDC target. If the US were to achieve the short-term NDC target, emissions reductions over the mid-term mitigation period could proceed on an orderly basis. Alternate scenarios of moderate and high rates of emissions over the short-term mitigation period result in disjointed pathways that would require abrupt implementation of drastic measures to rapidly cut emissions beginning in 2030. The US economy would incur vast loss of capital in the form of stranded assets across economic sectors if disjointed pathways were followed.

Emissions reduction pathways in the US under a 1.5°C scenario magnify any distortions to an orderly process of economic transition. Compensation for a short-term period of continued high emissions would require that the entire process of near complete decarbonization of the economy of the United States must occur within a 20-year window.

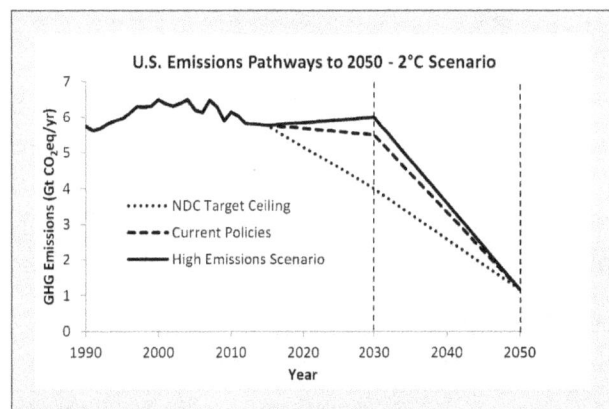

Figure 122. Emissions Trajectories for the US under a 2°C Surface Warming Scenario Extending to Mid Century. Low, moderate, and high emissions refer to possible outcomes during the short-term mitigation period to 2030.

As is the case for the US, other emissions-intensive advanced and emerging countries must progressively decarbonize their economies over the mid-term mitigation period. If high rates of emissions persist well past 2020, these countries will either fail to achieve mid-term targets or will incur the extreme costs and uncertainties of disjointed emissions reduction pathways.

Emissions from fossil fuel producing countries will decline with the drop in the demand for traditional energy products. Potentially, some countries could continue with domestic production and use of coal, oil, and natural gas. Pockets of carbon economies would undermine the global effort to avoid an over expenditure of the remaining emissions budget and would likely lead to future surface temperature elevations in excess of 2°C. By mid-century, rogue carbon-based economies could face strong international pressures including trade tariffs applied to high carbon products. Canada and Norway are examples of net fossil fuel exporting countries that are taking measures to decarbonize their economies outside of the oil and gas sector. With the decline in international demand for oil and gas during the mid-term mitigation period, Canada and Norway should be positioned to make the transition to low carbon economies.

As advanced and emerging economies progress through the mid-term mitigation period, the timing for the implementation of abatement measures will be based on the efficacy and the cost efficiency per unit of CO_2 emissions avoidance. Application of carbon capture and storage to cement manufacture is an example of a high cost abatement that would be introduced well after more cost-effective measures are in effect. Deep cuts in emissions from the aviation sub-sector are dependent on the availability of suitable low emissions second generation biofuels. Technical challenges and higher abatements costs may lead to residual emissions in some industry and transportation subsectors that could persist past mid-century. Regions with limited options for zero or very low emissions dispatchable power production may continue with some degree of fossil fuel use in the energy mix to produce electricity. In this circumstance, carbon capture and storage may be required to adequately control emissions from power production.

Under a 1.5°C scenario, the available carbon budget for emissions from the industry, energy, and transportation sectors is limited over the mid-term mitigation period. Arriving at less than 10 Gt/year of CO_2 emissions by mid-century will likely require rapid development and implementation of carbon capture and storage and sustainable production of low carbon liquid fuels for niche applications.

When compared with other advanced and emerging economies, EU countries have progressed further in the implementation of specific abatement measures. Over the mid-term mitigation period, the EU will likely retain a leadership position in developing and commercializing technologies to address technically challenging, more costly abatement opportunities.

Carbon pricing mechanisms likely will be required to control the size of the global herd of ruminant animals such that methane emissions and land use for agriculture are maintained within sustainable limits. GHG emissions from agriculture will continue past the mid-term mitigation period and ultimately must be balanced by carbon dioxide removal from the atmosphere.

Carbon pricing mechanisms to drive down emissions as cost-effectively as possible are likely to become commonplace during the mid-term mitigation period. An idealized scenario would see a global market for trading in emissions allowances. This could evolve by linking the growing collection of ETSs into a carbon club that would allow trading between systems. However, in comparing current ETSs, the inherent differences in GHG coverage, along with inequalities in carbon pricing, may impede the development of a global system. In North America, the broad-spectrum California-Ontario-Quebec ETS could grow to include Mexico and the remaining provinces of Canada and the states of the US. The well-established EU ETS, and the upcoming China ETS would appear to have similarities in that these systems cover emissions from power production and heavy industry. These two large cap and trade systems could become hubs for linkage of other small national ETSs. Potentially, these hubs could link to form a multinational carbon-trading club. The shared circumstance and objectives of the Climate Vulnerability Forum could result in another large ETS among developing countries. Remaining countries could initiate national ETSs that are designed for compatibility and linkage to specific large multinational trading systems. As emissions caps are ratcheted down on an annual basis within these systems, the price of carbon will continue to climb such that higher cost abatements including carbon capture and storage become economically viable. Over the mid-term mitigation period, it is difficult to envision successful, deep decarbonization of world economies in the absence of effective carbon pricing mechanisms around the globe.

7.3 Long-term Climate Change Mitigation Pathways (2050–2100)

Beyond mid-century under a less than 2°C scenario, global CO_2 emissions from all economic sectors will continue to be driven down to a floor level of residual emissions. To some extent, carbon dioxide will continue to be released from high value niche subsectors of industry and transportation that are lacking in practical abatement options. In addition, continued CO_2 emissions from the agriculture sector are unavoidable.

Between 2065 and 2080, residual CO_2 emissions from agriculture, industry and transportation will flat line at less than 5 Gt/year. Methane and nitrous oxide will continue to be released at a stable and predictable rate from the global agriculture sector. After all abatements are in effect, residual total GHG emissions from agriculture and other sources going forward to the end of the century and beyond will stabilize at a rate between 5 and 15 Gt of CO_2eq/year.

After 2050, programs of afforestation and reforestation will reach a practical limit that will be largely dependent on land use requirements for agriculture and energy crop production. The capacity of forest biomass to sequester atmospheric CO_2 will likely plateau at about 5 Gt/year but could be as high as 10 Gt/year.

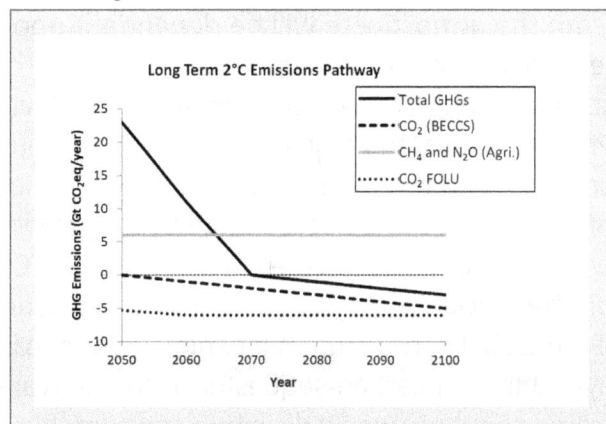

Figure 123. Potential Emissions Trajectories between 2050 and 2100 under a less than 2°C Scenario. Industrial carbon dioxide withdrawal begins in 2050 with initial roll out of bioenergy power plants equipped with carbon capture and storage (BECCS). Total carbon dioxide reduction from the combination of forest biomass and BECCS offsets the continued emissions of GHGs from agriculture and any residual emissions from other economic sectors. By the end of the century, the planet has transitioned to a controlled state of anthropogenic net negative emissions.

Under a 1.5°C pathway, the GHG emissions profile in the year 2050 will be similar to the 2065 emissions profile for the less than 2°C pathway. The timeline for transition to carbon neutral status under the 1.5 C pathway is advanced by about 15 years.

The complete process of climate change mitigation arriving at net zero accumulation of atmospheric greenhouse gases will likely require a global fleet of carbon capture equipped power plants that produce electricity from biomass combustion. The captured CO_2 that was removed from the atmosphere by the biomass would be transported by pipeline and permanently isolated in geological formations. The first BECCS power plants will likely come online by 2050 and by the end of the century the global fleet could provide 5–10 Gt/year of CO_2 storage capacity. These plants would be strategically located in regions of the world that would support rapid growth of energy crops grown on dedicated sustainably operated plantations. Carbon pricing mechanisms would be required such that total revenues from carbon storage plus power production would cover the complete costs of operating the biomass plantations and power plants, plus the costs of permanent storage of captured CO_2. Prior to the end of the century, the global capacity for carbon dioxide withdrawal from the combination of BECCS and forested areas could be designed to exceed residual GHG emissions. A globally managed system of BECCS power plants would allow for a gradual drawdown of the excess anthropogenic CO_2 that has accumulated in the atmosphere since James Watt advanced the design of the steam engine in 1776.

The final balance of processes that continue to emit GHGs and withdraw carbon dioxide from the atmosphere will be dependent upon the relative costs and capabilities of technologies available in the second half of the century. In the aviation subsector, aircraft could continue to be fueled by fossil fuel products or by second generation liquid biofuels. Potentially, the land requirements and complete costs of production of suitable biofuels could exceed the costs of constructing additional BECCS power plants as required to offset continued use of fossil fuels in the aviation subsector. The total carbon storage capacity required from the global fleet of BECCS power plants will be highly dependent on the efficiencies of future agricultural practices and on food consumption patterns. If the global herd of ruminant animals were to be cut well below current numbers, land availability for carbon-sequestering forested areas would increase, and GHG emissions from agriculture would drop by a considerable margin. These factors would reduce the future reliance on BECCS to offset ongoing emissions from agriculture.

Going forward to the next century and beyond, BECCS and perhaps other carbon dioxide removal technologies would be used to control atmospheric GHG concentrations within boundaries that will optimize average surface temperatures. The longer-term objective would be to transition beyond uncontrolled damaging anthropogenic climate change to anthropogenic climate control. By dialing in atmospheric CO_2 concentrates to target boundaries, average surface temperatures and thus the climate of Earth could be optimized for the preservation of ecosystems, natural habitats, food production systems and human infrastructure and endeavors.

7.4 The Year 2080

In 1960, after 2 years of direct sampling from the Mauna Loa observatory in Hawaii, David Keeling published definitive proof of progressive time dependent accumulation of CO_2 in the atmosphere. Continuous monitoring from Mauna Loa and other sites around the world in combination with the historical data set generated from ice core samples have confirmed that, from the onset of the Industrial Revolution, the activities of man have led to an increase in the concentration of atmospheric CO_2 from 280 to over 400 ppm. The greenhouse effect has been well understood since Svante Arrhenius's 1896 publication on the theoretical surface warming effects of atmospheric carbon dioxide. The 1°C increase in average temperatures since the beginnings of industrialization is the direct result of anthropogenic emissions of greenhouse gases. Until such time as the imbalance in the global carbon cycle is corrected, surface temperatures will continue to climb. Global warming will disrupt weather patterns, accelerate the melting of glaciers and icecaps, and damage natural ecosystems, habitats, human infrastructure, livelihoods and food production systems. The degree of damage will be in direct proportion to the magnitude of anthropogenic climate change.

In 1994, by enacting the United Nations Framework Convention on Climate Change, all countries unanimously committed to the goal of "preventing dangerous anthropogenic interference with the Earth's climate" by reducing greenhouse gas emissions. Since 1990, the annual rate of global GHG emissions has increased by 40% largely due to the fossil fuel-based industrialization of China and the failure of most advanced and emerging countries outside of the European Union to effectively curtail emissions.

By 2015, clear evidence of sea level rise, disrupted weather patterns and an increased frequency of severe weather events had emerged. Rapid rates of glacial melting, increased frequency and severity of floods, droughts, wildfires and heat waves were reported around the globe. On December 12, 2015, the 21st meeting of the Conference of Parties to the UNFCCC concluded with the universal agreement by all member states to limit future surface temperatures to well below 2°C and ideally 1.5°C above pre-industrial temperatures. The Paris Agreement came into force on November 4, 2016, 4 days prior to the election of Donald Trump as the future president of the United States.

At present, the aggregate of national ambitions to combat climate change is insufficient to meet the core objective of the Paris Agreement. The likely outcome of the implementation of current polices is a 3–4°C increase in average surface temperatures by the end of the century. The alternative is to strengthen national climate change action plans to further reduce domestic emissions and to facilitate access to lower cost international abatement opportunities. Cost-effective pathways to limit surface warming to well below 2°C would see countries progressing through low-cost, high-impact abatements to more challenging higher cost measures to further curtail emissions. An orderly process of decarbonization of economies would arrive at net zero GHG emissions sometime in middle of the second half of the century. By 2080, the transition to zero or net negative emissions would be completed. A 1.5°C pathway will require an immediate and sustained dedication of purpose from all nations to rapidly and aggressively implement abatements such that the aggregate effort arrives at net zero-emissions prior to 2060.

The simple fact is that we have less than a trillion tonnes of carbon dioxide in the remaining budget of emissions aligned with a 66% chance that end-of-century surface warming will be less than 2°C. Over-expending this budget will result in higher temperatures and will impose the extreme costs of climate change on future generations. Under-expending the budget will limit end-of-century surface warming to well below 2°C and minimize the damage and future costs of climate change. These are the facts and should be the guiding principle in global efforts to combat climate change.

By 2080, the overall impacts of climate change, based on an extrapolation of current policies will be somewhere between the catastrophic global outcome of a business as usual scenario in the absence of any mitigations and the manageable effects of a well below 2°C scenario. All regions will be impacted by climate change; however, with 3 to 4°C in surface warming, advanced economies will have sunk vast resources into climate change preparedness and adaptations. Infrastructure to limit damage from sea level rise, flooding, and storm surges would be commonplace. Costly measures to maintain water and food security would include relocation of farm lands and diversion of inland water flows. Measures to suppress wildfires and limit wildfire damage will be in place. The costs of climate change preparedness and adaptation in advanced countries will be well in excess of the costs that would have been incurred with an orderly decarbonization of global economies. Average surface temperature warming between 3 to 4°C will disproportionally impact highly vulnerable countries. By 2080, 30% of the global population of 10.8 billion will live in the countries of Sub-Saharan Africa.[498] These countries will not have sufficient capacity or, in many cases, options for preparedness and adaptation to floods, droughts, and severe weather events including extreme heat waves. Economic growth in these countries will have been stagnated by climate change. Crop failure and reduced agriculture productivity will have threatened food security, and flooding and droughts impacted water quality. Many countries will have become dependent on food aid to avoid malnutrition and possibly mass starvation. Low-lying coastal regions and small islands would be at risk to the damaging effects of sea level rise, storm surges, and coastal degradation. Climate change refugees, and widespread poverty and conflict within vulnerable regions are likely secondary outcomes of a global failure of mitigation. By 2080, the world would have become an uglier and meaner place with the disparities between wealthy and poor nations exacerbated by the disproportional impacts of climate change.

Limiting surface warming to well below 2°C, and ideally 1.5°C, will markedly limit the severity of climate change. The major benefit of effective mitigation will be an avoidance of the extreme costs of climate change damage that would have been imposed on future generations. Improved air quality, energy security, and poverty and conflict alleviation are secondary benefits to decarbonization of global economies. By necessity, Green

Revolution modern intensive agriculture practices will be common to all countries including Sub-Saharan African nations. Combustion of biomass for home heating and cooking purposes in least developed countries will have been relegated to the dustbin of archaic practices. This change alone will eradicate early deaths from poor indoor air quality. Near complete elimination of fossil fuel combustion across economic sectors will markedly improve outdoor air quality and improve the health status of citizens of many countries. The energy mix of renewables, hydro, and nuclear for the production of electricity along with electrification and improved energy efficiencies in the transportation, industry and buildings sectors, will result in energy independence for most countries. By 2080, the international trade of fossil fuel products will have declined to a trickle relative to peak demands in 2020. Energy independence, advances in agriculture, and efficiencies across economic sectors will contribute to the alleviation of global poverty and poverty-driven human conflict.

There is a brief window of opportunity to advance ambitions and strengthen policies to successfully mitigate future climate change. The expertise of the scientific community and climate change economists should be fully accessed by policy makers to develop effective climate change action plans. By establishing well-defined ceilings to annual emissions and a long-term budget for the total expenditure of emissions, the process of decarbonization can be completed in an orderly manner over the next 50 to 60 years. Industry will require clarity of intent for future planning such that the costs of transition and losses incurred through stranded assets are minimized.

The importance of this issue is such that a lack of awareness and inaction among political leaders equates to incompetence. Opposition to carbon pricing mechanisms and other policy measures designed to achieve cost-effective transitions to lower emissions practices is unfounded, and a form a climate change denial. The morality of political leadership that would deliberately and knowingly impose the extreme costs of climate change on future generations should be questioned. Incompetence, ignorance, and fear of change cannot be allowed to sway efforts to adequately curtail emissions.

The odds are in favour of my grandson living through the complete sequence of short, mid and long-term climate change mitigation periods. By 2080, my grandson would be 67 years of age, and the world into which he was born will have been radically transformed. The actions or inactions of decision-makers in positions of power at all levels of government during my grandson's childhood will define the extent of future climate change. My grandson's and subsequent generations will not be involved in the critical early phase decision-making process, but will either endure the cost and hardships imposed by a failure of mitigation or benefit from the transition to a sustainable future where greenhouse gas concentrations are maintained within safe parameters. Through action or inaction, a choice will be made as to the future climate of the planet. That choice is ours.

Epilogue

I do not believe in God, per say, or, for that matter, in any other form of faith-based reasoning. Inaction on the climate agenda is ultimately rooted in faith. Faith that immediate action is unnecessary, and that future generations will be better equipped to deal with the problem. Faith that the entire issue of anthropogenic climate change is a conspiracy invented by those opposed to free market economies or in a position to benefit from the implementation of climate change action plans. Faith that the free market system will make adjustments without the need for carbon pricing or other government interventions. Faith that climate change cannot be real, that somehow the scientists have got it wrong, that you cannot see emissions, and that the atmosphere is far too expansive to be influenced by the actions of man. Faith that we cannot afford to take action, that there is nothing that can be done, so why bother.

As a species we have, time and again, demonstrated a capacity for willful acts of wholesale destruction. We are most certainly capable of inaction or ineffectual delusional token action on the climate change agenda that would wreak havoc upon future generations.

We are also capable of much more than this. On July 4, 2012, scientists working with the Large Hadron Collider announced that they had detected a newly discovered particle with properties consistent with the theoretical Higgs boson. The LHC is by far the largest, most complex research instrument on Earth. The instrument was built by the European Organization for Nuclear Research over a 10-year period and completed in 2008 with an overall budget of 7.5 billion euros. Construction and operation of the LHC, was and is, based on a monumental effort of human cooperation involving 10,000 scientists and engineers from around the world and shared funding from the governments of Europe. Detection of the Higgs particle at the LHC represents a major scientific breakthrough in validating the Standard Model of subatomic particle physics.

As a species, we have demonstrated a capacity for inquiry and discovery such that we now are beginning to understand the origins of time, space, and matter. We have created stunning works of art, music, and theater that have explored the essence of the human condition and the beauty of the natural world.

Our capacity for comprehension and invention provides us with a viable cost-effective pathway to progressively decarbonize global economies over the next 50 years. We are fully capable of efficiently achieving this transition without over-expending the remaining budget of emissions in the less than 2°C surface warming scenario.

Our species has evolved such that we have the capabilities to change the composition of the atmosphere. We have an understanding of how these changes will impact climate systems and the environment of the planet. I find the full spectrum of human capabilities and the underlying responsibility to future generations to be deeply spiritual.

List of Figures

List of Tables

About the Book

The *Price of Carbon* could not have been written without the monumental efforts of the Intergovernmental Panel on Climate Change to compile the vast body of scientific literature on the subject of climate change into comprehensive reports. I am humbled by the magnitude and accomplishment of this work. In referencing IPCC reports, my goal was to further distill the findings and conclusions of this body of work within a narrative for a wider audience. Many of the images presented in the book were taken with permission from IPCC reports.

The science-based assessments of national climate action plans published by *Climate Action Tracker* are extensively referenced in the book. The various databases published by the World Bank have provided important statistics on economic performance, land use and population. The World Resources Institute detailed information on historical emissions of greenhouse gases allowed for a breakdown of sources of emissions on the basis of country and economic sector. All of these databases are available to the public on readily accessible websites. The wealth of information and quality of the presentation from organizations such as the Intergovernmental Panel on Climate Change, the International Energy Agency, Climate Action Tracker, the World Bank, the World Resources Institute and many other sources referenced in the book were critical to the writing of the *Price of Carbon*.

I am indebted to Bryan Maenz for endless hours of discussion on the subject and for his critical review of the manuscript. I would like to thank Katherine Kelly for her assistance in preparing the images contained in this book and Lynn Maenz for assistance in reviewing and editing the final proofs.

Finally, the professionalism and expertise of Tellwell Talent Inc. in all aspects of editing, design and marketing of this book is gratefully acknowledged.

Acronym List

°C – degrees Celsius

^{16}O – Common stable isotope of oxygen with an atomic weight of 16

^{18}O – Stable isotope of oxygen with an atomic weight of 18

$^{18}O/^{16}O$ – Ratio of isotopes of oxygen

ACES – American Clean Energy and Securities Act of 2009

AFOLU – Agriculture, Forestry and Other Land Use

BAU – Business as Usual

BCHP – Building Cooling, Heating and Power

BECCS – Bioenergy with Carbon Capture and Storage

BEV – Battery Electric Vehicles

BtL – Biomass to Liquid fuel

$CaCO_3$ – Calcium carbonate

CAT – Climate Action Tracker

CCS – Carbon Capture and Storage

CDM – Clean Development Mechanism

CDR – Carbon Dioxide Removal

CFCs – Chlorofluorocarbons (Freons)

CH_4 – Methane

CHP – Combined Heat and Power

CMA – Conference of Parties Meeting to the Paris Agreement

CO_2 – Carbon dioxide

COP – Conference of Parties (to the UNFCCC)

CPP – Clean Power Plan

CPS – Concentrated Solar Power

CVF – Climate Vulnerable Forum

CVLDC – Climate Vulnerable Least Develop Countries

DAC – Direct Air Capture of carbon dioxide

NDCs – Nationally Determined Contributions

DNA – Deoxyribonucleic Acid

DOE – US Department of Energy

E85 – Blend of 85% ethanol, 15% gasoline fuel

EAF – Electric Arc Furnace

EOR – Enhanced Oil Recovery

EPA – Environmental Protection Agency

ETSs – Emissions Trading Systems

FCPF – Forest Carbon Partnership Facility

FCV – Fuel Cell Vehicles

F-Gases – Fluorinated Gases

FOLU – Forestry and Other Land Use

GDP – Gross Domestic Product

Geo. – Geothermal power

GFEI – Global Fuel Economy Initiative

GHGs – Greenhouse Gases

Gt – Gigatonne (1 billion metric tonnes)

GTP – Global Temperature Potential

GW – Gigawatts (10^9 or 1 billion watts) electricity

GWP – Global Warming Potential

HCFCs – Halogenated chlorofluorocarbons

HDV – Heavy-Duty Vehicles

HVAC – Heating, Ventilation and Air Conditioning

ICE – Internal Combustion Engine

ICIs – International Cooperative Initiatives

IEA – International Energy Agency

INDCs – Intended Nationally Determined Contributions

IPCC – Intergovernmental Panel on Climate Change

IT – Information Technology

ITMOs – Internationally Transferred Mitigation Outcomes

IWG –Interagency Working Group

JIM – Joint Implementation Mechanism

km² – Square kilometres of land area

kWh – Kilowatts (10^3) of electricity per hour

LCOE – Levelized Cost of Electricity

LDV – Light-Duty Vehicles

LED – Light-Emitting Diodes

m/year – metre per year of glacial retreat

MAC – Marginal Abatement Costs

MBECS – Marine-Based Bioenergy with Carbon Storage

mm/year – millimetres per year of sea level rise

MMV – Measuring, Monitoring and Verification technologies applied to geological isolation of CO_2

MRV – Measurement, Reporting and Verification systems of forest monitoring

MSW – Municipal Solid Wastes

Mtoe – Millions of tonnes of oil equivalents as energy

MW – Megawatt (10^6 or 1 million watts) of electricity

N.R.C. – US National Research Council

N₂O – Nitrous oxide

NGCC – Natural Gas Combined Cycle power plants

NGOs – Non-Governmental Organizations

NZEBs – Near Zero-Emissions Buildings

O₃ – Ozone

OAE – Other Advanced Economies

OECD – Organization for Economic Cooperation and Development

OEE – Other Emerging Economies

PAT – Perform Achieve Trade energy efficiency trading system in India

PH – Passive House

ppm – parts per million

P-Tr – Permian-Triassic extinction event

RCP2.6 – Aggressive curtailment scenario with peak emissions by 2020

RCP4.5 – Intermediate curtailment scenario with peak emissions by 2040

RCP6.0 – Intermediate curtailment scenario with peak emission by 2080

RCP8.6 – Business as usual scenario with continued increased emissions

RCPs – Representative Concentration Pathways (Greenhouse gas emissions scenarios)

REDD – Reducing Emissions from Deforestation and Degradation

REDD+ – Reducing Emissions from Deforestation and Degradation plus Enhancement of Forest Carbon

RGGI – Regional Greenhouse Gas Initiative

RNA – Ribonucleic Acid

SC-CO₂ – Social Cost of Carbon Dioxide

SLR – Sea Level Rise

TPED – Total Primary Energy Demand

TPES – Total Primary Energy Supply

T$_s$ – Average surface temperature

TWh – Terawatts (10^{12} or 1 trillion watts) of electricity per hour

UN – United Nations

UNEP – United Nations Environment Programme

UNFCCC – United Nations Framework Convention on Climate Change

UREDD+ – Units of tradable emissions reduction within REDD and REDD+ programs

WAIS – Western Antarctic Ice Sheet

WHO – World Health Organization

ZEV Alliance – Zero-Emissions Vehicle Alliance

Suggested Readings

Intergovernmental Panel on Climate Change Fifth Assessment Reports. Available on-line. http://www.ipcc.ch/report/ar5/index.shtml.

Climate Change 2013: The Physical Science Basis. Contribution of Working Group I to the Fifth Assessment Report of the Intergovernmental Panel on Climate Change.

Climate Change 2014: Synthesis Report. Contribution of Working Groups I, II, and III to the Fifth Assessment Report of the Intergovernmental Panel on Climate Change.

Climate Change 2014: Impacts, Adaptation, and Vulnerability. Part A: Global and Sectoral Aspects. Contribution of Working Group II to the Fifth Assessment Report of the Intergovernmental Panel on Climate Change.

Climate Change 2014: Impacts, Adaptation, and Vulnerability. Part B: Regional Aspects. Contribution of Working Group II to the Fifth Assessment Report of the Intergovernmental Panel on Climate Change.

Climate Change 2014: Mitigation of Climate Change. Contribution of Working Group III to the Fifth Assessment Report of the Intergovernmental Panel on Climate Change.

World Resources Institute. http://www.wri.org/.
Climate Action Tracker. http://climateactiontracker.org/.
World Energy Outlook 2016. International Energy Agency. https://www.iea.org/

References

1   Encrenaz, T., Bibring, J. P., Blanc, M., Barucci, M. A., Roques, F., & Zarka, P. (2004). *The Solar System* (3rd Edition). Berlin. Springer.

2   Bonanno, A., Schlattl, H., & Paternò, L. (2008). The Age of the sun and the relativistic corrections in the EOS. *Astronomy and Astrophysics. 390*, 1155-1118.

3   History of earth. (2015). *Wikipedia.* Retrieved from https://en.wikipedia.org/wiki/History_of_Earth.

4   Beychok, M. (2014). Earth's atmospheric air. *The Encyclopedia of Earth.* Retrieved from http://www.eoearth.org/view/article/170977/.

5   Choi, C.Q. (2016, March 7). Life's building blocks form in replicated deep sea vents. *Astrobiology Magazine.* Retrieved from http://www.astrobio.net/origin-and-evolution-of-life/lifes-building-blocks-form-in-replicated-deep-sea-vents/.

6   Bell, E.A., Boehnke, P., Harrison, T.M., & Mao, W.L. (2015). Potentially biogenic carbon preserved in 4.1 billion-year-old zircon. *Proceedings of the National Academy of Sciences. 112*, 14519-14521.

7   Witze, A. (2016, August 13). Claims of earth's oldest fossils tantalize researchers. *Nature. News.* Retrieved from http://www.nature.com/news/claims-of-earth-s-oldest-fossils-tantalize-researchers-1.20506.

8   Woese, C., & Gogarten, J.P. (1999). When did eukaryotic cells (cells with Nuclei and other internal organelles) first evolve? What do we know about how they evolved from earlier life-forms? *Scientific America.* Retrieved from https://www.scientificamerican.com/article/when-did-eukaryotic-cells/.

9   Timeline of the evolutionary history of life. (2017). *Wikipedia.* Retrieved from https://en.wikipedia.org/wiki/Timeline_of_the_evolutionary_history_of_life#cite_note-27.

10   Buick, R. (2008). When did oxygenic photosynthesis evolve? *Philosophical Transactions of the Royal Society B Biological Sciences. 363*, 2731-2743.

11   What caused the snowball Earths? (2015). *Snowball Earth.* Retrieved from http://www.snowballearth.org/cause.html.

12   Ramesh, S. (2016). A brief history of Earth: how it all began. *The Wire.* Retrieved from https://thewire.in/40924/a-brief-history-of-the-earth-how-it-all-began/.

13   Sparling, B. (2001). The ozone layer. *NASA.* Retrieved from https://www.nas.nasa.gov/About/Education/Ozone/ozonelayer.html.

14   Marshall, M. (2010). The history of ice on Earth. *New Scientist.* Retrieved from https://www.newscientist.com/article/dn18949-the-history-of-ice-on-earth/.

15   Greenhouse and icehouse Earth. (2015). *Wikipedia.* Retrieved from https://en.wikipedia.org/wiki/Greenhouse_and_icehouse_Earth.

16   Ice age. (2015). *Wikipedia.* Retrieved from https://en.wikipedia.org/wiki/Ice_age.

17   Royer, D.L., Berner, R.A., Montañez, I.P., Tabor, N.J., & Beerling, D.J. (2004). CO_2 as a primary driver of Phanerozoic climate. *GSA Today. 14*, Retrieved from http://www.geosociety.org/gsatoday/archive/14/3/pdf/i1052-5173-14-3-4.pdf.

18   Cambrian explosion. (2015). *Wikipedia.* Retrieved from https://en.wikipedia.org/wiki/Cambrian_explosion.

19   Devonian. (2015). *Wikipedia.* Retrieved from https://en.wikipedia.org/wiki/Devonian.

20   Carboniferous. (2015). *Wikipedia.* Retrieved from https://en.wikipedia.org/wiki/Carboniferous.

21   Karoo ice age. (2015). *Wikipedia.* Retrieved from https://en.wikipedia.org/wiki/Karoo_Ice_Age.

22   Permian-Triassic extinction event. (2015). *Wikipedia.* Retrieved from https://en.wikipedia.org/wiki/Permian%E2%80%93Triassic_extinction_event.

23   Mesozoic. (2015). *Wikipedia.* Retrieved from https://en.wikipedia.org/wiki/Mesozoic.

24   Cretaceous-Paleogene extinction event. (2015). *Wikipedia.* Retrieved from https://en.wikipedia.org/wiki/Cretaceous%E2%80%93Paleogene_extinction_event.

25   Oxygen isotope ratio cycle. (2015). *Wikipedia.* Retrieved from https://en.wikipedia.org/wiki/Oxygen_isotope_ratio_cycle.

26   Formaminifera. (2015). *Wikipedia.* Retrieved from https://en.wikipedia.org/wiki/Foraminifera.

27   Hansen, J., Sato, M., Russell, G., & Kharecha, P. (2013). Climate sensitivity, sea level, atmospheric carbon dioxide. *Philosophical Transactions of the Royal Society A Mathematical, Physical and Engineering Sciences. 371,* 20120294. doi:10.1098/rsta.2012.0294.

28   Eocene. (2015). *Wikipedia.* Retrieved from https://en.wikipedia.org/wiki/Eocene.

29   Azolla event. (2015). *Wikipedia.* Retrieved from https://en.wikipedia.org/wiki/Azolla_event.

30   Haug, G.H. (2004, March 22). How the ismthus of Panama put ice in the arctic. *OceanuUS Magazine.* Retrieved from http://www.whoi.edu/oceanus/feature/how-the-isthmus-of-panama-put-ice-in-the-arctic.

31   Dome C. (2015). *Wikipedia.* Retrieved from https://en.wikipedia.org/wiki/Dome_C.

32   Kumar, A. (2015). Research at Concordia station. *European Space Agency. Chronicals from Concordia.* Retrieved from http://blogs.esa.int/concordia/research-on-planet-concordia/.

33   Concordia station. (2015). *Wikipedia.* Retrieved from https://en.wikipedia.org/wiki/Concordia_Station.

34   Ice core. (2015). *Wikipedia.* Retrieved from https://en.wikipedia.org/wiki/Ice_core.

35   Lüthi, D., Le Floch, M., Bereiter, B., Blunier, T., Barnola, J.-M., Siegenthaler, U., Raynaud, D., Jouzel, J., Fischer, H., Kawamura, K., & Stocker, T.F. (2008). High-resolution carbon dioxide concentration record 650,00–800,000 years before present. *Nature. 453,* 379-382.

36   Petit J.R., Jouzel J., Raynaud D., Barkov N.I., Barnola J.M., Basile I., Bender M., Chappellaz J., Davis J., Delaygue G., Delmotte M., Kotlyakov V.M., Legrand M., Lipenkov V., Lorius C., Pépin L., Ritz C., Saltzman E., & Stievenard M. (1999). Climate and atmospheric history of the past 420,000 years from the Vostok ice core, Antarctica. *Nature. 399,* 429-436.

37   Milutin Milanković. (2015). *Wikipedia.* Retrieved from https://en.wikipedia.org/wiki/Milutin_Milankovi%C4%87.

38   Orbital forcing. (2015). *Wikipedia.* Retrieved from https://en.wikipedia.org/wiki/Orbital_forcing.

39   Milankovitch cycles. (2015). *Wikipedia.* Retrieved from https://en.wikipedia.org/wiki/Milankovitch_cycles.

40   Holocene. (2015). *Wikipedia.* Retrieved from https://en.wikipedia.org/wiki/Holocene.

41   File:Holocene Temperature Variations.png. (2015). *Wikipedia.* Retrieved from https://commons.wikimedia.org/wiki/File:Holocene_Temperature_Variations.png.

42   Neoglaciation.(2015). *Wikipedia.* Retrieved from https://en.wikipedia.org/wiki/Neoglaciation.

43   Monnin, E. (2004). EPICA dome C ice core high resolution Holocene and transition CO₂ data. *National Climate Data Center.* Retrieved from http://www1.ncdc.noaa.gov/pub/data/paleo/icecore/antarctica/epica_domec/edc-co2.txt.

44   Crutzen, P.J., & Stoermer, E.F. (2000). The Anthropocene. *International Geosphere-Biosphere Programme. Newsletter 41.* Retrieved from http://www.igbp.net/download/18.316f18321323470177580001401/1376383088452/NL41.pdf.

45   PhyOrg. (2016, August 29). The Anthropocene is here: scientists. *PhyOrg.* Retrieved from https://phys.org/news/2016-08-anthropocene-scientists.html.

46   www.parliament.uk. (2015). An act declaring the rights and liberties of the subject and setting the succession of the crown. (1689). *Parliament of England, 1689.* Retrieved from www.parliament.uk/about/living-heritage/evolutionofparliament/parliamentaryauthority/revolution/collections1/collections-glorious-revolution/billofrights/.

47   Hatcher, J. (1993). *The history of the British coal industry: volume 1: before 1700: towards the age of coal.* New York. Oxford University Press.

48   Thomas Savery. (2015). *Wikipedia.* Retrieved from https://en.wikipedia.org/wiki/Thomas_Savery.

49   Thurston, R.H. (1886). *A History of the growth of the Steam Engine.* London. K. Paul, Trench, Trübner & co.

50   Rolt, L.T.C. & Allen J.S. (1977). *The steam engine of Thomas Newcomen.* New York. Science History Publications.

51   Rifkin, J. (2002). *The Hydrogen Economy.* New York. Tarcher/Penguin.

52   Kingsford, P.W. (2015). James Watt. *Encyclopedia Britannica.* Retrieved from http://www.britannica.com/biography/James-Watt.

53   Carnegie, A. (1905). *James Watt.* New York. Doubleday, Page.

54   Cast Iron. (2015). *Encyclopedia Britannica.* Retrieved from http://www.britannica.com/technology/cast-iron.

55   Wrought Iron. (2015) *Encyclopedia Britannica.* Retrieved from http://www.britannica.com/technology/wrought-iron.

56   Henry Cort. (2015). *Encyclopedia Britannica.* Retrieved from http://www.britannica.com/biography/Henry-Cort.

57   Industrial revolution. (2015). *Wikipedia*. Retrieved from https://en.wikipedia.org/wiki/Industrial_Revolution.

58   Puddling (metallurgy). (2015). *Wikipedia*. Retrieved from https://en.wikipedia.org/wiki/Puddling_(metallurgy).

59   Boylston, H.B. (1936). *An Introduction to the Metallurgy of Iron and Steel*. New York. J. Wiley and sons.

60   Spoerl, J.S. (2015). A brief history of iron and steel production. Retrieved from http://www.anselm.edu/homepage/dbanach/h-carnegie-steel.htm.

61   Henderson, L.J. (1917). *The Order of Nature*. Cambridge. Harvard University Press.

62   Hudson, P. (2008). *Chapter 2 – Industrial organization and structure.* In R. Floud (Ed.) The Cambridge Economic History of Modern Britain. Volume 1: Industrialization, 1700-1860. In Cambridge University Press.

63   Clark, G. (2007). *A Farewell to Alms: A Brief Economic History of the World*. Princeton University Press.

64   History of labor law in the United Kingdom. (2015). *Wikipedia*. Retrieved from https://en.wikipedia.org/wiki/History_of_labour_law_in_the_United_Kingdom.

65   United Nations. (2015). The world at 6 billion. (2015). *United Nations Department of Economic and Social Affairs. Population Division*. Retrieved from http://www.igbp.net/download/18.316f18321323470177580001401/1376383088452/NL41.pdf.

66   Nature (journal). (2015). *Wikipedia*. Retrieved from https://en.wikipedia.org/wiki/Nature_(journal).

67   National Academy of Sciences. (2015). Overview NAS History. *National Academy of Sciences*. Retrieved from http://www.nasonline.org/about-nas/history/.

68   Darwin. (1859). *The Origin of Species*. John Murray.

69   History of Coal Mining. (2015). *Wikipedia*. Retrieved from https://en.wikipedia.org/wiki/History_of_coal_mining.

70   Nasaw, D. (2006). *Andrew Carnegie*. The Penquin Group.

71   Arrhenius, S. (1896). On the influence of carbonic acid in the air upon the temperature of the ground. *Philosophical Magazine and Journal of Science. Series 5 Volume 41*, 237-276.

72   Chandler, A. 1987. The Paris Exposition Universelle of 1900. Retrieved from http://www.arthurchandler.com/paris-1900-exposition/. Oct. 2015

73   MacMillan, M. (2003). The War that Ended Peace. New York. Random House.

74   Siemens. (2015).Siemens history. *Siemens Ingenuity of Life*. Retrieved from https://www.siemens.com/history/en/news/1308_worlds-fairs.htm.

75   The early electric car site. (2015). *The Early Electric Car Site*. Retrieved from http://earlyelectric.com/timeline.html.

76   Ford Motor Company. (2015). *Wikipedia*. Retrieved from https://en.wikipedia.org/wiki/Ford_Motor_Company.

77   Wright Brothers. (2015). *Wikipedia*. Retrieved from https://en.wikipedia.org/wiki/Wright_brothers.

78   Weart, S. (2017). The discovery of global warming. *The Discovery of Global Warming*. Retrieved from http://history.aip.org/climate/co2.htm.

79   Revelle, R, & Suess H. E. (1957). Carbon Dioxide Exchange between Atmosphere and Ocean and the Question of an Increase of Atmospheric CO_2 During the Past Decades. *Tellus* **9**, 18-27.

80   Scripps Institution of Oceanography. (2015). Charles David Keeling biography. *Scripps CO_2 Program*. Retrieved from http://scrippsco2.ucsd.edu/history_legacy/charles_david_keeling_biography.

81   Keeling, C.D. (1960). The concentration and isotopic abundances of carbon dioxide in the atmosphere. *Tellus* **12**, 200-203.

82   Scripps Institution of Oceanography. (2017, January 31). The Keeling curve. Latest CO_2 reading January 31, 2017. *The Keeling Curve*. Retrieved from https://scripps.ucsd.edu/programs/keelingcurve/.

83   Monthly CO_2. (2015). *CO_2 Earth*. Retrieved from http://www.co2.earth/monthly-co2.

84   Keeling Curve. (2015). *Wikipedia*. Retrieved from https://en.wikipedia.org/wiki/Keeling_Curve.

85   Etheridge, D.M., Steel, L.P., Langenfelds, R.L., & Francey, R.J. (1998). Historical CO_2 record from the Law Dome DE08, DE08-2, and DSS ice cores. *Carbon Dioxide Information Analysis Center*. Retrieved from http://cdiac.ornl.gov/ftp/trends/co2/lawdome.combined.dat.

86   IPCC, 2014: Climate Change 2014: Synthesis Report. Contribution of Working Groups I, II, and III to the Fifth Assessment Report of the Intergovernmental Panel on Climate Change [Core Writing Team, R.K. Pacharui and L.A. Meyer (eds)]. IPCC, Geneva, Switzerland.

87   Myhre, G., D. Shindell, F.-M. Breon, W. Collins, J. Fuglestvedt, J. Huang, D. Koch, J.-F. Lamarque, D. Lee, B. Mendoza, T. Nakajima, A. Robock, G. Stephens, T. Takemura and H. Zhang, 2013: Anthropogenic and Natural Radiative Forcing. In: Climate Change 2013: The Physical Science Basis. Contribution of Working Group I to the Fifth Assessment Report of the Intergovernmental Panel on Climate Change [Stocker, T.F., D. Qin, G.-K. Plattner, M. Tignor, S.K. Allen, J. Boschung, A. Nauels, Y. Xia, V. Bex and P.M. Midgley (eds.)]. Cambridge University Press, Cambridge, United Kingdom and New York, NY, USA.

88 Ozone Secretariat. (2000). The Montreal protocol on substances that deplete the ozone layer. *United Nations Environment Programme.* Retrieved from http://ozone.unep.org/pdfs/Montreal-Protocol2000.pdf

89 Bullister, J.L. (2015). Atmospheric Histories (1765-2015) for CFC-11, CFC-12, CFC-13, CCl_4, SF_6 and N_2O. *Ocean CO₂ Carbon Dioxide Information Analysis Center.* Retrieved from http://cdiac.ornl.gov/oceans/CFC_ATM_Hist2015.html.

90 1883 eruption of Krakatoa. (2015). *Wikipedia.* Retrieved from https://en.wikipedia.org/wiki/1883_eruption_of_Krakatoa.

91 Climate change controversies. (2007). *The Royal Society.* Retrieved from https://royalsociety.org/~/media/Royal_Society_Content/policy/publications/2007/8031.pdf.

92 Ciais, P., C. Sabine, G. Bala, L. Bopp, V. Brovkin, J. Canadell, A. Chhabra, R. DeFries, J. Galloway, M. Heimann, C. Jones, C. Le Quere, R.B. Myneni, S. Piao and P. Thornton, 2013: Carbon and Other Biogeochemical Cycles. In: Climate Change 2013: The Physical Science Basis. Contribution of Working Group I to the Fifth Assessment Report of the Intergovernmental Panel on Climate Change [Stocker, T.F., D. Qin, G.-K. Plattner, M. Tignor, S.K. Allen, J Boschung, A. Nauels, Y. Xia, V. Bex and P.M. Midgley (eds.)]. Cambridge University Press, Cambridge, United Kingdom and New York, NY, USA.

93 Global Carbon Budget 2014. (2014). *Global Carbon Project.* Retrieved from http://www.globalcarbonproject.org/carbonbudget/14/hl-full.htm.

94 NASA. (2017). Global climate change. Vital signs of the planet. Facts. Global temperatures. *National Aeronautics and Space Administration.* Retrieved from http://climate.nasa.gov/vital-signs/global-temperature/.

95 Report of an Ad Hoc Study Group on Carbon Dioxide and Climate, Woods Hole, Massachusetts, July 23–27, 1979, to the Climate Research Board, Assembly of Mathematical and Physical Sciences, National Research Council. (1979). Carbon dioxide and climate: A scientific assessment. *National Research Council.* Washington, D.C.: The National Academies Press. ISBN 0-309-11910-3.

96 Smil, V. (2001). *Enriching the Earth. Fritz Haber, Carl Bosch and the Transformation of World Food Production.* London. The MIT Press.

97 Science heroes. (2015). *Science Heroes.* Retrieved from http://www.scienceheroes.com/index.php?option=com_content&view=article&id=258&Itemid=232.

98 Haber process. (2015). *Wikipedia.* Retrieved from https://en.wikipedia.org/wiki/Haber_process.

99 Green Revolution. (2015). *Wikipedia.* Retrieved from https://en.wikipedia.org/wiki/Green_Revolution.

100 Rowlatt, J. (2016, December 1). IR8: The miracle rice which saved millions of lives. *BBC News.* Retrieved from http://www.bbc.com/news/world-asia-india-38156350.

101 Workman, D. (2016, December 26). Rice Exports by Country. *World's Top Exports.* Retrieved from http://www.worldstopexports.com/rice-exports-country/.

102 Statistica. (2017). Leading soybean producing countries worldwide from 2012/13 to 2016/17 (in million metric tons). *The Statistical Portal.* Retrieved from https://www.statista.com/statistics/263926/soybean-production-in-selected-countries-since-1980/.

103 Top ten countries by agricultural exports. (2017). *Maps of the World.* Retrieved from http://www.mapsofworld.com/world-top-ten/world-top-ten-agricultural-exporters-map.html.

104 Norman Borlaug. (2015). *Wikipedia.* Retrieved from https://en.wikipedia.org/wiki/Norman_Borlaug.

105 United Nations. (2017). World population prospects 2017. *United Nations Department of Economic and Social Affairs.* Retrieved from https://esa.un.org/unpd/wpp/.

106 World Bank. (2015). Population, total. *The World Bank.* Retrieved from http://data.worldbank.org/indicator/SP.POP.TOTL?locations=IN. Accessed Nov. 2015

107 Smith P., M. Bustamante, H. Ahammad, H. Clark, H. Dong, E. A. Elsiddig, H. Haberl, R. Harper, J. House, M. Jafari, O. Masera, C. Mbow, N. H. Ravindranath, C. W. Rice, C. Robledo Abad, A. Romanovskaya, F. Sperling, and F. Tubiello, 2014: Agriculture, Forestry and Other Land Use (AFOLU). In: Climate Change 2014: Mitigation of Climate Change. Contribution of Working Group III to the Fifth Assessment Report of the Intergovernmental Panel on Climate Change [Edenhofer, O., R. Pichs-Madruga, Y. Sokona, E. Farahani, S. Kadner, K. Seyboth, A. Adler, I. Baum, S. Brunner, P. Eickemeier, B. Kriemann, J. Savolainen, S. Schlömer, C. von Stechow, T. Zwickel and J.C. Minx (eds.)]. Cambridge University Press, Cambridge, United Kingdom and New York, NY, USA.

108 CDIAC. (2017). Fossil fuel CO_2 emissions. *Carbon Dioxide Information Analysis Center.* Retrieved from http://cdiac.ornl.gov/trends/emis/overview_2013.html.

109 Marshall Plan. (2015). *Wikipedia.* Retrieved from https://en.wikipedia.org/wiki/Marshall_Plan.

110 The Marshall Plan, 1948. (2015). *Department of State. United States of America. Office of the Historian.* Retrieved from https://history.state.gov/milestones/1945-1952/marshall-plan.

111 Japanese post-war economic miracle. (2015). *Wikipedia.* Retrieved from https://en.wikipedia.org/wiki/Japanese_post-war_economic_miracle.

112 Bolt, J., Timmer, M., & van Zanden, J.L. (2014). "GDP per capita since 1820", in Jan Luiten van Zanden, et al. (Eds.), How Was Life?: Global Well-being since 1820, OECD Publishing. Retrieved from http://dx.doi.org/10.1787/9789264214262-7-en.

113 Energycarrier. H$_2$ the energy carrier. (2015). H$_2$*ydropole*. Retrieved from https://hydropole.ch/en/hydrogen/energycarrier/.

114 Ozone Secretariat. (2017). Montreal protocol – achievements to date and challenges ahead. *United Nations Environment Programme*. Retrieved from http://ozone.unep.org/en/focus/montreal-protocol-achievements-date-and-challenges-ahead.

115 IPCC. (2015). Organization. *Intergovernmental Panel on Climate Change*. Retrieved from http://www.ipcc.ch/organization/organization.shtml.

116 IPCC. (2015). Reports. *Intergovernmental Panel on Climate Change*. Retrieved from http://www.ipcc.ch/publications_and_data/publications_and_data_reports.shtml.

117 The Nobel Peace Prize 2007. (2015). *Nobelprize.org*. Retrieved from http://www.nobelprize.org/nobel_prizes/peace/laureates/2007/. Accessed Dec. 2015.

118 UNFCCC. (2015). First steps to a safer future: Introducing the United Nations Framework Convention on Climate Change. *United Nations Framework Convention on Climate Change*. Retrieved from https://unfccc.int/essential_background/convention/items/6036.php.

119 UNFCCC. (2015). Kyoto protocol. *United Nations Framework Convention on Climate Change*. Retrieved from https://unfccc.int/kyoto_protocol/items/2830.php.

120 Copenhagen Accord. (2015) *Wikipedia*. Retrieved from https://en.wikipedia.org/wiki/Copenhagen_Accord.

121 UNFCCC. (2014). Compilation of economy-wide emission reduction targets to be implemented by Parties included in Annex I to the convention. *United Nations Framework Convention on Climate Change*. Retrieved from http://unfccc.int/resource/docs/2014/sbsta/eng/inf06.pdf.

122 UNEP. (2015). The Emissions Gap Report 2015. *United Nations Environment Programme*. Nairobi.

123 UNFCCC. (2015). Durban: Towards full implementation of the UN Climate Change Convention. *United Nations Framework Convention on Climate Change*. Retrieved from http://unfccc.int/key_steps/durban_outcomes/items/6825.php.

124 WRI. (2014). Climate data explorer. *World Resources Institute*. Retrieved from http://cait.wri.org. Accessed April 2016.

125 OECD. (2015). List of OECD member countries – ratification of the convention on the OECD. *Organization of Economic Cooperation and Development*. Retrieved from http://www.oecd.org/about/membersandpartners/list-oecd-member-countries.htm.

126 Coal in China. (2015). *Wikipedia*. Retrieved from https://en.wikipedia.org/wiki/Coal_in_China.

127 IEA. (2014). CO$_2$ emissions from fuel combustion. *International Energy Agency*. 9 rue de la Fédération. 75739 Paris Cedex 15. France.

128 IPCC. (2006). *2006 IPCC Guidelines for National Greenhouse Gas Inventories*. Prepared by the National Greenhouse Gas Inventories Programme, Eggleston H.S., Buendia L., Miwa K., Ngara T., and Tanabe K. (Eds). Published: IGES, Japan.

129 IEA. (2016). *Key World Energy Statistics 2016*. International Energy Agency. 9 rue de la Fédération. 75739 Paris Cedex 15. France.

130 Coal Reserves. (2015). *BP Global*. Retrieved from http://www.bp.com/en/global/corporate/energy-economics/statistical-review-of-world-energy/coal/coal-reserves.html.

131 US EPA. (2015). The plain english guide to the clean air act. *United States Environmental Protection Agency*. Retrieved from https://www.epa.gov/sites/production/files/2015-08/documents/peg.pdf.

132 US EPA. (2015). Clean air act overview. *United States Environmental Protection Agency*. Retrieved from https://www.epa.gov/clean-air-act-overview/progress-cleaning-air-and-improving-peoples-health.

133 CATF. (2010). The toll from coal. *Clean Air Task Force*. Retrieved from http://www.catf.us/resources/publications/files/The_Toll_from_Coal.pdf.

134 WHO News Release. (2014, March 25). 7 million premature deaths annually linked to air pollution. World Health Organization. Retrieved from http://www.who.int/mediacentre/news/releases/2014/air-pollution/en/.

135 WHO Fact Sheet. (2016). Ambient (outdoor) air quality and health. *World Health Organization*. Retrieved from http://www.who.int/mediacentre/factsheets/fs313/en/.

136 United States Department of Labor. (2016). Coal fatalities for 1900 through 2016. *Mine Safety and Health Administration*. Retrieved from http://www.msha.gov/stats/centurystats/coalstats.asp.

137 Coal in China. (2015). *Wikipedia*. Retrieved from https://en.wikipedia.org/wiki/Coal_in_China.

138 Borenstein, S. (2008, March 14). Deaths per TWh for all energy sources: Rooftop solar power is actually more dangerous than Chernoby. *next BIG Future*. Retrieved from http://nextbigfuture.com/2008/03/deaths-per-twh-for-all-energy-sources.html.

139 European Commission. Press release. (2001). New research reveals the real cost of electricity in Europe. *European Commission. Research Directorate-General*. Retrieved from http://www.ier.uni-stuttgart.de/forschung/projektwebsites/newext/externen.pdf.

140   Environmental impact of the coal industry. (2015). *Wikipedia*. Retrieved from https://en.wikipedia.org/wiki/Environmental_impact_of_the_coal_industry.

141   IEA. (2013). Resources to reserves. *International Energy Agency*. IEA Publishing. Retrieved from https://www.iea.org/publications/freepublications/publication/Resources2013.pdf.

142   Deepwater Horizon oil spill. (2015). *Wikipedia*. Retrieved from https://en.wikipedia.org/wiki/Deepwater_Horizon_oil_spill.

143   Lac-Mégantic rail disaster. (2015). *Wikipedia*. Retrieved from https://en.wikipedia.org/wiki/Lac-M%C3%A9gantic_rail_disaster.

144   Exxon Valdez oil spill. (2015). *Wikipedia*. Retrieved from https://en.wikipedia.org/wiki/Exxon_Valdez_oil_spill.

145   IEA. (2016). Natural gas information. *International Energy Agency*. IEA Publishing. Retrieved from http://wds.iea.org/wds/pdf/Gas_documentation.pdf.

146   Hydroelectricity. (2015). *Wikipedia*. Retrieved from https://en.wikipedia.org/wiki/Hydroelectricity.

147   IAEA Report by the Director General (2014). Nuclear technology review 2014. *International Atomic Energy Agency*. Retrieved from https://www.iaea.org/About/Policy/GC/GC58/GC58InfDocuments/English/gc58inf-4_en.pdf.

148   Chernobyl accident appendix 1. Sequence of events. (2009). *World Nuclear Association*. Retrieved from http://www.world-nuclear.org/information-library/safety-and-security/safety-of-plants/appendices/chernobyl-accident-appendix-1-sequence-of-events.aspx.

149   Timeline. A chronology of events surrounding the Chernobly nuclear disaster. (2015) *The Chernobyl Gallery*. Retrieved from http://chernobylgallery.com/chernobyl-disaster/timeline/.

150   WHO. (2006). Health effects of the Chernobyl accident and special health care programs. *World Health Organization*. Geneva. WHO Press.

151   Chernobyl disaster. (2015). Wikipedia. Retrieved from https://en.wikipedia.org/wiki/Chernobyl_disaster.

152   International nuclear event scale. (2015). *Wikipedia*. Retrieved from https://en.wikipedia.org/wiki/International_Nuclear_Event_Scale.

153   Fukushima Daiichi nuclear disaster. (2015). *Wikipedia*. Retrieved from https://en.wikipedia.org/wiki/Fukushima_Daiichi_nuclear_disaster.

154   Comparison of Fukushima and Chernobyl nuclear accidents. (2015). *Wikipedia*. Retrieved from https://en.wikipedia.org/wiki/Comparison_of_Fukushima_and_Chernobyl_nuclear_accidents.

155   UNSCEAR. (2013). Volume 1. Report to the General Assembly. Scientific Annex A. Levels and effects of radiation exposure due to the nuclear accident after the 2011 great east-Japan earthquake and tsunami. *United Nations Scientific Committee on the Effects of Atomic Radiation*. New York. United Nations Publication.

156   Radioactive waste management. (2015). *World Nuclear Association*. Retrieved from http://www.world-nuclear.org/info/nuclear-fuel-cycle/nuclear-wastes/radioactive-waste-management/.

157   Radioactive waste. (2015). *Wikipedia*. Retrieved from https://en.wikipedia.org/wiki/Radioactive_waste.

158   Generation IV reactor. (2015). *Wikipedia*. Retrieved from https://en.wikipedia.org/wiki/Generation_IV_reactor.

159   2011 Tōhoku earthquake and tsunami. (2015). *Wikipedia*. Retrieved from https://en.wikipedia.org/wiki/2011_T%C5%8Dhoku_earthquake_and_tsunami.

160   Nuclear Power in France. (2017). *World Nuclear Association*. Retrieved from http://www.world-nuclear.org/information-library/country-profiles/countries-a-f/france.aspx./.

161   Statista (2015). Global electricity prices by select countries in 2015 (in U.S. dollar cents per kilowatt hour). *The Statistical Portal*. Retrieved from http://www.statista.com/statistics/263492/electricity-prices-in-selected-countries/.

162   Control of fire by early humans. (2016). *Wikipedia*. Retrieved from https://en.wikipedia.org/wiki/Control_of_fire_by_early_humans.

163   IEA. (2016). Topic renewable. Subtopic bioenergy. *International Energy Agency*. Retrieved from https://www.iea.org/topics/renewables/subtopics/bioenergy/.

164   Forest biomass and air emissions. (2016). *Washington State Department of Natural Resources*. Retrieved from http://file.dnr.wa.gov/publications/em_forest_biomass_and_air_emissions_factsheet_8.pdf.

165   Ontario Newsroom News. (2014, April 15). Creating cleaner air in Ontario. *Minister of Energy, Government of Ontario*. Retrieved from https://news.ontario.ca/mei/en/2014/04/creating-cleaner-air-in-ontario-1.html.

166   Schlömer S., T. Bruckner, L. Fulton, E. Hertwich, A. McKinnon, D. Perczyk, J. Roy, R. Schaeffer, R. Sims, P. Smith, and R. Wiser, 2014: Annex III: Technology-specific cost and performance parameters. In: Climate Change 2014: Mitigation of Climate Change. Contribution of Working Group III to the Fifth Assessment Report of the Intergovernmental Panel on Climate Change [Edenhofer, O., R. Pichs-Madruga, Y. Sokona, E. Farahani, S. Kadner, K. Seyboth, A. Adler, I. Baum, S. Brunner, P. Eickemeier, B. Kriemann, J. Savolainen, S. Schlömer, C. von Stechow, T. Zwickel and J.C. Minx (eds.)]. Cambridge University Press, Cambridge, United Kingdom and New York, NY, USA.

167 EC. (2016). Topic energy. Subtopic biomass. *European Commission.* Retrieved from
https://ec.europa.eu/energy/en/topics/renewable-energy/biomass.

168 IPCC. (2006). Solid Waste Disposal: Guidelines for National Greenhouse Gas Inventories. *Intergovernmental Panel on Climate Change.* Retrieved from http://www.ipcc-nggip.iges.or.jp/public/2006gl/pdf/5_Volume5/V5_3_Ch3_SWDS.pdf.

169 Dias De Oliveira, M.E., Vaughan, B.E., Rykiel Jr., & E.J. (2005). Ethanol as fuel: energy, carbon dioxide balances, and ecological footprint. *BioScience. 55,* 593-602.

170 Sapp, M. (2015, March 5) Brazil to Implement E27 on March 16. *Biofuels Digest.* Retrieved from
http://www.biofuelsdigest.com/bdigest/2015/03/05/brazil-to-implement-e27-on-march-16/.

171 IEA. (2010). Sustainable production of second-generation biofuels. *International Energy Agency.* IEA Publishing. Retrieved from https://www.iea.org/publications/freepublications/publication/second_generation_biofuels.pdf.

172 IEA. (2016). *World Energy Outlook 2016.* International Energy Agency. 9 rue de la Fédération. 75739 Paris Cedex 15, France.

173 Kutscher, C.F. (2007). Tackling climate change in the U.S. potential carbon emissions reductions
from energy efficiency and renewable energy by 2030. *American Solar Energy Society.* Retrieved from
http://www.ases.org/images/stories/file/ASES/climate_change.pdf.

174 IEA. (2014). Technology Roadmap. Solar Photovoltaic Energy. *International Energy Agency.* Retrieved from
http://www.iea.org/publications/freepublications/publication/TechnologyRoadmapSolarPhotovoltaicEnergy_2014edition.pdf.

175 Geothermal electricity. (2015). *Wikipedia.* Retrieved from https://en.wikipedia.org/wiki/Geothermal_electricity.

176 IGA. (2016). Electricity generation. Installed generating capacity. *International Geothermal Association.* Retrieved from https://www.geothermal-energy.org/electricity_generation/philippines.html.

177 Hartmann, D.L., A.M.G. Klein Tank, M. Rusticucci, L.V. Alexander, S. Bronnimann, Y. Charabi, F.J. Dentener, E.J. Dlugokencky, D.R. Easterling, A. Kaplan, B.J. Soden, P.W. Thorne, M. Wild and P.M. Zhai, 2013: Observations: Atmosphere and Surface. In: Climate Change 2013: The Physical Science Basis. Contribution of Working Group I to the Fifth Assessment Report of the Intergovernmental Panel on Climate Change [Stocker, T.F., D. Qin, G.-K. Plattner, M. Tignor, S.K. Allen, J. Boschung, A. Nauels, Y. Xia, V. Bex and P.M. Midgley (eds.)]. Cambridge University Press, Cambridge, United Kingdom and New York, NY, USA.

178 NOAA. (2017). Coral Reef Watch. *National Oceanic and Atmospheric Administration.*Retrieved from
https://coralreefwatch.noaa.gov/satellite/analyses_guidance/global_coral_bleaching_2014-17_status.php.

179 Porter, J.R., L. Xie, A.J. Challinor, K. Cochrane, S.M. Howden, M.M. Iqbal, D.B. Lobell, and M.I. Travasso, 2014: Food security and food production systems. In: *Climate Change 2014: Impacts, Adaptation, and Vulnerability. Part A: Global and Sectoral Aspects. Contribution of Working Group II to the Fifth Assessment Report of the Intergovernmental Panel on Climate Change* [Field, C.B., V.R. Barros, D.J. Dokken, K.J. Mach, M.D. Mastrandrea, T.E. Bilir, M. Chatterjee, K.L. Ebi, Y.O. Estrada, R.C. Genova, B. Girma, E.S. Kissel, A.N. Levy, S. MacCracken, P.R. Mastrandrea, and L.L. White (eds.)]. Cambridge University Press, Cambridge, United Kingdom and New York, NY, USA.

180 IPCC, 2013: Annex II: Climate System Scenario Tables [Prather, M., G. Flato, P. Friedlingstein, C. Jones, J.-F. Lamarque, H. Liao and P. Rasch (eds.)]. In: Climate Change 2013: The Physical Science Basis. Contribution of Working Group I to the Fifth Assessment Report of the Intergovernmental Panel on Climate Change [Stocker, T.F., D. Qin, G.-K. Plattner, M. Tignor, S.K. Allen, J. Boschung, A. Nauels, Y. Xia, V. Bex and P.M. Midgley (eds.)]. Cambridge University Press, Cambridge, United Kingdom and New York, NY, USA.

181 Collins, M., R. Knutti, J. Arblaster, J.-L. Dufresne, T. Fichefet, P. Friedlingstein, X. Gao, W.J. Gutowski, T. Johns, G. Krinner, M. Shongwe, C. Tebaldi, A.J. Weaver and M. Wehner, 2013: Long-term Climate Change: Projections, Commitments and Irreversibility. In: Climate Change 2013: The Physical Science Basis. Contribution of Working Group I to the Fifth Assessment Report of the Intergovernmental Panel on Climate Change [Stocker, T.F., D. Qin, G.-K. Plattner, M. Tignor, S.K. Allen, J. Boschung, A. Nauels, Y. Xia, V. Bex and P.M. Midgley (eds.)]. Cambridge University Press, Cambridge, United Kingdom and New York, NY, USA.

182 Balog, J. (2009). Extreme Ice Now: Vanishing Glaciers and Changing Climate: A Progress Report. *National Geographic Society.* Peguin Random House.

183 Vaughan, D.G., J.C. Comiso, I. Allison, J. Carrasco, G. Kaser, R. Kwok, P. Mote, T. Murray, F. Paul, J. Ren, E. Rignot, O. Solomina, K. Steffen and T. Zhang, 2013: Observations: Cryosphere. In: Climate Change 2013: The Physical Science Basis. Contribution of Working Group I to the Fifth Assessment Report of the Intergovernmental Panel on Climate Change [Stocker, T.F., D. Qin, G.-K. Plattner, M. Tignor, S.K. Allen, J. Boschung, A. Nauels, Y. Xia, V. Bex and P.M. Midgley (eds.)]. Cambridge University Press, Cambridge, United Kingdom and New York, NY, USA.

184 Stocker, T.F., D. Qin, G.-K. Plattner, L.V. Alexander, S.K. Allen, N.L. Bindoff, F.-M. Bréon, J.A. Church, U. Cubasch, S. Emori, P. Forster, P. Friedlingstein, N. Gillett, J.M. Gregory, D.L. Hartmann, E. Jansen, B. Kirtman, R. Knutti, K. Krishna Kumar, P. Lemke, J. Marotzke, V. Masson-Delmotte, G.A. Meehl, I.I. Mokhov, S. Piao, V. Ramaswamy, D. Randall, M. Rhein, M. Rojas, C. Sabine, D. Shindell, L.D. Talley, D.G. Vaughan and S.-P. Xie, 2013: Technical Summary. In: Climate Change 2013: The Physical Science Basis. Contribution of Working Group I to the Fifth Assessment Report of the Intergovernmental Panel on Climate Change [Stocker, T.F., D. Qin, G.-K. Plattner, M. Tignor, S.K. Allen, J. Boschung, A. Nauels, Y. Xia, V. Bex and P.M. Midgley (eds.)]. Cambridge University Press, Cambridge,United Kingdom and New York, NY, USA.

185 Glaciers online. (2017). Vadret da Morteratsch. *Swissedduc.ch.* Retrieved from
http://www.swisseduc.ch/glaciers/morteratsch/index-en.html.

186 Lythe, M.B., & Vaughan, D.G. (2001). BEDMAP: A new ice thickness and subglacial topographic model of Antarctica. *Journal of Geological Science. 106*, 11335-11351.

187 Witze, A. (2015). Antarctic coast meltdown could trigger ice-sheet collapse. *Nature doi:10.1038/nature.2015.18688 10.1038/nature.2015.18688.*

188 Church, J.A., P.U. Clark, A. Cazenave, J.M. Gregory, S. Jevrejeva, A. Levermann, M.A. Merrifield, G.A. Milne, R.S. Nerem, P.D. Nunn, A.J. Payne, W.T. Pfeffer, D. Stammer and A.S. Unnikrishnan, 2013: Sea Level Change. In: Climate Change 2013: The Physical Science Basis. Contribution of Working Group I to the Fifth Assessment Report of the Intergovernmental Panel on Climate Change [Stocker, T.F., D. Qin, G.-K. Plattner, M. Tignor, S.K. Allen, J. Boschung, A. Nauels, Y. Xia, V. Bex and P.M. Midgley (eds.)]. Cambridge University Press, Cambridge, United Kingdom and New York, NY, USA.

189 Strauss, B.H., Kulp, S., & Levermann, A. (2015). Mapping Choices: Carbon, Climate and Rising Seas, Our Global Legacy. *Climate Central.* Retrieved from http://sealevel.climatecentral.org/uploads/research/Global-Mapping-Choices-Report.pdf.

190 IPCC, 2014: Summary for policymakers. In: Climate Change 2014: Impacts, Adaptation, and Vulnerability. Part A: Global and Sectoral Aspects. Contribution of Working Group II to the Fifth Assessment Report of the Intergovernmental Panel on Climate Change [Field, C.B., V.R. Barros, D.J. Dokken, K.J. Mach, M.D. Mastrandrea, T.E. Bilir, M. Chatterjee, K.L. Ebi, Y.O. Estrada, R.C. Genova, B. Girma, E.S. Kissel, A.N. Levy, S. MacCracken, P.R. Mastrandrea, and L.L. White (eds.)]. Cambridge University Press, Cambridge, United Kingdom and New York, NY, USA, pp. 1-32.

191 Rhein, M., S.R. Rintoul, S. Aoki, E. Campos, D. Chambers, R.A. Feely, S. Gulev, G.C. Johnson, S.A. Josey, A. Kostianoy, C. Mauritzen, D. Roemmich, L.D. Talley and F. Wang, 2013: Observations: Ocean. In: Climate Change 2013: The Physical Science Basis. Contribution of Working Group I to the Fifth Assessment Report of the Intergovernmental Panel on Climate Change [Stocker, T.F., D. Qin, G.-K. Plattner, M. Tignor, S.K. Allen, J. Boschung, A. Nauels, Y. Xia, V. Bex and P.M. Midgley (eds.)]. Cambridge University Press, Cambridge, United Kingdom and New York, NY, USA.

192 Solubility pump. (2016). *Wikipedia.* Retrieved from https://en.wikipedia.org/wiki/Solubility_pump.

193 Biological pump. (2016). *Wikipedia.* Retrieved from https://en.wikipedia.org/wiki/Biological_pump.

194 White Cliffs of Dover. (2016). *Wikipedia.* Retrieved from https://en.wikipedia.org/wiki/White_Cliffs_of_Dover.

195 Pörtner, H.-O., D.M. Karl, P.W. Boyd, W.W.L. Cheung, S.E. Lluch-Cota, Y. Nojiri, D.N. Schmidt, and P.O. Zavialov, 2014: Ocean systems. In: Climate Change 2014: Impacts, Adaptation, and Vulnerability. Part A: Global and Sectoral Aspects. Contribution of Working Group II to the Fifth Assessment Report of the Intergovernmental Panel on Climate Change [Field, C.B., V.R. Barros, D.J. Dokken, K.J. Mach, M.D. Mastrandrea, T.E. Bilir, M. Chatterjee, K.L. Ebi, Y.O. Estrada, R.C. Genova, B. Girma, E.S. Kissel, A.N. Levy, S. MacCracken, P.R. Mastrandrea, and L.L. White (eds.)]. Cambridge University Press, Cambridge, United Kingdom and New York, NY, USA, pp. 411-484.

196 Larsen, J.N., O.A. Anisimov, A. Constable, A.B. Hollowed, N. Maynard, P. Prestrud, T.D. Prowse, and J.M.R. Stone, 2014: Polar regions. In: Climate Change 2014: Impacts, Adaptation, and Vulnerability. Part B: Regional Aspects. Contribution of Working Group II to the Fifth Assessment Report of the Intergovernmental Panel on Climate Change [Barros, V.R., C.B. Field, D.J. Dokken, M.D. Mastrandrea, K.J. Mach, T.E. Bilir, M. Chatterjee, K.L. Ebi, Y.O. Estrada, R.C. Genova, B. Girma, E.S. Kissel, A.N. Levy, S. MacCracken, P.R. Mastrandrea, and L.L. White (eds.)]. Cambridge University Press, Cambridge, United Kingdom and New York, NY, USA, pp. 1567-1612.

197 Benaršek, N, Feely, R.A., Reum, J.C.P., Peterson, B., Menkel, J., Alin, S.R., & Hales, B. (2014). *Limacina helicina* shell dissolution as an indicator of declining habitat suitability owing to ocean acidification in the California Current Ecosystem. *Proc. Roy. Soc. 281* 20140123.Retrieved from http://dx.doi.org/10.1098/rspb.2014.0123.

198 NOAA News Report. (2014). NOAA-led researchers discover ocean acidity is dissolving shells of tiny snails off the U.S. West Coast. *National Ocean and Atmospheric Administration.* Retrieved from http://www.noaanews.noaa.gov/stories2014/20140430_oceanacidification.html.

199 Hoegh-Guldberg, O., R. Cai, E.S. Poloczanska, P.G. Brewer, S. Sundby, K. Hilmi, V.J. Fabry, and S. Jung, 2014: The Ocean. In: Climate Change 2014: Impacts, Adaptation, and Vulnerability. Part B: Regional Aspects. Contribution of Working Group II to the Fifth Assessment Report of the Intergovernmental Panel on Climate Change[Barros, V.R., C.B. Field, D.J. Dokken, M.D. Mastrandrea, K.J. Mach, T.E. Bilir, M. Chatterjee, K.L. Ebi, Y.O. Estrada, R.C. Genova, B. Girma, E.S. Kissel, A.N. Levy, S. MacCracken, P.R. Mastrandrea, and L.L. White (eds.)]. Cambridge University Press, Cambridge, United Kingdom and New York, NY, USA, pp. 1655-1731.

200 Countries of Africa. (2013). *One World Nations Online.* Retrieved from http://www.nationsonline.org/oneworld/africa.htm.

201 World Bank. (2016). Poverty and Equity. Sub-Saharan Africa. *The World Bank.* Retrieved from http://povertydata.worldbank.org/poverty/region/SSA. Accessed Jan 2016.

202 Niang, I., O.C. Ruppel, M.A. Abdrabo, A. Essel, C. Lennard, J. Padgham, and P. Urquhart, 2014: Africa. In: Climate Change *2014:* Impacts, Adaptation, and Vulnerability. Part B: Regional Aspects. Contribution of Working Group II to the Fifth Assessment Report of the Intergovernmental Panel on Climate Change [Barros, V.R., C.B. Field, D.J. Dokken, M.D. Mastrandrea, K.J. Mach, T.E. Bilir, M. Chatterjee, K.L. Ebi, Y.O. Estrada, R.C. Genova, B. Girma, E.S. Kissel, A.N. Levy, S. MacCracken, P.R. Mastrandrea, and L.L. White (eds.)]. Cambridge University Press, Cambridge, United Kingdom and New York, NY, USA, pp 1199-1265.

203 UN. (2015). Statistical annex. Human development report 2015. *United Nations Development Programme.* Retrieved from http://hdr.undp.org/sites/default/files/hdr_2015_statistical_annex.pdf.

204 World Bank. (2016). Data. Sub-saharan Africa (developing only). *The World Bank.* Retrieved from http://data.worldbank.org/region/SSA.

205 World Bank. (2016). Data. GDP growth (annual %). *The World Bank.* Retrieved from http://data.worldbank.org/indicator/NY.GDP.MKTP.KD.ZG/countries/GH-zf?display=graph.

206 World Bank. (2016). Africa. fact sheet: The World Bank and agriculture in Africa. *The World Bank.* Retrieved from http://web.worldbank.org/WBSITE/EXTERNAL/COUNTRIES/AFRICAEXT/0,,contentMDK:21935583~pagePK:146736~piPK:146830~theSitePK:258644,00.html.

207 Fauna of Africa. (2016). *Wikipedia.* Retrieved from https://en.wikipedia.org/wiki/Fauna_of_Africa.

208 Keenan, R. J., Reams, G.A., Achard, F., V. de Frietas, J., Grainger, A., & Lindquist, E. (2015). Dynamics of global forest area: Results from the FAO Global Forest Resources Assessment 2015. *Forest Ecology and Management. 352,* 9-20.

209 FAO. (2010). Global forest resources assessment 2010. FAO forestry paper 163. *Food and Agriculture Organization of the United Nations.* Retrieved from http://www.fao.org/docrep/013/i1757e/i1757e.pdf

210 UN. (2015). World Population Prospects, the 2015 Revision. *United Nations Department of Economics and Social Affairs, Population Division.* Retrieved from http://esa.un.org/unpd/wpp/Download/Probabilistic/Population/. Accessed Feb 2016.

211 Magrin, G.O., J.A. Marengo, J.-P. Boulanger, M.S. Buckeridge, E. Castellanos, G. Poveda, F.R. Scarano, and S. Vicuña, 2014: Central and South America. In: Climate Change 2014: Impacts, Adaptation, and Vulnerability. Part B: Regional Aspects. Contribution of Working Group II to the Fifth Assessment Report of the Intergovernmental Panel on Climate Change [Barros, V.R., C.B. Field, D.J. Dokken, M.D. Mastrandrea, K.J. Mach, T.E. Bilir, M. Chatterjee, K.L. Ebi, Y.O. Estrada, R.C. Genova, B. Girma, E.S. Kissel, A.N. Levy, S. MacCracken, P.R. Mastrandrea, and L.L. White(eds.)]. Cambridge University Press, Cambridge, United Kingdom and New York, NY, USA, pp. 1499-1566.

212 Kovats, R.S., R. Valentini, L.M. Bouwer, E. Georgopoulou, D. Jacob, E. Martin, M. Rounsevell, and J.-F. Soussana, 2014: Europe. In: Climate Change 2014: Impacts, Adaptation, and Vulnerability. Part B: Regional Aspects. Contribution of Working Group II to the Fifth Assessment Report of the Intergovernmental Panel on Climate Change [Barros, V.R., C.B. Field, D.J. Dokken, M.D. Mastrandrea, K.J. Mach, T.E. Bilir, M. Chatterjee, K.L. Ebi, Y.O. Estrada, R.C. Genova, B. Girma, E.S. Kissel, A.N. Levy, S. MacCracken, P.R. Mastrandrea, and L.L. White eds.)]. Cambridge University Press, Cambridge, United Kingdom and New York, NY, USA, pp. 1267-1326.

213 World Bank. (2016). GDP per capita (current USD). *The World Bank.* Retrieved from http://data.worldbank.org/indicator/NY.GDP.PCAP.CD.

214 World Bank. (2016). Data. Agricultural land (% of land area). *The World Bank.* Retrieved from http://data.worldbank.org/indicator/AG.LND.AGRI.ZS.

215 Rating Countries. (2016). *Climate Action Tracker.* Retrieved from http://climateactiontracker.org/countries.html.

216 World Bank. (2016). GDP (current USD). *The World Bank.* Retrieved from http://data.worldbank.org/indicator/NY.GDP.MKTP.CD.

217 Hijioka, Y., E. Lin, J.J. Pereira, R.T. Corlett, X. Cui, G.E. Insarov, R.D. Lasco, E. Lindgren, and A. Surjan, 2014: Asia. In: Climate Change 2014: Impacts, Adaptation, and Vulnerability. Part B: Regional Aspects. Contribution of Working Group II to the Fifth Assessment Report of the Intergovernmental Panel on Climate Change[Barros, V.R., C.B. Field, D.J. Dokken, M.D. Mastrandrea, K.J. Mach, T.E. Bilir, M. Chatterjee, K.L. Ebi, Y.O. Estrada, R.C. Genova, B. Girma, E.S. Kissel, A.N. Levy, S. MacCracken, P.R. Mastrandrea, and L.L. White(eds.)]. Cambridge University Press, Cambridge, United Kingdom and New York, NY, USA, pp. 1327-1370.

218 UN. (2009). Vital Forest Graphics. 2009. *United Nations Environment Programme.* Retrieved from http://www.grida.no/_res/site/file/publications/vital_forest_graphics.pdf pp45.

219 Burke, L., Selig, E., & Spalding, M. (2002). Reefs at Risk in Southeast Asia. *World Resources Institute.* Retrieved from http://www.wri.org/sites/default/files/pdf/rrseasia_full.pdf. pp 8.

220 Coral reefs. (2016). *World Conservation Union and the World Wildlife Federation.* Retrieved from http://www.ioseaturtles.org/Education/coralreefbooklet.pdf.

221 Wilkinson, C. (2008). Status of coral eefs of the world: 2008. *Global Coral Reef Monitoring Network.* Retrieved from http://cmsdata.iucn.org/downloads/coral_status2004_exec_summ.pdf.

222 Fuchs, R.J. (2010). Cities at Risk: Asia's Coastal cities in an age of Climate Change. *Asia Pacific Issues, No 96.* East-West Center, Honolulu.

223 Miracle on the Han River. (2016). *Wikipedia.* Retrieved from https://en.wikipedia.org/wiki/Miracle_on_the_Han_River.

224 Reisinger, A., R.L. Kitching, F. Chiew, L. Hughes, P.C.D. Newton, S.S. Schuster, A. Tait, and P. Whetton, 2014: Australasia. In: Climate Change 2014: Impacts, Adaptation, and Vulnerability. Part B: Regional Aspects. Contribution of Working Group II to the Fifth Assessment Report of the Intergovernmental Panel on ClimateChange [Barros, V.R., C.B. Field, D.J. Dokken, M.D. Mastrandrea, K.J. Mach, T.E. Bilir, M. Chatterjee, K.L. Ebi, Y.O. Estrada, R.C. Genova, B. Girma, E.S. Kissel, A.N. Levy, S. MacCracken, P.R. Mastrandrea, and L.L. White(eds.)]. Cam bridge University Press, Cambridge, United Kingdom and New York, NY, USA, pp. 1371-1438.

225 Inland waters. Water Storage. (2010). *Australian Bureau of Statistics. Government of Australia.* Retrieved from http://www.abs.gov.au/ausstats/abs@.nsf/Lookup/by%20Subject/1370.0~2010~Chapter~Water%20storage%20(6.3.6.2).

226 Media release. Australia Bureau of Statistics. (2006, November 28). Drought drives down water consumption. *Australian Bureau of Statistics. Government of Australia.* Retrieved from http://www.abs.gov.au/ausstats/abs@.nsf/mediareleasesbyTopic/7EF75E1ECCA37DE3CA2577E700158AD8?OpenDocument.

227 Great barrier reef. (2016). *Wikipedia.* Retrieved from https://en.wikipedia.org/wiki/Great_Barrier_Reef.

228 Current world population. (2016). *Worldometers.* Retrieved from http://www.worldometers.info/world-population/.

229 National Corn Growers Association. (2015). Corn usage by segment. *World of Corn.* Retrieved from http://www.worldofcorn.com/#corn-usage-by-segment. Accessed Feb 2016.

230 FOA. (2016). Livestock Primary. *Food and Agriculture Organization of the United Nations* Retrieved from http://www.fao.org/faostat/en/#data/QL/visualize.

231 USDA. (2015). Agricultural Trade. *United States Department of Agriculture. Economic Research Service.* Retrieved from http://www.ers.usda.gov/data-products/ag-and-food-statistics-charting-the-essentials/agricultural-trade.aspx.

232 Water in California. (2016). *Wikipedia.* Retrieved from https://en.wikipedia.org/wiki/Water_in_California.

233 Water Stress in the U.S. (2016). *U.S. Global Change Research Program.* Retrieved from http://www.globalchange.gov/browse/multimedia/water-stress-us.

234 Asad, M., & Dinar, A. (2006). The role of water policy in Mexico. *World Bank. Latin Amercian and Caribbean Region.* October 2006. Number 95. Retrieved from http://documents.worldbank.org/curated/en/358891468050661935/pdf/382050ENGLISH01ve0950Water01PUBLIC1.pdf.

235 Romero-Lankao, P., J.B. Smith, D.J. Davidson, N.S. Diffenbaugh, P.L. Kinney, P. Kirshen, P. Kovacs, and L. Villers Ruiz, 2014: North America. In: Climate Change 2014: Impacts, Adaptation, and Vulnerability. Part B: Regional Aspects. Contribution of Working Group II to the Fifth Assessment Report of the Intergovernmental Panel on Climate Change [Barros, V.R., C.B. Field, D.J. Dokken, M.D. Mastrandrea, K.J. Mach, T.E. Bilir, M. Chatterjee, K.L. Ebi, Y.O. Estrada, R.C. Genova, B. Girma, E.S. Kissel, A.N. Levy, S. MacCracken, P.R. Mastrandrea, and L.L. White (eds.)]. Cambridge University Press, Cambridge, United Kingdom and New York, NY, USA, pp. 1439-1498.

236 World Bank. (2016). Agriculture, value added (% of GDP). *The World Bank.* Retrieved from http://data.worldbank.org/indicator/NY.GDP.MKTP.CD.

237 EIA. (2016). International Energy Statistics. *U.S. Energy Information Administration.* Retrieved from http://www.eia.gov/cfapps/ipdbproject/IEDIndex3.cfm?tid=5&pid=53&aid=1.

238 Amazon basin. (2016). *Wikipedia.* Retrieved from https://en.wikipedia.org/wiki/Amazon_basin.

239 EurekAlert! The Global Source for Science News. Press Release. (2013, October 17). Field Museum scientists estimated 16,000 tree species in the Amazon. American Association for the Advancement of Science. Retrieved from https://www.eurekalert.org/pub_releases/2013-10/fm-fms101413.php.

240 Azevedo-Ramos, C. (2008). Sustainable development and challenging deforestation in the Brazilian Amazon: the good, the bad and the ugly. *Unasylva. 230*(59) 12-16.

241 FAOSTAT. (2016). Rankings. Countries by commodity. *Food and Agriculture Organization of the United Nations. Statistics Division.* Retrieved from http://faostat3.fao.org/browse/rankings/countries_by_commodity/.

242 Ethanol fuel in Brazil. (2016). *Wikipedia.* Retrieved from https://en.wikipedia.org/wiki/Ethanol_fuel_in_Brazil.

243 Sao Paulo (state). (2016). *Wikipedia.* Retrieved from https://en.wikipedia.org/wiki/S%C3%A3o_Paulo_(state).

244 Sao Paulo. (2016). *Wikipedia.* Retrieved from https://en.wikipedia.org/wiki/S%C3%A3o_Paulo.

245 AOSIS. (2016). About AOSIS. *Alliance of Small Island States.* Retrieved from http://aosis.org/about/.

246 Nurse, L.A., R.F. McLean, J. Agard, L.P. Briguglio, V. Duvat-Magnan, N. Pelesikoti, E. Tompkins, and A. Webb, 2014: Small islands. In: Climate Change 2014: Impacts, Adaptation, and Vulnerability. Part B: Regional Aspects. Contribution of Working Group II to the Fifth Assessment Report of the Intergovernmental Panel on Climate Change [Barros, V.R., C.B. Field, D.J. Dokken, M.D. Mastrandrea, K.J. Mach, T.E. Bilir, M. Chatterjee, K.L. Ebi, Y.O. Estrada, R.C. Genova, B. Girma, E.S. Kissel, A.N. Levy, S. MacCracken, P.R. Mastrandrea, and L.L. White (eds.)]. Cambridge University Press, Cambridge, United Kingdom and New York, NY, USA, pp. 1613-1654.

247 Maldives. (2016). *Wikipedia.* Retrieved from https://en.wikipedia.org/wiki/Maldives.

248 United Nations. (2016). Maldives. Climate change adaptation. *United Nations Development Programme.* Retrieved from http://www.adaptation-undp.org/explore/maldives.

249 SIDS. (2016). Malé Declaration on the human dimension of global climate change. *Small Island Developing States.* Retrieved from http://www.ciel.org/Publications/Male_Declaration_Nov07.pdf.

250 United Nations. (2016). Report of the Special Rapporteur on the issue of human rights obligations relating to the enjoyment of a safe, clean, healthy and sustainable environment. *Human Rights Council of the United Nations.* Human Rights Council. Thirty-first Session. Report 52. A/HRC/31/52.

251 Electricity production by sources in France 1960-2012. (2014). *Bluenomics.* Retrieved from https://www.bluenomics.com/sector

252 United Nations Framework Convention on Climate Change. (1992). *United Nations.* Retrieved from http://unfccc.int/files/essential_background/background_publications_htmlpdf/application/pdf/conveng.pdf.

253 WRI. (2016). Climate data explorer. Pre-2020 pledges map. *World Resources Institute.* Retrieved from http://cait.wri.org.

254 WRI. (2016). Climate data explorer. Paris contributions map. *World Resources Institute.* Retrieved from http://cait.wri.org/indc/#/. March 2016.

255 Cebr. (2016, December 26). Cebr's World Economic League Table. *Center for Economic and Business Research.* Retrieved from https://www.cebr.com/reports/welt-2016/.

256 Fransen, T., Song, R., Stolle, & F. Henderson, G. (2015). A closer look at china's new climate plan (INDC). *World Resource Institute.* Retrieved from http://www.wri.org/blog/2015/07/closer-look-chinas-new-climate-plan-indc.

257 What is CAT? (2016). *Climate Action Tracker.* Retrieved from http://climateactiontracker.org/about.html.

258 November 2015 Paris attacks. (2016). *Wikipedia.* Retrieved from https://en.wikipedia.org/wiki/November_2015_Paris_attacks.

259 UNFCCC. (2015). Conference of Parties. Provisional list of participants.*United Nations Framework Convention on Climate Change.* Retrieved from http://unfccc.int/resource/docs/2015/cop21/eng/misc02p01.pdf.

260 Fabius, L. (2015). Paris 2015 | COP 21- Speech by Laurent Fabius, French Minister of Foreign Affairs and International Development President of COP21 (30 November, 2015). *France Diplomatie.* Retrieved from http://www.diplomatie.gouv.fr/en/french-foreign-policy/climate/2015-paris-climate-conference-cop21/article/paris-2015-cop-21-speech-by-laurent-fabius-french-minister-of-foreign-affairs.

261 Loeak, H.E.C.J. (2015). H.E. Christopher J. Loeak President of the Republic of the Marshall Islands opening of the UN Framework Convention on Climate Change's 21st Conference of Parties Paris, France 30 November 2015. *United Nations Framework Convention on Climate Change.* Retrieved from https://unfccc.int/files/meetings/paris_nov_2015/application/pdf/cop21cmp11_leaders_event_marshall_islands.pdf.

262 Hollande, F. (2015). COP21 opening speech: President François Hollande. *République Française.* Retrieved from http://www.ambafrance-rsa.org/COP21-opening-speech-President-Francois-Hollande.

263 Cañete, M.A. (2016). EU climate commissioner: How we formed the high ambition coalition. *Business Green.* Retrieved from http://www.businessgreen.com/bg/opinion/2439215/eu-climate-commissioner-how-we-formed-the-high-ambition-coalition.

264 Harvey, F. (2015, December 14). COP21: UN climate change conference Paris. Paris climate change agreement: the world's greatest diplomatic success. *The Guardian.* Retrieved from http://www.theguardian.com/environment/2015/dec/13/paris-climate-deal-cop-diplomacy-developing-united-nations.

265 UNFCCC. (2015). Adoption of the Paris agreement. *United Nations Framework Convention on Climate Change.* FCCC/CP/2015/L.9/Rev.1 Retrieved from https://unfccc.int/resource/docs/2015/cop21/eng/l09r01.pdf. .

266 UNFCCC. (2015). The Paris agreement. *United Nations Framework Convention on Climate Change.* Retrieved from http://unfccc.int/paris_agreement/items/9485.php.

267 Clarke L., K. Jiang, K. Akimoto, M. Babiker, G. Blanford, K. Fisher-Vanden, J.-C. Hourcade, V. Krey, E. Kriegler, A. Löschel,D. McCollum, S. Paltsev, S. Rose, P. R. Shukla, M. Tavoni, B. C. C. van der Zwaan, and D.P. van Vuuren, 2014. Assessing Transformation Pathways. In: Climate Change 2014: Mitigation of Climate Change. Contribution of Working Group IIIto the Fifth Assessment Report of the Intergovernmental Panel on Climate Change [Edenhofer, O., R. Pichs-Madruga,Y. Sokona, E. Farahani, S. Kadner, K. Seyboth, A. Adler, I. Baum, S. Brunner, P. Eickemeier, B. Kriemann, J. Savolainen, S.Schlömer, C. von Stechow, T. Zwickel and J.C. Minx (eds.)]. Cambridge University Press, Cambridge, United Kingdom and New York, NY, USA.

268 United States Department of State. (2010). Appendix 1. Annex 1 Parties. United States of America. *United Nations Framework Convention on Climate Change.* Retrieved from http://unfccc.int/files/meetings/cop_15/copenhagen_accord/application/pdf/unitedstatescphaccord_app.1.pdf.

269 IPCC, 2014: Summary for Policymakers. In: Climate Change 2014: Mitigation of Climate Change. Contribution of Working Group III to the Fifth Assessment Report of the Intergovernmental Panel on Climate Change [Edenhofer, O., R. Pichs-Madruga, Y. Sokona, E. Farahani, S. Kadner, K. Seyboth, A. Adler, I. Baum, S. Brunner, P. Eickemeier, B. Kriemann, J. Savolainen, S. Schlomer, C. von Stechow, T. Zwickel and J.C. Minx (eds.)]. Cambridge University Press, Cambridge, United Kingdom and New York, NY, USA.

270 Edenhofer O., R. Pichs-Madruga, Y. Sokona, S. Kadner, J. C. Minx, S. Brunner, S. Agrawala, G. Baiocchi, I. A. Bashmakov, G. Blanco, J. Broome, T. Bruckner, M. Bustamante, L. Clarke, M. Conte Grand, F. Creutzig, X. Cruz-Núñez, S. Dhakal, N. K. Dubash, P. Eickemeier, E. Farahani, M. Fischedick, M. Fleurbaey, R. Gerlagh, L. Gómez-Echeverri, S. Gupta, J. Harnisch, K. Jiang, F. Jotzo, S. Kartha, S. Klasen, C. Kolstad, V. Krey, H. Kunreuther, O. Lucon, O. Masera, Y. Mulugetta, R. B. Norgaard, A. Patt, N. H. Ravindranath, K. Riahi, J. Roy, A. Sagar, R. Schaeffer, S. Schlömer, K. C. Seto, K. Seyboth, R. Sims, P. Smith, E. Somanathan, R. Stavins, C. von Stechow, T. Sterner, T. Sugiyama, S. Suh, D. Ürge-Vorsatz, K. Urama, A. Venables, D. G. Victor, E.Weber, D. Zhou, J. Zou, and T. Zwickel, 2014. Technical Summary. In: Climate Change 2014: Mitigation of Climate Change. Contribution of Working Group III to the Fifth Assessment Report of the Intergovernmental Panel on Climate Change[Edenhofer, O., R. Pichs-Madruga, Y. Sokona, E. Farahani, S. Kadner, K. Seyboth, A. Adler, I. Baum, S. Brunner, P. Eickemeier, B. Kriemann, J. Savolainen, S. Schlömer, C. von Stechow, T. Zwickel and J. C. Minx (eds.)]. Cambridge University Press, Cambridge,United Kingdom and New York, NY, USA.

271   Bruckner T., I. A. Bashmakov, Y. Mulugetta, H. Chum, A. de la Vega Navarro, J. Edmonds, A. Faaij, B. Fungtammasan, A. Garg, E. Hertwich, D. Honnery, D. Infield, M. Kainuma, S. Khennas, S. Kim, H. B. Nimir, K. Riahi, N. Strachan, R. Wiser, and X. Zhang, 2014: Energy Systems. In: Climate Change 2014: Mitigation of Climate Change. Contribution of Working Group III to the Fifth Assessment Report of the Intergovernmental Panel on Climate Change [Edenhofer, O., R. Pichs-Madruga, Y. Sokona, E. Farahani, S. Kadner, K. Seyboth, A. Adler, I. Baum, S. Brunner, P. Eickemeier, B. Kriemann, J. Savolainen, S. Schlömer, C. von Stechow, T. Zwickel and J.C. Minx (eds.)]. Cambridge University Press, Cambridge, United Kingdom and New York, NY, USA.

272   Victor D. G., D. Zhou, E. H. M. Ahmed, P. K. Dadhich, J. G. J. Olivier, H-H. Rogner, K. Sheikho, and M. Yamaguchi, 2014: Introductory Chapter. In: Climate Change 2014: Mitigation of Climate Change. Contribution of Working Group III to the Fifth Assessment Report of the Intergovernmental Panel on Climate Change [Edenhofer, O., R. Pichs-Madruga, Y. Sokona, E. Farahani, S. Kadner, K. Seyboth, A. Adler, I. Baum, S. Brunner, P. Eickemeier, B. Kriemann, J. Savolainen, S. Schlömer, C. von Stechow, T. Zwickel and J.C. Minx (eds.)]. Cambridge University Press, Cambridge, United Kingdom and New York, NY, USA.

273   Bluestein, J., Mallya, H., Yandoli, L., Polchert, M. Amarin, & N. (2015). Methane emissions from the oil and gas industry: "Making Sense of the Noise". *IFC International.* Retrieved from https://www.icf.com/-/media/files/.../methane_emissions_from_oil_gas_industry.pdf.

274   World Bank. (2014). Time to end routine gas flairing. *The World Bank.* Retrieved from http://www.worldbank.org/en/news/feature/2014/07/15/gas-flaring-reduction-takes-center-stage-at-global-event.

275   Linnit, C. (2016,March 16). Canada-U.S. plan to nearly halve methane emissions could be huge deal for the climate. *DESMOGCANADA.* Retrieved from http://www.desmog.ca/2016/03/16/canada-u-s-plan-nearly-halve-methane-emissions-could-be-huge-deal-climate.

276   World Bank. (2016). Global gas flaring reduction partnership (GGFR). *The World Bank.* Retrieved from http://www.worldbank.org/en/programs/gasflaringreduction.

277   Kutscher, C.F. (2007). Tackling climate change in the U.S. Potential carbon emissions reductions from energy efficiency and renewable energy by 2030. American Solar Energy Society. Retrieved from https://web.archive.org/web/20081126220129/

278   Neslen, A. (2016, January 18). Demark broke world record for wind power in 2015. *The Guardian.* Retrieved from https://www.theguardian.com/environment/2016/jan/18/denmark-broke-world-record-for-wind-power-in-2015

279   EIA. (2015). Levelized cost and levelized avoided cost of new generation resources in the annual energy outlook. *U.S. Energy Information Administration.* Retrieved from https://www.eia.gov/forecasts/aeo/pdf/electricity_generation.pdf.

280   Matek, B. (2016). 2016 Annual U.S. & global geothermal power production report. *Geothermal Energy Association.* Retrieved from http://geo-energy.org/reports/2016/2016%20Annual%20US%20Global%20Geothermal%20Power%20Production.pdf

281   Statista. (2015). Global electricity prices by select countries in 2015 (in U.S. dollar cents per kilowatt hour). *The Statistical Portal.* Retrieved from http://www.statista.com/statistics/263492/electricity-prices-in-selected-countries/.

282   ECIU. (2015). Germany's energy transition. *Energy and Climate Intelligence Unit.* Retrieved from http://eciu.net/briefings/international-perspectives/germanys-energy-transition.

283   MIT. (2016). Carbon capture & sequestration technologies. LaBarge fact sheet: Carbon dioxide capture and storage project. *Massachusetts Institute of Technology.* MIT CC&ST Program. Retrieved from https://sequestration.mit.edu/index.html.

284   Leung, D.Y.C., Caramanna, G., & Mercedes Maroto-Valer, M. (2014). An overview of current status of carbon dioxide capture and storage technologies. *Renewable and Sustainable Energy Reviews. 39,* 426-443.

285   Consoli, C. (2016). The global CCS instituted releases the first global storage portfolio. *Global CCS Institute.* Retrieved from http://www.globalccsinstitute.com/insights/authors/ChrisConsoli/2016/03/08/global-ccs-institute-releases-first-global-storage-portfolio?author=Mjg5ODg%3D.

286   IEA. (2006). Energy technology essentials 2006. CO_2 capture and storage. *International Energy Agency.* Retrieved from https://www.iea.org/publications/freepublications/publication/essentials1.pdf.

287   EIA. (2013). Updated capital cost estimates for utility scale electricity generating plants. *U.S. Energy Information Administration.* Retrieved from http://www.eia.gov/forecasts/capitalcost/pdf/updated_capcost.pdf.

288   Rubin, E.S., & Zhai, H. (2012). The cost of carbon capture and storage for natural gas combined cycle power plants. *Environ. Sci. Technol. 46,* 3076-3084.

289   World Population. (2016). *Google.* Retrieved from https://www.google.ca/search?sourceid=navclient&aq=&oq=world+po&ie=UTF-8&rlz=1T4LENP_enCA511CA496&q=world+population+1990&gs_l=hp..6.0i131j0l4j41l3.0.0.3.565530.USRgy-jFwgU.

290   FAO. (2015). Livestock and landscapes. *Food and Agriculture Organization of the United Nations.* Retrieved from http://www.fao.org/docrep/018/ar591e/ar591e.pdf.

291   International Partnership on Mitigation and MRV. (2016). Implementation prevention and control policies for reducing deforestation. *Partnership on Transparency.* Retrieved from https://mitigationpartnership.net/gpa/implementing-prevention-and-control-policies-reducing-deforestation.

292   Butler, R. (2016). Calculating deforestation figures for the Amazon. *Mongabay.com.* Retrieved from http://rainforests.mongabay.com/amazon/deforestation_calculations.html.

293   Assunção, J. & Chiavari, J. (2015). Towards efficient land use in Brazil. *The New Climate Economy.* Retrieved from http://2015.newclimateeconomy.report/wp-content/uploads/2015/09/Towards-Efficient-Land-Use-Brazil.pdf.

294 FAOSTAT. (2016). Emissions – land use. *Food and Agriculture Organization of the United Nations Statistical Division.* Retrieved from http://faostat3.fao.org/browse/G2/*/E.

295 UNFCCC. (2005). Reducing emissions from deforestation and forest degradation in developing countries: approaches to stimulate action. *United Nations Framework Convention on Climate Change.* Conference of Parties 11th Session. Montreal. Retrieved from http://unfccc.int/resource/docs/2005/cop11/eng/misc01.pdf.

296 UNFCCC. (2007). Report of the Conference of Parties on its thirteenth session *United Nations Framework Convention on Climate Change.* Conference of Parties 13th Session. Bali. Retrieved from http://unfccc.int/resource/docs/2007/cop13/eng/06a01.pdf.

297 FOA. (2016). Partner countries. *UN-REDD Programme.* Retrieved from http://www.un-redd.org/partner-countries.

298 FCPF. (2015). About FCPC. *Forest Carbon Partnership Facility.* Retrieved from https://www.forestcarbonpartnership.org/.

299 World Bank. (2013). Costa Rica first to negotiate sale of forestry carbon credits. *The World Bank.* Retrieved from http://www.worldbank.org/en/news/press-release/2013/09/10/creditos-por-reduccion-de-carbono. Accessed May 2016.

300 Nield, D. (2017, January 3). Costa Rica went 250 days in 2016 without burning any fossil fuels. *Science Alert.* Retrieved from http://www.sciencealert.com/costa-rica-went-250-days-in-2016-without-burning-any-fossil-fuels.

301 Stehfest, E., Bouwman, L., van Vuuren, D.P., den Elzen, M.G.J., Eickhout, B., & Kabut, P. (2009). Climate benefits of changing diet. *Climate Change. 95*, 83-102.

302 Thornton, P.K. (2010). Livestock production: recent trends, future prospects. *Philosophical Transactions of the Royal Society B Biological Sciences. 365*, 2853-2867.

303 FAO. (2011). Food wastage footprint & climate change. *Food and Agriculture Organization.* Retrieved from http://www.fao.org/3/a-bb144e.pdf.

304 Chum, H., A. Faaij, J. Moreira, G. Berndes, P. Dhamija, H. Dong, B. Gabrielle, A. Goss Eng, W. Lucht, M. Mapako, O. Masera Cerutti, T. McIntyre, T. Minowa, K. Pingoud, 2011: Bioenergy. In IPCC Special Report on Renewable Energy Sources and Climate Change Mitigation [O. Edenhofer, R. Pichs-Madruga, Y. Sokona, K. Seyboth, P. Matschoss, S. Kadner, T. Zwickel, P. Eickemeier, G. Hansen, S. Schlomer, C. von Stechow (eds)], Cambridge University Press, Cambridge, United Kingdom and New York, NY, USA.

305 WHO. (2016). Indoor air pollution. *World Health Organization.* Retrieved from http://www.who.int/indoorair/en/.

306 IEA. (2016). WEO 2016 Energy access database. *International Energy Agency.*Retrieved from http://www.worldenergyoutlook.org/resources/energydevelopment/energyaccessdatabase/.

307 Coal plant conversion projects. (2015). *Sourcewatch.* Retrieved from http://www.sourcewatch.org/index.php/Coal_plant_conversion_projects.

308 Fischedick M., J. Roy, A. Abdel-Aziz, A. Acquaye, J. M. Allwood, J.-P. Ceron, Y. Geng, H. Kheshgi, A. Lanza, D. Perczyk, L. Price, E. Santalla, C. Sheinbaum, and K. Tanaka, 2014: Industry. In: Climate Change 2014: Mitigation of Climate Change. Contribution of Working Group III to the Fifth Assessment Report of the Intergovernmental Panel on Climate Change [Edenhofer, O., R. Pichs-Madruga, Y. Sokona, E. Farahani, S. Kadner, K. Seyboth, A. Adler, I. Baum, S. Brunner, P. Eickemeier, B. Kriemann, J. Savolainen, S. Schlömer, C. von Stechow, T. Zwickel and J.C. Minx (eds.)]. Cambridge University Press, Cambridge, United Kingdom and New York, NY, USA.

309 Sims R., R. Schaeffer, F. Creutzig, X. Cruz-Núñez, M. D'Agosto, D. Dimitriu, M. J. Figueroa Meza, L. Fulton, S. Kobayashi, O. Lah, A. McKinnon, P. Newman, M. Ouyang, J. J. Schauer, D. Sperling, and G. Tiwari, 2014: Transport. In: Climate Change 2014: Mitigation of Climate Change. Contribution of Working Group III to the Fifth Assessment Report of the Intergovernmental Panel on Climate Change [Edenhofer, O., R. Pichs-Madruga, Y. Sokona, E. Farahani, S. Kadner, K. Seyboth, A. Adler, I. Baum, S. Brunner, P. Eickemeier, B. Kriemann, J. Savolainen, S. Schlömer, C. von Stechow, T. Zwickel and J.C. Minx (eds.)]. Cambridge University Press, Cambridge, United Kingdom and New York, NY, USA.

310 GFEI. (2016). About GFEI. *Global Fuel Economy Initiative.* Retrieved from http://www.globalfueleconomy.org/.

311 U.S. EPA. (2015). Light-duty automotive technology, carbon dioxide emissions, and fuel economy trends: 1975-2015. *United States Environmental Protection Agency.* (EPA). Retrieved from https://www3.epa.gov/otaq/fetrends.htm.

312 EC. (2011). White Paper. Roadmap to a single European transport area – Toward a competi-tive and resource efficient transportation system. *European Commission.* Retrieved from http://eur-lex.europa.eu/legal-content/EN/TXT/PDF/?uri=CELEX:52011DC0144&from=EN

313 The Week. (2016, June 6). Tesla Model 3: Prices, specs and rumors. *The Week.* Retrieved from http://www.theweek.co.uk/tesla-model-3/70320/tesla-model-3-owners-wont-get-free-supercharging.

314 Supercharger. (2016) *Tesla.* Retrieved from https://www.teslamotors.com/en_CA/supercharger.

315 Ogden, J.M. (1999). Prospects for building a hydrogen energy infrastructure. *Annual Review of Energy and the Environment. 24*, 227–279.

316 ZEV Alliance. (2015). International ZEV Alliance announcement. Rational. *Zero Emissions Vehicle Alliance.* Retrieved from https://www.scribd.com/doc/292065952/ZEV-Alliance-COP21-Announcement-3-Dec-2015.

317 Toyota environmental challenge 2050. (2016). *Toyota.* Retrieved from http://www.toyota-global.com/sustainability/environment/challenge2050/.

318 Nissan News. (2017, January 5). PRESS KIT: Nissan Intelligent Mobility at CES. *Nissan.* Retrieved from http://nissannews.com/en-US/nissan/usa/channels/us-nissan-2017-ces/releases/press-kit-nissan-intelligent-mobility-at-ces.

319 Nissan Motor Corporation. (2016). Long-term goal for reducing CO_2. *Nissan.* Retrieved from http://www.nissan-global.com/EN/TECHNOLOGY/NISSAN/VALUE/. Accessed June 2016.

320 Lucon O., D. Urge-Vorsatz, A. Zain Ahmed, H. Akbari, P. Bertoldi, L. F. Cabeza, N. Eyre, A. Gadgil, L. D. D. Harvey, Y. Jiang, E. Liphoto, S. Mirasgedis, S. Murakami, J. Parikh, C. Pyke, and M. V. Vilarino, 2014: Buildings. In: Climate Change 2014: Mitigation of Climate Change. Contribution of Working Group III to the Fifth Assessment Report of the Intergovernmental Panel on Climate Change [Edenhofer, O., R. Pichs-Madruga, Y. Sokona, E. Farahani, S. Kadner, K. Seyboth, A. Adler, I. Baum, S. Brunner, P. Eickemeier, B. Kriemann, J. Savolainen, S. Schlomer, C. von Stechow, T. Zwickel and J.C. Minx (eds.)]. Cambridge University Press, Cambridge, United Kingdom and New York, NY, USA.

321 PHI. (2016). Passive house requirements. *Passive House Institute* Retrieved from http://passiv.de/en/index.html.

322 iPHA. (2016). The global Passive House platform. *The International Passive House Association.* Retrieved from http://www.passivehouse-international.org/index.php?page_id=65.

323 Promotion of European Passive Houses. (2006). Energy Saving Potential. *Intelligent Energy Europe.* Retrieved from http://erg.ucd.ie/pep/pdf/Energy_Saving_Potential_2.pdf.

324 European Commission Joint Research Center Science for Policy Report. (2016). Synthesis report on the national plans for nearly zero energy buildings (NZEBs). *JRC Science Hub.* Retrieved from http://publications.jrc.ec.europa.eu/repository/bitstream/JRC97408/reqno_jrc97408_online%20nzeb%20 report%281%29.pdf.

325 Ministry of Ecology, Sustainable Development and Energy, France. (2015). Energy transition for green growth act. *République Française.* Retrieved from http://www.developpement-durable.gouv.fr/IMG/pdf/14123-8-éGB_loi-TE-mode-emploi_DEF_light.pdf.

326 World Bank. (2016). End Poverty in South Asia. *The World Bank.* Retrieved from http://blogs.worldbank.org/endpovertyinsouthasia/upgrading-informal-housing-step-right-direction.

327 World Bank. (2016). Access to electricity. *The World Bank.* Retrieved from http://data.worldbank.org/indicator/EG.ELC.ACCS.ZS.

328 iPHA. (2016). Passive house legislation. *The International Passive House Association.* Retrieved from http://www.passivehouse-international.org/index.php?page_id=176.

329 NYC Built to Last. (2014). One City Built to Last. *City of New York.* Retrieved from http://www.nyc.gov/html/builttolast/pages/home/home.shtml.

330 National Academies of Sciences Engineering Medicine. (1979) .*Carbon Dioxide and Climate: A Scientific Assessment.* Washington, D.C.: The National Academies Press.

331 Burns, J. (2009, Auguest 27). Artificial trees to cut carbon. *BBC News.* Retrieved from http://news.bbc.co.uk/2/hi/science/nature/8223528.stm

332 Hughes, A.D., Black, K.D., Campbell, I., Davidson, K., Kelly, M.S., & Stanley, M.S. (2012). Does seaweed offer a solution for bioenergy with biological carbon capture and storage? *Greenhouse Gas Sci. Technol. 2,* 402-407.

333 Keith, D. (2013). *A Case for Climate Engineering.* The MIT Press. Cambridge MA.

334 EC. (2013). Emissions trading dystem (EU ETS). Phases 1 and 2 (2005-2012). *European Commission.* Retrieved from http://ec.europa.eu/clima/policies/ets/pre2013/index_en.htm.

335 EC. (2016). Emissions trading system (EU ETS). *European Commission.* Retrieved from http://ec.europa.eu/clima/policies/ets/index_en.htm.

336 EC. (2016). Emissions trading system (EU ETS) Revisions for phase 4 (2021-2030). *European Commission.* Retrieved from http://ec.europa.eu/clima/policies/ets/revision/index_en.htm.

337 EC. (2016). The EU emissions trading system (EU ETS). Fact sheet. European Union. *European Commission.* Retrieved from http://ec.europa.eu/clima/publications/docs/factsheet_ets_en.pdf.

338 Swartz, J. (2016). China's national emissions trading system. *ICTSD Global Platform on Climate Change, Trade and Sustainable Energy. International Center for Trade and Sustainable Development.* Retrieved from http://www.ieta.org/resources/China/ Chinas_National_ETS_Implications_for_Carbon_Markets_and_Trade_ICTSD_March2016_Jeff_Swartz.pdf.

339 ICAP. (2016). Emissions trading worldwide: Status report 2016. *International Carbon Action Partnership.* Retrieved from https://icapcarbonaction.com/images/StatusReport2016/ICAP_Status_Report_2016_Online.pdf.

340 ICAP. (2016). Ontario passes ETS legislation. *International Carbon Action Partnership.* Retrieved from https://icapcarbonaction.com/en/news-archive/394-ontario-passes-ets-legislation.

341 Australian Government. (2016). The safeguard mechan- ism – Overview. *Department of Energy and the Environment.* Retrieved from https://www.environment.gov.au/climate-change/emissions-reduction-fund/publications/factsheet-erf-safeguard-mechanism.

342 ICAP. (2016, August 17). Mexico to launch pilot ETS. *International Carbon Action Partnership*. Retrieved from https://icapcarbonaction.com/en/news-archive/407-mexico-to-launch-pilot-ets.

343 Zwick, S. (2016). Building on Paris, countries assemble the carbon markets of tomorrow. *Ecosystems Marketplace*. Retrieved from http://www.ecosystemmarketplace.com/articles/building-on-paris-countries-assemble-the-carbon-markets-of-tomorrow/.

344 Zwick, S. (2016). The road from Paris: Green lights, speed bumps, and the future of carbon markets. *Ecosystems Marketplace*. Retrieved from http://www.ecosystemmarketplace.com/articles/green-lights-and-speed-bumps-on-road-to-markets-under-paris-agreement/.

345 World Bank and Ecofys. (2016). Carbon pricing watch 2016. *The World Bank and Ecofys*. Retrieved from http://www.ecofys.com/files/files/world-bank-group_ecofys-carbon-pricing-watch_160525.pdf.

346 Melo, R. (2013). A quick look at Sweden's carbon tax. *Real Melo*. Retrieved from http://blogs.ubc.ca/realmelo/2013/03/06/a-quick-look-at-swedens-carbon-tax/.

347 Andersson, M. & Lövin, I. (2016). Sweden: Decoupling GDP Growth from CO_2 emissions is possible. *The World Bank*. Development in a Changing Climate. Retrieved from http://blogs.worldbank.org/climatechange/sweden-decoupling-gdp-growth-co2-emissions-possible.

348 World Bank and Ecofys. (2017). Carbon pricing watch 2017. *The World Bank*. Retrieved from https://openknowledge.worldbank.org/bitstream/handle/10986/26565/9781464811296.pdf?sequence=4&isAllowed=y.

349 Bae, C. (2013). Denmark's carbon tax policy. Retrieved from http://blogs.ubc.ca/cindybae/2013/02/07/denmarks-carbon-tax-policy/. .

350 Big rise in Swiss carbon tax from 1 January. (2015). *Le News*. Retrieved from http://lenews.ch/2015/12/29/big-rises-in-swiss-carbon-tax-from-1-january-2016/.

351 EurActiv. (2015, July 27). France targets carbon tax in energy transition law. *Journal de l'environement*. Retrieved from https://www.euractiv.com/section/sustainable-dev/news/france-targets-carbon-tax-in-energy-transition-law/.

352 Carbon Tax Center. (2016). Where carbon is taxed. *Carbon Tax Center. Pricing carbon efficiently and equitably.* Retrieved from http://www.carbontax.org/where-carbon-is-taxed/#Ireland.

353 Alberta Government. (2016). Carbon levy and rebates. *Government of Alberta*. Retrieved from http://www.alberta.ca/climate-carbon-pricing.aspx.

354 France sets carbon price floor. (2016). *The Guardian*. Retrieved from https://www.theguardian.com/environment/2016/may/17/france-sets-carbon-price-floor.

355 Vivid Economics with Ecofys. (2013). Carbon leakage prospects under Phase III of the EU ETS, report prepared for DECC. *vivideconomics ECOFYS*. Retrieved from https://www.gov.uk/government/uploads/system/uploads/attachment_data/file/318893/carbon_leakage_prospects_under_phase_III_eu_ets_beyond.pdf. Accessed Sept. 2016.

356 EC. (2016). Carbon Leakage. *European Commission Climate Action*. Retrieved from http://ec.europa.eu/clima/policies/ets/allowances/leakage/index_en.htm.

357 C2ES. (2016). Summary of California's Cap and Trade Program. *Center for Climate Change and Energy Solutions*. Retrieved from http://www.c2es.org/us-states-regions/action/california/cap-trade-regulation#sub8.

358 Gupta S., J. Harnisch, D. C. Barua, L. Chingambo, P. Frankel, R. J. Garrido Vázquez, L. Gómez-Echeverri, E. Haites, Y. Huang, R. Kopp, B. Lefèvre, H. Machado-Filho, and E. Massetti. 2014. Cross-cutting Investment and Finance Issues. In: Climate Change 2014: Mitigation of Climate Change. Contribution of Working Group III to the Fifth Assessment Report of the Intergovernmental Panel on Climate Change [Edenhofer, O., R. Pichs-Madruga, Y. Sokona, E. Farahani, S. Kadner, K. Seyboth, A. Adler, I. Baum, S. Brunner, P. Eickemeier, B. Kriemann, J. Savolainen, S. Schlömer, C. von Stechow, T. Zwickel and J.C. Minx (eds.)]. Cambridge University Press, Cambridge, United Kingdom and New York, NY, USA.

359 The latest information on climate funds. (2016). *Climate Funds Update*. Retrieved from http://www.climatefundsupdate.org/.

360 Climate, energy and China's 13th five-year plan in graphics. (2016, March 18). *Chinadialogue*. Retrieved from https://www.chinadialogue.net/article/show/single/en/8734-Climate-energy-and-China-s-13th-Five-Year-Plan-in-graphics.

361 US EPA. (2016). 2016. Clean power plan for existing power plants. *United States Environmental Protection Agency*. Retrieved from https://www.epa.gov/cleanpowerplan/clean-power-plan-existing-power-plants.

362 RGGI. (2016). RGGI, Inc. *Regional Greenhouse Gas Initiative*. Retrieved from https://www.rggi.org/.

363 Magill, B. (2016, April 12). The suit against the clean power plan, explained. *Climate Central*. Retrieved from http://www.climatecentral.org/news/the-suit-against-the-clean-power-plan-explained-20234.

364 The White House. (2015). President Obama's climate action plan. 2nd Anniversary Progress Report. *The White House Washington*. Retrieved from https://obamawhitehouse.archives.gov/sites/default/files/docs/cap_progress_report_final_w_cover.pdf

365 Kolbert, E. (2017, June 1). Au Revoir: Trump exits the Paris climate agreement. Daily Comment. *The New Yorker*. Retrieved from http://www.newyorker.com/news/daily-comment/au-revoir-trump-exits-the-paris-climate-accord.

366 Upton, J. (2016). 3 ways Trump could abandon the Paris climate pact. *Climate Central*. Retrieved from http://www.climatecentral.org/news/trump-could-abandon-paris-climate-agreement-20711.

367 C2ES. (2016). California global warming solutions act (AB32). *Center for Climate Change and Energy Solutions*. Retrieved from http://www.c2es.org/us-states-regions/action/california/ab32. Accessed Sept. 2016.

368 CA Gov. (2015, April 29). Governor Brown establishes most ambitious greenhouse gas reduction target in North America. *Office of Governor Edmund G. Brown Jr.* Retrieved from https://www.gov.ca.gov/news.php?id=18938.

369 CA Gov. (2016). California climate change. *Office of Governor Edmund G. Brown Jr.* Retrieved from http://climatechange.ca.gov/.

370 US EPA. (2016). State energy CO_2 emissions. *United States Environmental Protection Agency*. Retrieved from https://www.epa.gov/statelocalclimate/state-energy-co2-emissions.

371 C2ES. (2016). Climate action plans. *Center for Climate Change and Energy Solutions*. Retrieved from http://www.c2es.org/us-states-regions/policy-maps/climate-action-plans.

372 New York State climate action council. (2010). Climate action plan interim report. *Center for Climate Change and Energy Solutions*. Retrieved from http://www.c2es.org/docUploads/states/climate-action-plan/ny_climate_action_plan_report_2010.pdf.

373 Governor's action team on energy and climate change. (2008). Florida's energy and climate change action plan. *Center for Climate Change and Energy Solutions*. Retrieved from http://www.c2es.org/docUploads/states/climate-action-plan/florida_ccactionplan_oct2008.pdf.

374 Mayors climate protection center. (2014). The U.S. mayors climate protection agreement. *The United States Conference of Mayors*. Retrieved from http://usmayors.org/climateagreement/Final%20USCM%202014%20Mayors%20Climate%20Protection%20Agreement.pdf.

375 Washington Governor, Jay Inslee. (2017, June 1). Insell, New York Governor Cuomo, and California Governor Brown announce formation of the United States Climate Allicance. *Office of the Governor*. Retrieved from http://governor.wa.gov/news-media/inslee-new-york-governor-cuomo-and-california-governor-brown-announce-formation-united.

376 United States Climate Alliance. (2017). *Wikipedia*. Retrieved from https://en.wikipedia.org/wiki/United_States_Climate_Alliance.

377 EU. (2009). Presidency conclusions. *Council of the European Union*. Retrieved from https://www.consilium.europa.eu/uedocs/cms_data/docs/pressdata/en/ec/108622.pdf

378 EC. (2011). A Roadmap for moving to a competitive low carbon economy in 2050. *European Commission*. Retrieved from http://eur-lex.europa.eu/legal-content/EN/TXT/PDF/?uri=CELEX:52011DC0112&from=EN.

379 EC. (2016). Climate action. EU action. Effort sharing decision. *European Commission*. Retrieved from http://ec.europa.eu/clima/policies/effort/index_en.htm.

380 EC. (2016). Climate action. EU action. Emissions trading system (EU ETS). Revisions for phase 4 (2021-2030). *European Commission*. Retrieved from http://ec.europa.eu/clima/policies/ets/revision/index_en.htm.

381 EC. (2016). Climate action. EU action. Climate strategies and targets. 2030 climate & energy framework. *European Commission*. Retrieved from http://ec.europa.eu/clima/policies/strategies/2030/index_en.htm. Accessed Oct. 2016.

382 E.C. (2016). Energy. Buildings. *European Commission*. Retrieved from http://ec.europa.eu/energy/en/topics/energy-efficiency/buildings.

383 E.C. (2016). Climate action. EU action. Reducing CO_2 emissions from passenger cars. *European Commission*. Retrieved from http://ec.europa.eu/clima/policies/transport/vehicles/cars/index_en.htm.

384 Morris, C.M. & Pehnt, M. (2014). Energy transition. The German energiewende. Key findings. *Energytransition*. Retrieved from http://energytransition.de/2012/10/key-findings/.

385 Wettengel, J. & Amelang, S. (2016). German ministry avoids concrete targets in weakened climate action plan. *Chinadialogue*. Retrieved from https://www.chinadialogue.net/article/show/single/en/9250-German-ministry-avoids-concrete-targets-in-weakened-Climate-Action-Plan.

386 Wheeler, B., & Hunt, A. (2017, June 17). Brexit: All you need to know about the UK leaving the EU. *BBC News*. Retrieved from http://www.bbc.com/news/uk-politics-32810887.

387 Gov.UK. (2010). Beyond Copenhagen: The UK Government's international climate change action plan. *Department of Energy and Climate Change*. Retrieved from https://www.gov.uk/government/uploads/system/uploads/attachment_data/file/238437/7850.pdf.

388 Ministry of Ecology, Sustainable Development and Energy. (2015). Energy transition for green growth act. *République Française*. Retrieved from http://www2.developpement-durable.gouv.fr/IMG/pdf/16172-GB_loi-TE-les-actions_DEF_light.pdf

389 Donat, L., Velten, E.K., Prahl, A., & Zane, E.B. (2014). Assessment of climate change policies in the context of the European Semester. Country Report: Italy. *Ecologic Institute*. Retrieved from http://ecologic.eu/sites/files/publication/2014/countryreport_it_ecologicelareon_jan2014_0.pdf

390 C2ES. (2016). India's climate and energy policies. *Center for Climate and Energy Solutions.* Retrieved from https://www.c2es.org/international/key-country-policies/india.

391 Press information bureau government of India cabinet. (2015 June 17). Revision of cumulative targets under the national solar mission from 20,000 MW by 2021-22 to 100,000 MW. *Governemet of India Cabinet.* Retrieved from http://pib.nic.in/newsite/PrintRelease.aspx?relid=122566.

392 C2ES. (2016). India and climate change. *Center for Climate Change and Energy Solutions.* Retrieved from http://www.c2es.org/international/key-country-policies/india.

393 Vaidyula, M., Rittenhouse, K., Sopher, P., Francis, D., & Swartz, J. (2015). India an emissions trading case study. *International Emissions Trading Association.* Retrieved from http://www.ieta.org/resources/Resources/Case_Studies_Worlds_Carbon_Markets/india_case_study_may2015.pdf.

394 Moarif, S. (2012). Market-based climate mitigation policies in emerging economies. *Center for Climate and Energy Solutions.* Retrieved from http://www.c2es.org/docUploads/market-based-climate-mitigation-policies-emerging-economies.pdf.

395 Government of India. (2015). National mission on sustainable habitat. *Ministry of Urban Development.* Retrieved from http://moud.gov.in/policies/NMSH.

396 Government of India. (2015). Smart Cities Mission. *Ministry of Urban Development.* Retrieved from http://smartcities.gov.in/.

397 Press information bureau government of India cabinet. (2016). Compensatory afforestation fund bill, 2016 passed by Rajya Sabha. *Government of India Cabinet.* Retrieved from http://pib.nic.in/newsite/mbErel.aspx?relid=14793.

398 The Soviet of the Republics of the Supreme Soviet of the USSR. (1991). Declaration № 142-H. *Wikisource.* Retrieved from https://en.wikisource.org/wiki/Declaration_no._142-N_of_the_Soviet_of_the_Republics_of_the_Supreme_Soviet_of_the_USSR.

399 IEA. (2014). Russia 2014. Energy policies beyond IEA countries. *International Energy Agency.* Retrieved from https://www.iea.org/publications/freepublications/publication/Russia_2014.pdf.

400 Summary of GHG Emissions from Russian Federation. (2016). *United Nations Climate Change Secretariat.* Retrieved from https://unfccc.int/files/ghg_emissions_data/application/pdf/rus_ghg_profile.pdf.

401 Ministry of Energy of the Russian Federation. (2010). Energy strategy 2030. *Government of the Russian Federation.* Retrieved from http://www.energystrategy.ru/projects/docs/ES-2030_(Eng).pdf.

402 IPP. (2011). RU-3:Federal law on energy conservation and energy efficiency. *Institute for Industrial Productivity.* Retrieved from http://iepd.iipnetwork.org/policy/federal-law-energy-conservation-and-energy-efficiency. .

403 World Bank. (2016). Environment. Table 3.6. World development Indicators: Energy production and use. *The World Bank.* Retrieved from http://wdi.worldbank.org/table/3.6.

404 Indonesia forest information and data. (2011). *Mongabay.* Retrieved from http://rainforests.mongabay.com/deforestation/2000/Indonesia.htm.

405 LSE. (2015). Indonesia. Approach to climate change. *London School of Economics and Political Science and Grantham Research Institute on Climate Change and the Environment.* Retrieved from http://www.lse.ac.uk/GranthamInstitute/country-profiles/indonesia/

406 REDD+ taskforce (Indonesia). (2016). *The REDD Desk.* Retrieved from http://theredddesk.org/countries/actors/redd-taskforce-indonesia.

407 Tharakan, P. (2015). Summary of Indonesia's energy sector assessment. *Asian Development Bank.* Retrieved from https://www.adb.org/sites/default/files/publication/178039/ino-paper-09-2015.pdf.

408 Indonesia. (2016). *Global Forest Watch.* Retrieved from http://www.globalforestwatch.org/country/IDN.

409 UNFCCC. (2015). Federal Republic of Brazil. 2015. Intended nationally determined contribution. *United Nations Framework Convention on Climate Change.* Retrieved from http://www4.unfccc.int/submissions/INDC/Published%20 Documents/Brazil/1/BRAZIL%20iNDC%20english%20FINAL.pdf.

410 GFC Task Force. (2014). Contributions to the national REDD+ strategy: A proposal for allocation between states and the union. *Governors' Climate Greenhouse Gas Taskforce.* Retrieved from http://www.idesam.org.br/wp-content/uploads/2014/02/gcf-contribuicoes-para-estrategia-nacionalde-redd-english.pdf.

411 Marcelo, M. M., Lima, L., & Leonel, D.A. (2014). Low carbon agriculture in Brazil: The environmental and trade impact of current farm Policies; issue paper No. 54; *International Centre for Trade and Sustainable Development.* Geneva, Switzerland, Retrieved from http://www.ictsd.org/sites/default/files/research/Low-Carbon%20Agriculture%20in%20Brazil.pdf.

412 Deckers, D. (2016). Low carbon agriculture plan. *Ministry of Agriculture, Livestock and Food Supply.* Retrieved from http://www.nwf.org/pdf/REDD-Workshop/DeckersBrazilianMinistrySanDiego2011.pdf.

413 IEA. (2016). Brazil (partner country). *International Energy Agency.* Retrieved from https://www.iea.org/countries/non-membercountries/brazil/.

414 LSE. (2007). National Energy Plan (PNE 2030). *London School of Economics and Political Science and Grantham Research Institute on Climate Change and the Environment.* Retrieved from http://www.lse.ac.uk/GranthamInstitute/law/national-energy-plan-2030-pne-2030/.

415   Nachmany, M., Fankhauser, S., Davidová, J., Kingsmill, N., Landesman, T. Roppongi, H., Schleifer, P., Setzer, J., Sharman, A., Stolle Singleton, C., Sundaresan, J. & Townshend, T. (2015). Climate change legislation in Brazil. *London School of Economics and Political Science and Grantham Research Institute on Climate Change and the Environment.* Retrieved from http://www.lse.ac.uk/GranthamInstitute/wp-content/uploads/2015/05/BRAZIL.pdf.

416   Lane, J. (2016). Biofuels Mandates Around the World: 2016. *BiofuelsDigest.* Retrieved from http://www.biofuelsdigest.com/bdigest/2016/01/03/biofuels-mandates-around-the-world-2016/.

417   Dahan, L., Rittenhouse, K., Sopher, P., Francis, D., Swartz, J., & De Clara, S. (2015). Brazil. The world's carbon markets: a case study to emissions trading. *International Emissions Trading Association.* Retrieved from http://www.ieta.org/resources/Resources/Case_Studies_Worlds_Carbon_Markets/brazil_case_study_may2015.pdf.

418   Kuramochi, T. (2014). GHG mitigation in Japan: an overview of the current policy landscape. *World Resource Institute.* Retrieved from http://www.wri.org/sites/default/files/wri_workingpaper_japan_final_ck_6_11_14.pdf.

419   NBR. (2016). Energy mix in Japan – before and after Fukushima. *The National Bureau of Asian Research.* Retrieved from http://www.nbr.org/downloads/pdfs/eta/PES_2013_handout_kihara.pdf.

420   IEE Japan. (2010). Japan energy brief. No. 7: May 2010. *Institute of Energy Economics, Japan.* Retrieved from https://eneken.ieej.or.jp/en/jeb/1005.pdf.

421   Arima, J. (2011). Japan's energy policy – pre and post Fukushima. *World Energy Insight* Retrieved from https://www.worldenergy.org/wp-content/uploads/2012/10/PUB_World-Energy-Insight_2011_WEC.pdf.

422   UNSCEAR 2013 Report. (2013). *Volume 1. Report to the General Assembly. Scientific Annex A. Levels and effects of radiation exposure due to the nuclear accident after the 2011 great east-Japan Earthquake and tsunami.* New York. United Nations Publication,

423   IEEFA. (2016). IEEFA Japan briefing: Japan's energy transformation. *Institute for Energy Economics and Financial Analysis.* Retrieved from http://ieefa.org/wp-content/uploads/2016/03/Japan-Energy-Brief.pdf.

424   UNFCCC. (2015). Submission of Japan's intended nationally determined contribution (INDC). INDCs as communicated by parties. *United Nations Framework Convention on Climate Change.* Retrieved from http://www4.unfccc.int/Submissions/INDC/Published%20Documents/Japan/1/20150717_Japan's%20INDC.pdf.

425   Ahmad, F.M. (2016). Technological innovation, sustainable development, and post-Paris voluntary cooperation – a closer look at Japan's joint credit mechanism. *Center for Climate Change and Energy Solutions.* Retrieved from http://www.c2es.org/newsroom/articles/technological-innovation-sustainable-development-post-paris-voluntary-cooperation-closer-look-japans-joint.

426   Ministry of Economy, Trade and Industry. (2015). Japan's energy plan. *Agency for Natural Resources and Energy.* Retrieved from http://www.enecho.meti.go.jp/en/category/brochures/pdf/energy_plan_2015.pdf.

427   IEA. (2016). Feed-in tariff for electricity generation from renewable energy. *International Energy Agency.* Retrieved from http://www.iea.org/policiesandmeasures/pams/japan/name-30660-en.php.

428   ICAP. (2016). Japan – Tokyo cap-and-trade program. *International Carbon Partnership.* Retrieved from https://icapcarbonaction.com/en/?option=com_etsmap&task=export&format=pdf&layout=list&systems%5B%5D=51.

429   ICAP. (2016). Japan – Saitama target setting emissions trading system. *International Carbon Partnership.* Retrieved from https://icapcarbonaction.com/en/?option=com_etsmap&task=export&format=pdf&layout=list&systems%5B%5D=84.

430   UNFCCC. (2016). Status of Ratification of the Kyoto protocol. *United Nations Framework Convention on Climate Change.* Retrieved from http://unfccc.int/kyoto_protocol/status_of_ratification/items/2613.php.

431   Government of Canada. (2015). Greenhouse gas emissions by economic sector. *Environment and Climate Change Canada.* Retrieved from https://www.ec.gc.ca/indicateurs-indicators/default.asp?lang=en&n=F60DB708-1.

432   Government of Canada. (2015). Greenhouse gas emissions by province and territory. *Environment and Climate Change Canada.* Retrieved from https://www.ec.gc.ca/indicateurs-indicators/default.asp?lang=en&n=18F3BB9C-1.

433   Government of Canada. (2016). Population by year, by province and territory. *Statistics Canada.* Retrieved from http://www.stats.gov.nl.ca/statistics/population/PDF/Annual_Pop_Prov.PDF.

434   Government of Ontario. (2016). Ontario's five year climate change action plan 2016-2020. *Ministry of the Environment and Climate Change.* Retrieved from http://www.applications.ene.gov.on.ca/ccap/products/CCAP_ENGLISH.pdf.

435   Electricity generated in Ontario. (2016). *Canadian Nuclear Society.* Retrieved from https://www.cns-snc.ca/media/ontarioelectricity/ontarioelectricity.html.

436   Government of Canada. (2015). Electric power generation. *Statistics Canada.* Retrieved from http://www5.statcan.gc.ca/cansim/a37.

437   Government of British Columbia. (2014). Climate action in British Columbia. 2014 Progress report. *Ministry of the Environment British Columbia.* Retrieved from http://www2.gov.bc.ca/assets/gov/environment/climate-change/policy-legislation-and-responses/2014-progress-to-targets.pdf.

438   Oil and gas production by Canadian province. (2014). *SaskWind.* Retrieved from http://www.saskwind.ca/oil-gas-by-province/.

439 Evans, R.L. & Bryant, T. (2013). Greenhouse gas emissions from the Canadian oil and gas sector. *Trottier Energy Futures Project.* Retrieved from https://www.cae-acg.ca/wp-content/uploads/2014/01/Greenhouse-Gas-Emissions-from-the-Canadian-Oil-and-Gas-Sector.pdf.

440 Alberta Government. (2015). Climate leadership plan. *Environment and Natural Resources.* Retrieved from http://www.alberta.ca/climate-leadership-plan.aspx.

441 Government of Canada. (2016). The pan-Canadian framework on clean growth and climate change. *Environment and Natural Resources.* Retrieved from https://www.canada.ca/en/services/environment/weather/climatechange/pan-canadian-framework.html

442 UNFCCC. (2015). Submission of Mexico's intended nationally determined contribution (INDC). INDCs as communicated by parties. *United Nations Framework Convention on Climate Change.* Retrieved from http://www4.unfccc.int/submissions/INDC/Published%20Documents/Mexico/1/MEXICO%20INDC%2003.30.2015.pdf.

443 Avila, A. (2012). Mexican general law on climate change serves as a model for mitigation and on the ground action. *Center for Clean Air Supply.* Retrieved from http://ccap.org/mexican-general-law-on-climate-change-serves-as-a-model-for-mitigation-and-on-the-ground-action/.

444 Federal Government of Mexico. (2013). National climate change strategy 10-20-40 vision. *Ministry of the Environment and Natural Resources.* Retrieved from https://mitigationpartnership.net/sites/default/files/encc_englishversion.pdf.

445 World Bank. (2016). Poverty and equity. Nigeria. *The World Bank.* Retrieved from http://povertydata.worldbank.org/poverty/country/NGA.

446 World Bank. (2016). Employment in agriculture (% of total employment). *The World Bank.* Retrieved from http://data.worldbank.org/indicator/SL.AGR.EMPL.ZS?end=2007&locations=NG&start=1983&view=chart.

447 FOA. (2010). Global forest resources assessment 2010 country report Nigeria. *Food and Agriculture Organization of the United Nations.* Retrieved from http://www.fao.org/forestry/20406-0d1f56d9ee7a6fd2079bcd520715362c3.pdf.

448 Olagunju, T.E. (2015). Drought, desertification and the Nigerian environment: A review. *Journal of Ecology and the Natural Environment. 7*, pp 196-209.

449 UNFCCC. (2015). Submission of Nigeria's intended nationally determined contribution (INDC). INDCs as communicated by parties. *United Nations Framework Convention on Climate Change.* Retrieved from http://www4.unfccc.int/submissions/INDC/Published%20Documents/Nigeria/1/Approved%20Nigeria's%20INDC_271115.pdf.

450 Australian Government. (2015). Australian energy update 2015. *Department of Industry and Science. Office of the Chief Economist.* Retrieved from http://www.industry.gov.au/Office-of-the-Chief-Economist/Publications/Documents/aes/2015-australian-energy-statistics.pdf.

451 Rocha, M., Hare, B., Parra, P., Cantzler, J., Höhne, N., Jeffery, L., Alexander, R., Wong, L. Wouters, K. & Blok, K. (2015). Australia set to overshoot its 2030 target by a large margin. *Climate Action Tracker.* Retrieved from http://climateactiontracker.org/assets/publications/briefing_papers/Australia.pdf.

452 UNFCCC. (2015). Submission of Australia's intended nationally determined contribution (INDC). INDCs as communicated by parties. *United Nations Framework Convention on Climate Change.* Retrieved from http://www4.unfccc.int/submissions/INDC/Published%20Documents/Australia/1/Australias%20Intended%20Nationally%20Determined%20Contribution%20to%20a%20new%20Climate%20Change%20Agreement%20-%20August%202015.pdf.

453 Australian Government. (2015). Carbon pricing mechanism. *Clean Energy Regulator.* Retrieved from http://www.cleanenergyregulator.gov.au/Infohub/CPM/About-the-mechanism.

454 Australian Government. (2016). Emissions reduction fund. *Department of the Environment and Energy.* Retrieved from https://www.environment.gov.au/climate-change/emissions-reduction-fund/about.

455 Australian Government. (2011). Clean energy act 2011. *Federal Register of Legislation.* Retrieved from https://www.legislation.gov.au/Details/C2011A00131.

456 About C40 Cities. (2016). *C40 Cities.* Retrieved from http://www.c40.org/about.

457 Deadline 2020. How cities can get the job done. (2016). *C40 Cities and ARUP.* Retrieved from http://www.c40.org/researches/deadline-2020.

458 Barboza, T. (2014, September 22). L.A, Houston, Philadelphia mayors vow more action on climate change. *Los Angeles Times.* Retrieved from http://www.latimes.com/science/sciencenow/.

459 Zero Zone Inc. et al., v. United States Department of Energy et al. (2016). United States Court of Appeals for the Seventh Circuit. Retrieved from http://media.ca7.uscourts.gov/cgi-bin/rssExec.pl?Submit=Display&Path=Y2016/D08-08/C:14-2159:J:Ripple:aut:T:fnOp:N:1807496:S:0.

460 US EPA (2016). Summary of executive order 12866 – regulatory planning and review. *United States Environmental Protection Agency.* Retrieved from https://www.epa.gov/laws-regulations/summary-executive-order-12866-regulatory-planning-and-review.

461 Interagency Working Group on Social Cost of Greenhouse Gases, United States Government. (2016). Technical support document: Technical update of the social cost of carbon for regulatory impact

analysis - under executive order 12866. *United States Environmental Protection Agency.* Retrieved from https://www.whitehouse.gov/sites/default/files/omb/inforeg/scc_tsd_final_clean_8_26_16.pdf.

462 Moore, F. C. & Diaz, D. B. (2015). Temperature impacts on economic growth warrant stringent mitigation policy. *Nature Climate Change. 5*, 127-131.

463 SP. (2011). The value of carbon in decision-making. *Sustainable Prosperity.* Retrieved from http://www.sustainableprosperity.ca/sites/default/files/publications/files/The%20Value%20of%20Carbon%20in%20 Decision-Making_0.pdf.

464 BNEF. (2010). Carbon markets – North America – research note. *Bloomberg New Energy Finance.* (January 2010). bnef.com/ WhitePapers/download/25.

465 Kesicki, F. (2010). Marginal abatement cost curves for policy making – expert-based vs. model-derived curves. *UCL Energy Institute.* Retrieved from http://www.homepages.ucl.ac.uk/~ucft347/Kesicki_MACC.pdf.

466 Aden, N. (2016). The roads to decoupling: 21 countries are reducing carbon emissions while growing GDP. *World Resource Institute.* Retrieved from http://www.wri.org/blog/2016/04/roads-decoupling-21-countries-are-reducing-carbon-emissions-while-growing-gdp.

467 Chateau, J., Saint-Martin, A., & Manfredi, T. (2011). Employment impacts of climate change mitigations policies in OECD: a general-equilibrium perspective. *OECD. Environment Working Paper 32.* Retrieved from http://www.oecd.org/officialdocuments/publicdisplaydocumentpdf/?cote=ENV/WKP(2011)2&docLanguage=En.

468 Government of Canada. (2016). Canada's greenhouse gas emissions: developments, prospects and reductions. *Office of the Parliamentary Budget Officer.* Retrieved from http://www.pbo-dpb.gc.ca/web/default/files/Documents/Reports/2016/ClimateChange/PBO_Climate_Change_EN.pdf.

469 Stern, N. (2006). *The Economics of Climate Change.* Cambridge University Press.

470 Nakhooda, S., & Norman, M. (2014). Climate finance: is it making a difference? A review of the effectiveness of multilateral climate funds. *Overseas Development Institute.* Retrieved from https://www.odi.org/sites/odi.org.uk/files/odi-assets/publications-opinion-files/9359.pdf.

471 OECD. (2015). "Climate finance in 2013-14 and the USD 100 billion goal", a report by the OECD and development (OECD) in collaboration with Climate Policy Initiative (CPI). *Organization of Economic Cooperation and Development.* Retrieved from http://www.oecd.org/environment/cc/OECD-CPI-Climate-Finance-Report.htm.

472 World Bank. 2016. World's largest concentrated solar plant opened in Morocco. *The World Bank.* Retrieved from http://www.worldbank.org/en/news/press-release/2016/02/04/worlds-largest-concentrated-solar-plant-opened-in-morocco.

473 C2ES. (2016). The American clean energy and security act (Waxman-Markey bill). *Center for Climate and Energy Solutions.* Retrieved from http://www.c2es.org/federal/congress/111/acesa.

474 Weiss, D.J. (2010). Anatomy of a senate climate bill death. *Center for American Progress.* Retrieved from https://www.americanprogress.org/issues/green/news/2010/10/12/8569/anatomy-of-a-senate-climate-bill-death/.

475 IEA. (2016). Key world energy statistics. *International Energy Agency.* Retrieved from https://www.iea.org/publications/freepublications/publication/KeyWorld2016.pd.

476 Trump, D.J. (2012, November 6) *Twitter.* Retrieved from https://twitter.com/realdonaldtrump/status/265895292191248385?lang=en.

477 Associated Press. (2016, May 27) Donald Trump vows to cancel Paris agree- ment and stop all payments to UN Climate change fund. *The Telegraph.* Retrieved from http://www.telegraph.co.uk/news/2016/05/27/donald-trump-ill-cancel-paris-climate-agreement-and-stop-all-pay/.

478 Roth, S. (2016, December 9). Scott Pruitt, Trump's EPA pick, rejects climate change science and fights for fossil fuels. *USA Today.* Retrieved from https://www.usatoday.com/story/news/environment/2016/12/09/ scott-pruitt-trumps-epa-pick-rejects-climate-science-and-fights-fossil-fuels/95200658/.

479 New York State. (2016, August 1). Governor Cuomo announces establishment of clean energy standard that mandates 50 percent renewable by 2030. *Pressroom, Govenor Andrew M. Cuomo.* Retrieved from https://www.governor.ny.gov/news/ governor-cuomo-announces-establishment-clean-energy-standard-mandates-50-percent-renewables.

480 Pyper, J. (2015). New York State pledges to start talks on a North American carbon market. *Greentech Media.* Retrieved from https://www.greentechmedia.com/articles/read/ new-york-state-pledges-to-start-talks-on-a-north-american-carbon-market.

481 EIA. (2017, January 19). Texas state profile and energy estimates. *United States Energy Information Administration.* Retrieved from https://www.eia.gov/state/?sid=TX.

482 The White House. Office of the Press Secretary. (2015, November 6). Statement by the President on the Keystone XL Pipeline. *The White House President Barack Obama.* Retrieved from https://www.whitehouse.gov/the-press-office/2015/11/06/statement-president-keystone-xl-pipeline.

483 EASAC. (2016). Greenhouse gas footprints of different oil feedstocks. *European Academies Science Advisory Council.* Retrieved from http://www.easac.eu/fileadmin/PDF_s/reports_statements/EASAC_16_GGF_Pantone432_Web_final.pdf.

484 About the science. (2016). *350.org.* Retrieved from https://350.org/about/science/.

485 CVF. (2016). CVF participating countries. *Climate Vulnerable Forum*. Retrieved from http://www.thecvf.org/.

486 Hilton, I. (2016, November 21). With Trump, China emerges as global leader on climate. *Environment 360. Reporting, Analysis, Opinion and Debate*. Retrieved from http://e360.yale.edu/feature/with_trump_china_stands_along_as_global_climate_leader/3057/.

487 GWEC. (2016). Global statistics. *Global Wind Energy Council*. Retrieved from http://www.gwec.net/global-figures/graphs/.

488 IEA. (2016). 2015 Snapshot of global photovoltaic markets. *International Energy Agency*. Retrieved from http://www.iea-pvps.org/fileadmin/dam/public/report/PICS/IEA-PVPS_-__A_Snapshot_of_Global_PV_-_1992-2015_-_Final_2_02.pdf.

489 IEA. (2015). Energy and Climate Change. World Energy Outlook Special Report. *International Energy Agency*. Retrieved from https://www.iea.org/publications/freepublications/publication/WEO2015SpecialReportonEnergyandClimateChange.pdf.

490 Chung, J. (2016). South Korea to shut 10 ageing coal-fired power plants by 2025. *Rueters*. Retrieved from http://www.reuters.com/article/us-southkorea-coal-idUSKCN0ZM06A.

491 Buckley, T. (2016). Cancellation of 4 ultra mega power plants underscores India's commitment to transition. *Institute for Energy Economics and Financial Analysis*. Retrieved from http://ieefa.org/ieefa-asia-note-cancellation-4-ultra-mega-power-plants-underscores-indias-commitment-transition%E2%80%A8%E2%80%A8/.

492 Government of India. (2016). Draft national electricity plan. (Volume 1). *Ministry of Power. Central Electricity Authority*. Retrieved from http://www.cea.nic.in/reports/committee/nep/nep_dec.pdf.

493 Brodie, C. (2017, May 23). India will sell only electric cars within the next 13 years. *World Economic Formum*. Retrieved from https://www.weforum.org/agenda/2017/05/india-electric-car-sales-only-2030/.

494 Dewan Nasional Perubahan Iklim, Indonesia. (2010). Indonesia's greenhouse gas abatement cost curve. *National Council on Climate Change*. Retrieved from http://www.mmechanisms.org/document/country/IDN/Indonesia_ghg_cost_curve_english.pdf.

495 Krisnawati, H., Imanuddin, R., Adinugroho, W.C. & Hutabarat, S. (2015). National inventory of greenhouse gas emissions and removals on Indonesia's forests and peatlands. *Research, Development and Innovation Agency of the Ministry of Environment and Forestry*. Bogor, Indonesia.

496 CVF. (2016). The Marrakech Communique. *Climate Vulnerable Forum*. Retrieved from http://www.thecvf.org/wp-content/uploads/2016/11/CVF-Marrakech-Communique-for-Adoption.pdf

497 PRB. (2013). World population data sheet 2013. *Population Reference Bureau*. Retrieved from http://www.prb.org/publications/datasheets/2013/2013-world-population-data-sheet/data-sheet.aspx.

498 Sub-Saharan Africa 2080. (2016). *United Nations Department of Economic and Social Affairs, Population Division*. Retrieved from https://populationpyramid.net/sub-saharan-africa/2080/.

Index

www.ingramcontent.com/pod-product-compliance
Lightning Source LLC
Chambersburg PA
CBHW081052220326
41598CB00038B/7065